Augmented Reality, Virtual Reality & Semantic 3D Reconstruction

Augmented Reality, Virtual Reality & Semantic 3D Reconstruction

Editors

Zhihan Lv
Jing-Yan Wang
Neeraj Kumar
Jaime Lloret

MDPI • Basel • Beijing • Wuhan • Barcelona • Belgrade • Manchester • Tokyo • Cluj • Tianjin

Editors

Zhihan Lv
Uppsala University
Sweden

Jing-Yan Wang
PEGASUS FZ LLC
United Arab Emirates

Neeraj Kumar
Deemed University
India

Jaime Lloret
Universitat Politecnica de Valencia
Spain

Editorial Office
MDPI
St. Alban-Anlage 66
4052 Basel, Switzerland

This is a reprint of articles from the Special Issue published online in the open access journal *Applied Sciences* (ISSN 2076-3417) (available at: https://www.mdpi.com/journal/applsci/special_issues/AR_VR).

For citation purposes, cite each article independently as indicated on the article page online and as indicated below:

LastName, A.A.; LastName, B.B.; LastName, C.C. Article Title. *Journal Name* **Year**, *Volume Number*, Page Range.

ISBN 978-3-0365-6061-8 (Hbk)
ISBN 978-3-0365-6062-5 (PDF)

Contents

applied sciences

MDPI

Editorial

Special Issue on "Augmented Reality, Virtual Reality & Semantic 3D Reconstruction"

Zhihan Lv [1,*], Jing-Yan Wang [2], Neeraj Kumar [3,4] and Jaime Lloret [5]

1 School of Data Science and Software Engineering, Qingdao University, Qingdao 266000, China
2 PEGASUS FZ LLC, Abu Dhabi 51133, United Arab Emirates; jywang.ieee@gmail.com
3 Computer Science and Engineering, Thapar Institute of Engineering and Technology, Deemed University, Patiala 147004, Punjab, India; neeraj.kumar@thapar.edu
4 School of Computer Science, University of Petroleum and Energy Studies, Dehradun 248001, Uttarakhand, India
5 Department of Communications, Polytechnic University of Valencia, 46022 Valencia, Spain; jlloret@dcom.upv.es
* Correspondence: lvzhihan@gmail.com

Citation: Lv, Z.; Wang, J.-Y.; Kumar, N.; Lloret, J. Special Issue on "Augmented Reality, Virtual Reality & Semantic 3D Reconstruction". *Appl. Sci.* **2021**, *11*, 8590. https://doi.org/10.3390/app11188590

Received: 28 August 2021
Accepted: 9 September 2021
Published: 16 September 2021

Publisher's Note: MDPI stays neutral with regard to jurisdictional claims in published maps and institutional affiliations.

1. Introduction

Augmented Reality is a key technology that will facilitate a major paradigm shift in the way users interact with data and has only just recently been recognized as a viable solution for solving many critical needs. Enter augmented reality (AR) technology, which can be used to visualize data from hundreds of sensors simultaneously, overlaying relevant and actionable information over your environment through a headset. Semantic 3D reconstruction makes AR technology much more promising, with much more semantic information. Although, there are several methods currently available as post-processing approaches to extract semantic information from the reconstructed 3D models, the obtained results are uncertainty, and are evenly incorrect. Thus, it is necessary to explore or develop a novel 3D reconstruction approach to automatic recover 3D geometry models and obtained semantic information in simultaneous.

The rapid advent of deep learning brought new opportunities to the field of semantic 3D reconstruction from photo collections. Deep learning-based methods are not only able to extract semantic information but can also be used to enhance some fundamental techniques in semantic 3D reconstruction: those fundamental techniques include feature matching or tracking, stereo matching, camera pose estimation, and multiview stereo. Moreover, deep learning techniques can be used to extract priors from photo collections, the obtained information in turn can improve the quality of 3D reconstruction.

The aim of this Special Issue is to provide a platform for researchers to share innovative work in the field of semantic 3D reconstruction, virtual reality, and augmented reality, including deep learning-based approaches to 3D reconstruction, and software platforms of deep learning for virtual reality and augmented reality.

2. Augmented Reality, Virtual Reality and Semantic 3D Reconstruction

As highly immersive virtual reality (VR) content, 360° video allows users to observe all viewpoints within the desired direction from the position where the video is recorded. In 360° video content, virtual objects are inserted into recorded real scenes to provide a higher sense of immersion. Lee et al. [1] propose a new method for previsualization and 3D composition that overcomes the limitations of existing methods. This system achieves real-time position tracking of the attached camera using a ZED camera and a stereovision sensor, and real-time stabilization using a Kalman filter. The proposed system shows high time efficiency and accurate 3D composition.

Dynamic hand gesture recognition based on one-shot learning requires full assimilation of the motion features from a few annotated data. However, how to effectively extract the spatio-temporal features of the hand gestures remains a challenging issue. Ma et al. [2] propose a skeleton-based dynamic hand gesture recognition using an enhanced network (GREN) based on one-shot learning by improving the memory-augmented neural network, which can rapidly assimilate the motion features of dynamic hand gestures. Besides, the network effectively combines and stores the shared features between dissimilar classes, which lowers the prediction error caused by unnecessary hyperparameters updating, and improves the recognition accuracy with the increase of categories. The experimental results demonstrate that the GREN network is feasible for skeleton-based dynamic hand gesture recognition based on one-shot learning.

Human cognitive processes in wayfinding may differ depending on the time taken to accept visual information in environments. Kim [3] investigated users' wayfinding processes using eye-tracking experiments, simulating a complex cultural space to analyze human visual movements in perception and the cognitive processes through visual perception responses. The results show that the methods for analyzing the gaze data may vary in terms of processing, analysis, and scope of the data depending on the purpose of the virtual reality experiments. Further, they demonstrate the importance of what purpose statements are given to the subject during the experiment and the possibility of a technical approach being used for the interpretation of spatial information.

Ref. [4] report concerns a study of the impact of a semi-immersive VR system in a group of 25 children in a kindergarten context. The children were involved in several different games and activity types. Their reactions and behaviors were recorded through observation grids addressing task comprehension, participation and enjoyment, interaction and cooperation, conflict, strategic behaviors, and adult-directed questions concerning the activity, the device or general help requests. The grids were compiled at the initial, intermediate and final timepoint during each session. The results show that the activities are easy to understand, enjoyable, and stimulate strategic behaviors, interaction and cooperation, while they do not elicit the need for many explanations. These results are discussed within a neuroconstructivist educational framework and the suitability of semi-immersive, virtual-reality-based activities for cognitive empowerment and rehabilitation purposes is discussed.

As a classical method widely used in 3D reconstruction tasks, the multisource Photometric Stereo can obtain more accurate 3D reconstruction results compared with the basic Photometric Stereo, but its complex calibration and solution process reduces the efficiency of this algorithm. Wang et al. [5] propose a multisource Photometric Stereo 3D reconstruction method based on the fully convolutional network (FCN). The experimental results show that their method has a good effect on solving the main problems faced by the classical method.

The Diagnosis of Attention Deficit/Hyperactivity Disorder (ADHD) requires an exhaustive and objective assessment in order to design an intervention that is adapted to the peculiarities of the patients. The authors of [6] aimed to determine if the most commonly used ADHD observation scale—the Evaluation of Attention Deficit and Hyperactivity (EDAH) scale—is able to predict performance in a Continuous Performance Test based on Virtual Reality (VR-CPT). The findings may partially explain why the impulsive–hyperactive and the combined presentations of ADHD might be considered as unique and qualitatively different subcategories of ADHD. These results also highlighted the importance of measuring not only the observable behaviors of ADHD individuals, but also the scores in performance tests that are attained by the patients themselves.

Image matching techniques offer valuable opportunities for the construction industry. Sabzevar et al. [7] developed and evaluated an orientation and positioning approach that decreased the variation in camera viewpoints and image transformation on construction sites. The results show that images captured while using this approach had less image transformation in contrast to images not captured using this approach.

Super-resolution reconstruction is an increasingly important area in computer vision. To alleviate the problems that super-resolution reconstruction models based on generative adversarial networks are difficult to train and contain artifacts in reconstruction results, Jiang and Li [8] presented a TSRGAN model which was based on generative adversarial networks. The author redefined the generator network and discriminator network. The experimental results show that the method made the average Peak Signal to Noise Ratio of reconstructed images reach 27.99 dB and the average Structural Similarity Index reach 0.778 without losing too much speed, which was superior to other comparison algorithms in objective evaluation index. What is more, TSRGAN significantly improved subjective visual evaluations. Experimental results prove the effectiveness and superiority of TSRGAN algorithm.

As virtual reality (VR) and the corresponding 3D documentation and modelling technologies evolve into increasingly powerful and established tools for numerous applications in architecture, monument preservation, conservation/restoration and the presentation of cultural heritage, new methods for creating information-rich interactive 3D environments are increasingly in demand. In [9], the authors describe the development of an immersive virtual reality application for the Imperial Cathedral in Königslutter. A specialized technical workflow was developed to build the virtual environment in Unreal Engine 4 (UE4) and integrate the panorama photographs. A simple mechanic was developed using the native UE4 node-based programming language to switch between these two modes of visualization.

Semantic modeling is a challenging task that has received widespread attention in recent years. With the help of mini Unmanned Aerial Vehicles (UAVs), multiview high-resolution aerial images of large-scale scenes can be conveniently collected. In [10], Wei et al. propose a semantic Multi-View Stereo (MVS) method to reconstruct 3D semantic models from 2D images. The graph-based semantic fusion procedure and refinement based on local and global information can suppress and reduce the reprojection error. In the work by Zha et al. [11] a group of images captured from an eye-in-hand vision system carried on a robotic manipulator are segmented by deep learning and geometric features and create a semantic 3D reconstruction using a map stitching method. The results demonstrate that the quality of segmented images and the precision of semantic 3D reconstruction are effectively improved by their method.

Consumer depth cameras bring about cheap and fast acquisition of 3D models. However, the precision and resolution of these consumer depth cameras cannot satisfy the requirements of some 3D face applications. Zhang et al. [12] present a super-resolution method for reconstructing a high resolution 3D face model from a low resolution 3D face model acquired from a consumer depth camera. They evaluated the method both qualitatively and quantitatively, and the experimental results validate their method.

Personalized production is moving the progress of industrial automation forward, and demanding new tools for improving the decision-making of the operators. In [13], the author presents a new, projection-based augmented reality system for assisting operators during electronic component assembly processes. The paper describes both the hardware and software solutions, and depicts the results obtained during a usability test with the new system.

Lip reading recognition is a new technology in the field of human–computer interaction. It is particularly important in a noisy environment and within the hearing-impaired population. This information is a visual language that benefits from Augmented Reality (AR). Wen and Lu [14] implemented the mobile end lip-reading recognition system based on Raspberry Pi for the first time, and the recognition application has reached the latest level of their research. Proved by experimental results, their model has fewer parameters and lower complexity. The accuracy of the model in the test dataset is 86.5%.

Augmented reality (AR) has evolved hand in hand with advances in technology, and today is considered as an emerging technique in its own right. The aim of the study in [15] was to analyze students' perceptions of how useful AR is in the school environment. During the study, a teaching proposal using AR related to the content of some curricular areas was put forward in the framework of the 3P learning model. The participants' perceptions of this technique were analyzed according to each variable, both overall and by gender, via a questionnaire. The initial results indicate that this technique is, according to the students, useful for teaching the curriculum. The conclusion is that AR can increase students' motivation and enthusiasm while enhancing teaching and learning at the same time.

Recently, associations between the release of scents to the visual content of the scenario has been studied. Alraddadi [16] proposed an approach that combines audio and visual contents to automatically trigger scents through an olfactory device using deep learning techniques. The proposed approach can be applied to different virtual environments as long as scents can be associated with visual and auditory content.

Pham et al. [17] develop a construction hazard investigation system leveraging object anatomization on an Interactive Augmented Photoreality platform (iAPR). A prototype is developed and evaluated objectively through interactive system trials with educators, construction professionals, and learners. The findings demonstrate that the iAPR platform has significant pedagogic methods to improve learners' construction hazard investigation knowledge and skills, which improve safety performance.

Feature tracking in image collections significantly affects the efficiency and accuracy of Structure from Motion (SFM). Insufficient correspondences may cause errors. In [18], the author presents a Superpixel-based feature tracking method for structure from motion. The experimental results show that the proposed method achieves better performance with respect to the state of the art methods.

The present study in [19] focuses on determining the performance and scientific production of augmented reality in higher education (ARHE). A total of 552 scientific publications on the Web of Science (WoS) have been analyzed. The results show that scientific productions on ARHE are not abundant; the main limitation of the study is that the results only reveal the status of this issue in the WoS database.

Author Contributions: Z.L., J.-Y.W., N.K. and J.L. all participated in the review of the papers published in this special issue to the same extent, and jointly supervised the quality of the papers. Z.L. was responsible for writing the relevant editorial content. All authors have read and agreed to the published version of the manuscript.

Funding: This research received no external funding.

Acknowledgments: We would like to thank all the authors, the dedicated referees, the editor team of Applied Sciences for their valuable contributions, making this Special Issue a success.

Conflicts of Interest: The authors declare no conflict of interest.

References

1. Lee, J.; Ha, S.; Gentet, P.; Hwang, L.; Kwon, S.; Lee, S. A Novel Real-Time Virtual 3D Object Composition Method for 360° Video. *Appl. Sci.* **2020**, *10*, 8679. [CrossRef]
2. Ma, C.; Zhang, S.; Wang, A.; Qi, Y.; Chen, G. Skeleton-Based Dynamic Hand Gesture Recognition Using an Enhanced Network with One-Shot Learning. *Appl. Sci.* **2020**, *10*, 3680. [CrossRef]
3. Kim, J.Y.; Kim, M.J. Exploring Visual Perceptions of Spatial Information for Wayfinding in Virtual Reality Environments. *Appl. Sci.* **2020**, *10*, 3461. [CrossRef]
4. Lorusso, M.L.; Travellini, S.; Giorgetti, M.; Negrini, P.; Reni, G.; Biffi, E. Semi-Immersive Virtual Reality as a Tool to Improve Cognitive and Social Abilities in Preschool Children. *Appl. Sci.* **2020**, *10*, 2948. [CrossRef]
5. Wang, R.; Wang, X.; He, D.; Wang, L.; Xu, K. FCN-Based 3D Reconstruction with Multi-Source Photometric Stereo. *Appl. Sci.* **2020**, *10*, 2914. [CrossRef]
6. Areces, D.; Rodríguez, C.; García, T.; Cueli, M. Is an ADHD Observation-Scale Based on DSM Criteria Able to Predict Performance in a Virtual Reality Continuous Performance Test? *Appl. Sci.* **2020**, *10*, 2409. [CrossRef]
7. Foroughi Sabzevar, M.; Gheisari, M.; Lo, J. Development and Assessment of a Sensor-Based Orientation and Positioning Approach for Decreasing Variation in Camera Viewpoints and Image Transformations at Construction Sites. *Appl. Sci.* **2020**, *10*, 2305. [CrossRef]
8. Jiang, Y.; Li, J. Generative Adversarial Network for Image Super-Resolution Combining Texture Loss. *Appl. Sci.* **2020**, *10*, 1729. [CrossRef]
9. Walmsley, A.P.; Kersten, T.P. The Imperial Cathedral in Königslutter (Germany) as an Immersive Experience in Virtual Reality with Integrated 360° Panoramic Photography. *Appl. Sci.* **2020**, *10*, 1517. [CrossRef]

10. Wei, Z.; Wang, Y.; Yi, H.; Chen, Y.; Wang, G. Semantic 3D Reconstruction with Learning MVS and 2D Segmentation of Aerial Images. *Appl. Sci.* **2020**, *10*, 1275. [CrossRef]
11. Zha, F.; Fu, Y.; Wang, P.; Guo, W.; Li, M.; Wang, X.; Cai, H. Semantic 3D Reconstruction for Robotic Manipulators with an Eye-In-Hand Vision System. *Appl. Sci.* **2020**, *10*, 1183. [CrossRef]
12. Zhang, F.; Zhao, J.; Wang, L.; Duan, F. 3D Face Model Super-Resolution Based on Radial Curve Estimation. *Appl. Sci.* **2020**, *10*, 1047. [CrossRef]
13. Ojer, M.; Alvarez, H.; Serrano, I.; Saiz, F.A.; Barandiaran, I.; Aguinaga, D.; Querejeta, L.; Alejandro, D. Projection-Based Augmented Reality Assistance for Manual Electronic Component Assembly Processes. *Appl. Sci.* **2020**, *10*, 796. [CrossRef]
14. Wen, J.; Lu, Y. Automatic Lip Reading System Based on a Fusion Lightweight Neural Network with Raspberry Pi. *Appl. Sci.* **2019**, *9*, 5432. [CrossRef]
15. López-García, A.; Miralles-Martínez, P.; Maquilón, J. Design, Application and Effectiveness of an Innovative Augmented Reality Teaching Proposal through 3P Model. *Appl. Sci.* **2019**, *9*, 5426. [CrossRef]
16. Alraddadi, S.; Alqurashi, F.; Tsaramirsis, G.; Al Luhaybi, A.; Buhari, M.S. Aroma Release of Olfactory Displays Based on Audio-Visual Content. *Appl. Sci.* **2019**, *9*, 4866. [CrossRef]
17. Pham, H.C.; Dao, N.-N.; Cho, S.; Nguyen, P.T.; Pham-Hang, A.-T. Construction Hazard Investigation Leveraging Object Anatomization on an Augmented Photoreality Platform. *Appl. Sci.* **2019**, *9*, 4477. [CrossRef]
18. Cao, M.; Jia, W.; Lv, Z.; Zheng, L.; Liu, X. Superpixel-Based Feature Tracking for Structure from Motion. *Appl. Sci.* **2019**, *9*, 2961. [CrossRef]
19. López Belmonte, J.; Moreno-Guerrero, A.-J.; López Núñez, J.A.; Pozo Sánchez, S. Analysis of the Productive, Structural, and Dynamic Development of Augmented Reality in Higher Education Research on the Web of Science. *Appl. Sci.* **2019**, *9*, 5306. [CrossRef]

applied sciences

MDPI

Article

A Novel Real-Time Virtual 3D Object Composition Method for 360° Video

Jaehyun Lee [1], Sungjae Ha [2], Philippe Gentet [2], Leehwan Hwang [1], Soonchul Kwon [3] and Seunghyun Lee [4,*]

1. Department of Plasma Bio-Display, Kwangwoon University, Seoul 01897, Korea; noa6142@kw.ac.kr (J.L.); hlh2143@kw.ac.kr (L.H.)
2. Spatial Computing Convergence Center, Kwangwoon University, Seoul 01897, Korea; sungjae@kw.ac.kr (S.H.); pgentet@kw.ac.kr (P.G.)
3. Graduate School of Smart Convergence, Kwangwoon University, Seoul 01897, Korea; ksc0226@kw.ac.kr
4. Ingenium College, Kwangwoon University, Seoul 01897, Korea

* Correspondence: shlee@kw.ac.kr; Tel.: +82-940-5290

Received: 26 October 2020; Accepted: 30 November 2020; Published: 4 December 2020

Abstract: As highly immersive virtual reality (VR) content, 360° video allows users to observe all viewpoints within the desired direction from the position where the video is recorded. In 360° video content, virtual objects are inserted into recorded real scenes to provide a higher sense of immersion. These techniques are called 3D composition. For a realistic 3D composition in a 360° video, it is important to obtain the internal (focal length) and external (position and rotation) parameters from a 360° camera. Traditional methods estimate the trajectory of a camera by extracting the feature point from the recorded video. However, incorrect results may occur owing to stitching errors from a 360° camera attached to several high-resolution cameras for the stitching process, and a large amount of time is spent on feature tracking owing to the high-resolution of the video. We propose a new method for pre-visualization and 3D composition that overcomes the limitations of existing methods. This system achieves real-time position tracking of the attached camera using a ZED camera and a stereo-vision sensor, and real-time stabilization using a Kalman filter. The proposed system shows high time efficiency and accurate 3D composition.

Keywords: virtual reality; 3D composition; pre-visualization; stereo vision; 360° video

1. Introduction

Three-hundred-and-sixty-degree video is receiving attention as highly immersive virtual reality (VR) content, where users can observe all viewpoints in their desired direction from the fixed position where the video is recorded, through the intentions of the videographer (who dictates environment position and height). Such video has been used to create highly realistic virtual environments not only in the media industry, including the capture of live performances, movies, and broadcasting, but also in education and games. It can provide a higher sense of immersion to users through the insertion of a computer-graphics-based virtual object, and subsequent user interaction with this inserted virtual object. These techniques have become essential elements for VR content. Typical examples include synthesizing virtual characters or objects in VR movies or displaying information markers in a 3D virtual space. This technique of inserting virtual objects into 360° video is called 3D composition.

In general, 360° video is viewed by wearing a head-mounted display (HMD). Many people experience physical discomfort and symptoms such as headaches, disorientation, and nausea when they wear an HMD [1]. This is VR motion sickness. One of the reasons this occurs is due to the user receiving insufficient updates regarding sensory information from the vestibular system [2]. When 360° video content includes fast camera movement, visual information keeps changing but the user's actual body position is fixed, which causes motion sickness. For this reason, most 360° video clips are taken from a fixed position. Synthesizing a virtual object into a fixed 360° video clip does not require a long processing time. The clip can be inserted at the desired position from the center of the camera. There have recently been various types of VR content used in film, education, and tourism which include stable movements filmed using special drones or cars. In the case of a 360° video clip including camera motion, a process of synchronizing the motion of the Red-Green-Blue (RGB) camera (actual camera) and a virtual camera is applied for the 3D composition. This process works by extracting internal (focal length) and external (position and rotation) parameters from the RGB camera used to capture a real scene [3,4]. From these parameters, we can retrieve the motion of the RGB camera, and this is called camera tracking [5]. The traditional 3D composition method estimates the trajectory of the camera by analyzing the feature points of each frame from the captured images. This method has a disadvantage in that the video resolution and camera-tracking processing times are proportional, and the composition results can only be confirmed after several processes (e.g., recording and camera tracking).

In this paper, we propose a novel method using stereo vision that can extract a depth map in real-time for 3D composition, rather than the traditional method using captured images.

2. Background Theory and Related Studies

2.1. 3D Composition

For a realistic 3D composition, it is mandatory that the RGB camera in the real space and the virtual camera in the virtual space have the same viewpoint. In the traditional method, the internal and external parameters can be estimated by searching the feature points from bright spots and dark spots and analyzing the feature point correspondence between each frame. Typical examples of this include simultaneous localization and mapping (SLAM) [6–8] and structure-from-motion (SfM) [9,10]. The external parameters extracted by these algorithms can be linked with virtual cameras in various 3D programs such as 3D Max and Maya, as applied in video production, and the Unity 3D and Unreal engines for game production. Figure 1 shows a traditional 3D composition method.

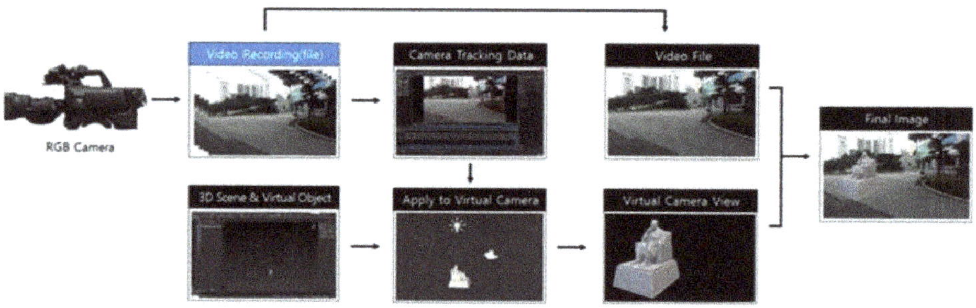

Figure 1. The traditional process using camera-tracking software (Boujou, After Effects) for creating a 3D composition by extracting feature points and estimating camera trajectory from video frames. The blue box shows the recording step (production) and the black boxes show the post-recording steps (post-production).

In general, the 3D composition method tends to depend on the camera-tracking result. Therefore, if the camera-tracking process fails the video must be reshot, which wastes time and money. In previous studies, we reported that various factors may lead to the failure of camera tracking, including an occlusion by a person or object, and motion blur caused by fast camera movement [11]. However, this is more likely to occur in a 2D video shot with relatively numerous camera movements. For a 360° video clip there is a low possibility of camera tracking failures from such factors because stable camera movements are applied to prevent user motion sickness when wearing an HMD. Nevertheless, there is a factor that has not yet been mentioned, caused by a difference in the production processes between 2D and 360° video. In 360° video more than two cameras are used for capturing each different camera view, and after recording in real-time a 360° panoramic view is created through a matching process called "stitching", which overlaps parts from each video clip [12]. During this stitching process errors can occur as a result of inaccurate matching due to lens distortion. These errors interfere with the tracking of the feature points in a 360° video clip containing camera movement. As a result, accurate 3D composition is hindered, and human resources are wasted. Figure 2 shows such stitching errors.

Figure 2. Errors of stitching in a 360° video.

There have been various studies undertaken with the aim of solving this problem. Most of them use a method of applying camera tracking to perspective views of a 360° video clip before the stitching process. One such method proposed by Michiels et al. uses a perspective view from one of the 360° camera rigs to obtain an undistorted image for eliminating the stitching errors [13]. In addition, Huang et al. proposed a method for obtaining stable tracking results, which uses an image correction by overlapping the point where the distortion occurs with the position difference between frames [14]. Furthermore, tracking algorithms for spherical images such as spherical scale invariant feature transform (SSIFT) [15] and spherical oriented fast and rotated brief (SPHORB) have been developed [16]. These methods can reduce the stitching errors caused by a misplaced feature point, but basically, it is progressed from the recorded video. In addition, most 360° video clips have a high resolution of more than 4K, which means a significant amount of time is consumed in camera tracking.

2.2. Stereo Vision

Representative algorithms for estimating the location of a device in real space and generating a map of the surrounding environment are simultaneous localization and mapping (SLAM) [4–6] and visual inertial odometry (VIO) [17,18]. SLAM and VIO can be applied to different types of sensors such as stereo vision, time-of-flight (ToF), and lidar, depending on the environment. Among them, stereo vision uses two cameras to extract the depth map and calculate the three-dimensional position of the feature point to calculate the relative motion. It has the advantage of being relatively inexpensive when compared with lidar and it can measure a wider distance than ToF [19].

In this paper, we used a ZED, which was developed by Stereo Lab [20]. A ZED is a stereo-vision device which uses the SLAM algorithm to provide various software tools, a software development kit (SDK) to generate 3D environment mapping and point clouds from real scenes for estimating position tracking in real-time. Various studies have been conducted on the accuracy of ZED. Ibragimov et al. investigated various Robot Operating System (ROS)-based visual SLAM methods and analyzed their feasibility for a mobile robot application in a homogeneous indoor environment. It was verified that the odometry errors of the ZED are as low as those of lidar [21]. In addition, Alapetite et al. compared the ZED with OptiTrack to analyze its accuracy [22].

In this study, we used the real-time positional tracking value of the ZED as the external parameter value of a mounted 360° camera. In addition, we converted the extracted data into a script suitable for a 3D program (e.g., 3D Max, Maya, Unity) to create a virtual camera.

2.3. Related Studies

There have been various studies relating to the 3D composition of virtual objects in 360° video clips. These studies are based on VR, augmented reality, and mixed reality (MR). Focusing on research on 360° video, Rhee et al. implemented a real-time lighting and material expression of virtual objects, according to the positional change reconstructing the camera trajectory from the captured 360° video [23]. Furthermore, the proposed MR360 is used to synthesize virtual objects with real background images. However, it is based on a fixed 360° video, and thus it differs from our proposed method, which contains camera movement [24].

Similarly, Tarko et al. implemented real-time 3D composition using the Unity game engine through a stabilization process after camera tracking [25]. However, camera tracking was based on the captured image. Here, real-time indicates a real-time composition in a 3D program after the tracking process, not during the recording step. Our proposed method is a real-time composition performed at the same time as the video recording.

We recently proposed a novel system that uses Microsoft HoloLens to track positions precisely for match-moving techniques [11] and studied a virtual camera for making motion-graphics using transformed data from the ZED [26]. In this paper, we propose a stabilized 3D composition system and a pre-visualization system using the ZED based on these previous studies.

3. Proposed System and Experiment

In this paper, we propose a novel system that uses ZED stereo vision to track the trajectory precisely for 3D composition in a 360° video. The proposed system also includes a pre-visualization system that can be confirmed to result from a 3D composition while recording the 360° video. Figure 3 shows the complete workflow of the proposed system.

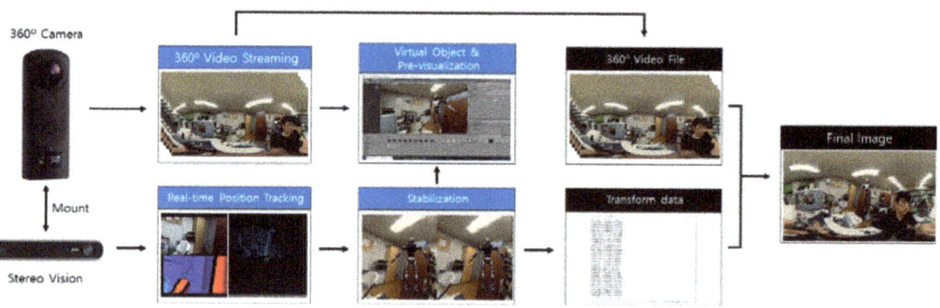

Figure 3. The proposed system workflow using stereo vision for extracting the external parameters of the 360° cameras, which were mounted together. The blue boxes show the recording step (production) and the black boxes show the post-recording steps (post-production).

3.1. Real-Time 3D Composition Using Stereo Vision

In our proposed system, we use a 360° camera "Z1", developed by Ricoh-theta [27]. Z1 can record in 4K (3840 × 2160). It can also use real-time video streaming with stitching to 3D programs such as Unity and Unreal. This 360° camera system and the ZED mounted on a rig are connected to a PC through a USB 3.0 port. In addition, the ZED is configured such that it faces the same direction as the front of the 360° camera. The 360° camera is used to record the images of the real scene, and at the same time, the ZED extracts the external parameter by generating a depth map in real-time. The ZED generates the initial value of position data (0,0,0) when the program starts, so the difference in the physical distance between the ZED and Z1 is not considered. Figure 4 shows the rig-mounted 360° camera and the ZED.

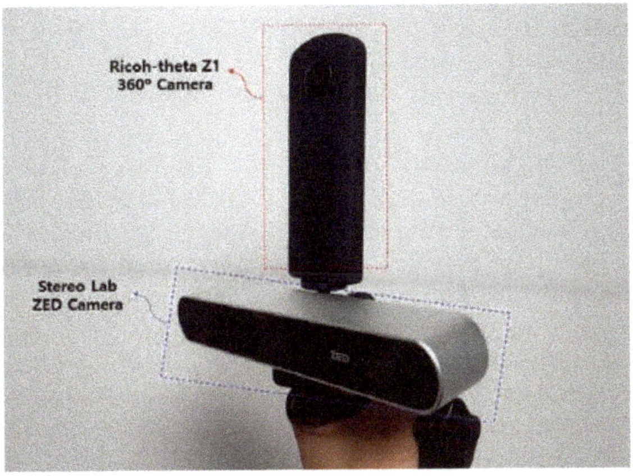

Figure 4. 360° camera and stereo-vision ZED camera.

The extraction and saving of external stereo-vision parameters are applied within Unity 3D, which is used for simultaneous processing with a pre-visualization system to confirm the composition result. For our method, we propose a stabilization process for external parameters in order to obtain better performance from the noise that generally contains stereo vision. The external parameters extracted from the ZED are saved as new data through a linear Kalman filter in real-time.

The Kalman filter is an algorithm that was developed by Kalman during the early 1960s [28,29]. It is used to track the optimal value by removing the noise included in the measured value using the prior and prediction states. It consists of a prediction step and an update step. In the prediction step, an expected value is calculated when the input value is received, according to the prior estimated value. In the update step, an accurate value is calculated based on the prior predicted value and the actual measured value. In other words, a correct value is derived by repeatedly applying the prediction and update steps. It is suitable for real-time processing because it makes predictions based on the immediately preceding data, rather than all previous data [30–32].

The trajectory data stabilized through the Kalman filter can be saved in various formats for application to 3D programs during post-production. In this paper, we saved the data using the 3ds Max file scripting language (.ms) to create a virtual camera in 3ds Max. Figure 5a shows the 3ds Max script file and Figure 5b shows the 3D composition in the 3ds Max program.

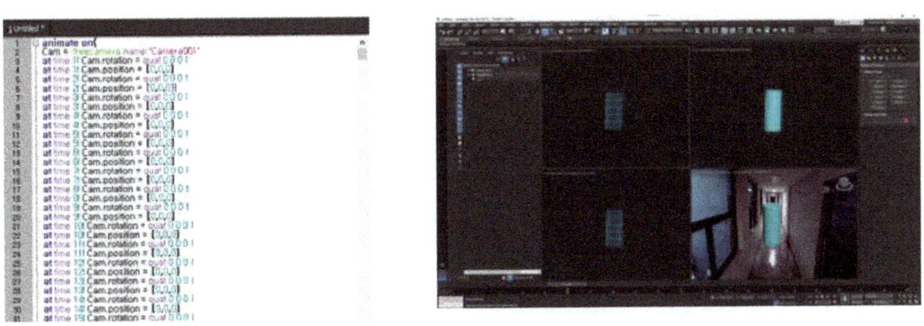

(a) (b)

Figure 5. 3D composition process in 3ds Max: (**a**) Max script and (**b**) 3ds Max scene.

To measure the accuracy of the camera trajectory with a Kalman filter, the traditional tracking method using an RGB camera was set to the ground truth, in order to compare the applied Kalman filter and raw data of the camera trajectory. The use of the traditional tracking method as a ground truth—even if it is not the best—allows us to show that the proposed method has the same camera trajectory accuracy as the traditional method.

3.2. Pre-Visualization

The purpose of the pre-visualization system is to confirm the composition result while recording the 360° video. For this purpose, we connect the 360° camera and stereo-vision ZED to a PC through a USB 3.0 port to send a video signal and trajectory data within the 3D program. In this study, we used the Unity game engine, which synchronizes the external parameters using the virtual camera from the ZED and generates a 360° virtual space for streaming the 360° camera video feed of the texture of a spherical object in real-time. The spherical object is set to 2.5 m in radius so as not to interfere with the placement of the virtual object. It also follows the virtual camera. It streams the video feed at 4K resolution at 60 fps, with a delay of 0.212 s. If the frame rate and time code do not match, the 3D composition will fail. To avoid this, the update function in Unity is set to 60 updates per second using FixedUpdate which has a static update rate, and a 0.212 s delay is given to the ZED data to match the time code.

The pre-visualization system uses simple 3D objects such as a box, cylinder, and a human-shaped figure. The real-time lighting and texture composition mentioned in various studies can be applied to our proposed method, although the purpose of our system is to confirm the possibility of such composition, and not perfect its application. Therefore, our system does not consider real-time lighting and texture composition techniques. Figure 6 shows the pre-visualization system and a simple 3D object.

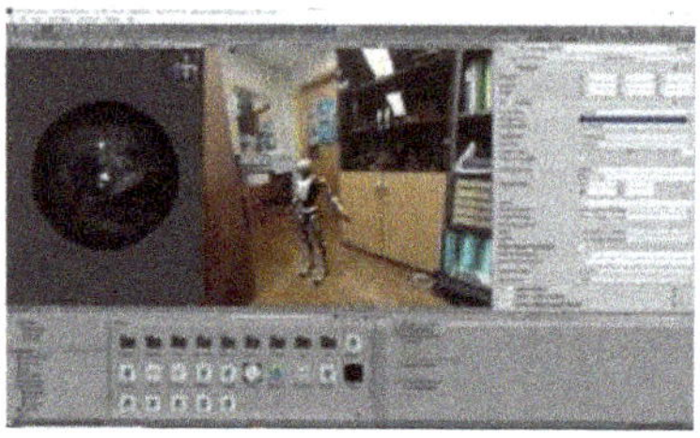

(**a**)

(**b**)

Figure 6. Pre-visualization system: (**a**) pre-visualization Unity scene and (**b**) simple 3D object.

4. Experimental Results

In our proposed system, in order to measure the camera trajectory and verify the composition of the pre-visualization system, we recorded two different 360° video clips, indoors and outdoors. The scenes were captured for duration of 26 s and 19 s at rate of 60 fps. Figure 7 shows the 360° images recorded.

(**a**)

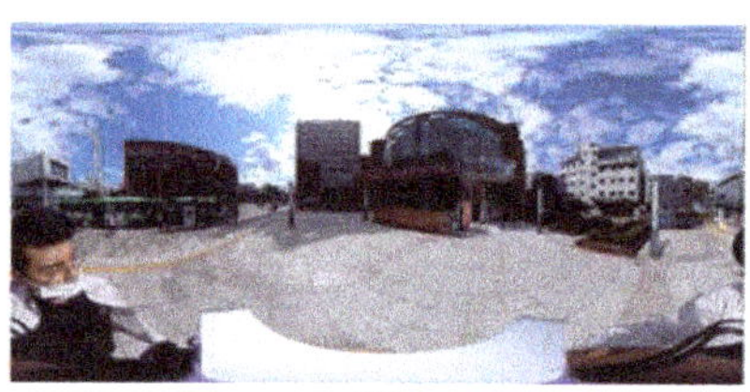

(**b**)

Figure 7. Experimental space for recording of 360° video: (**a**) indoor and (**b**) outdoor.

4.1. Camera Trajectory

The camera trajectory experiment was undertaken to show the efficiency of the proposed system through comparison with the traditional method of extracting camera trajectory, and additionally to show the improved accuracy of camera trajectory using the Kalman filter. Therefore, the proposed system and an RGB camera were used simultaneously for extracting each camera trajectory. The camera trajectory of the traditional method was set as the ground truth. For various camera movements, we used only hands without special equipment such as a stabilizer. Figure 8a,c shows the camera trajectory extracted from the ZED in comparison with the ground truth, which was recorded using the RGB camera. Figure 8b,d shows the camera trajectory extracted from the ZED with a Kalman filter in comparison with the ground truth. The deviations in percentage error calculated for both raw trajectory data and trajectory data with a Kalman filter, in comparison with the ground truth, are shown in Table 1. From Figure 8 and Table 1, it can be seen that the camera trajectory extracted from the ZED with a Kalman filter is mostly aligned with the ground truth, with a percentage error of less than 3.1%. In addition, the raw camera trajectory data extracted from the ZED is also mostly aligned with the ground truth. However, position X indoors shows a percentage error of 11.8%. By contrast, the Kalman filter shows a percentage error of 2.6%, which is less than that of the raw data.

As a result, it can be seen that the data extracted from the ground truth using the traditional method and the stereo-vision approach do not show a significant difference. This indicates that the proposed method achieved significant results for real-time composition. However, as can be seen in Table 1, the trajectory data following application of the Kalman filter show a lower difference from the ground truth when compared to the traditional method for all data. This indicates that applying the Kalman filter is more effective in preventing noise in the stereo-vision sensor and obtaining stable data.

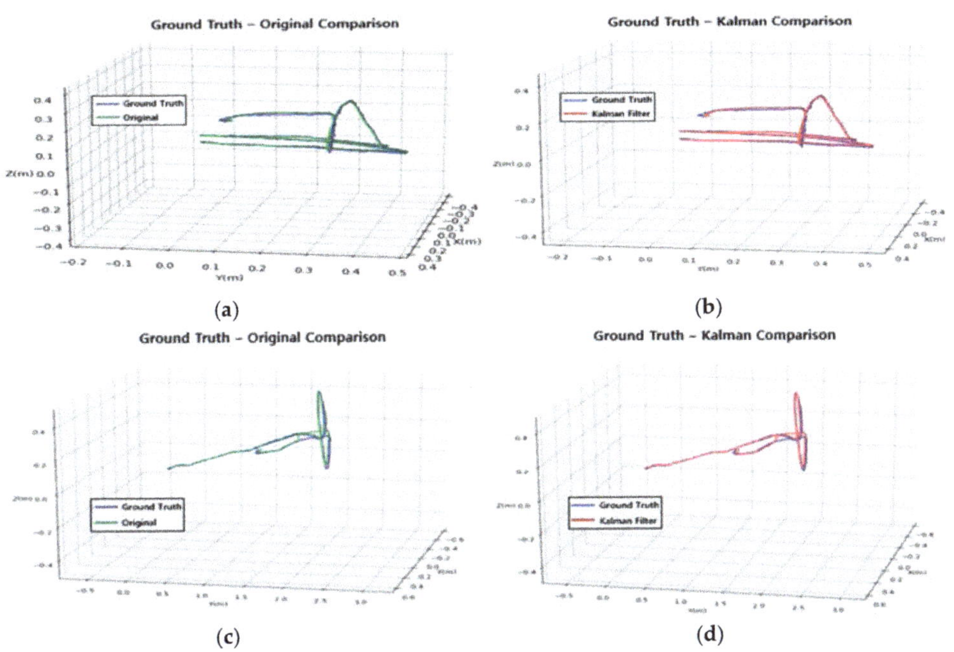

Figure 8. Accuracy evaluation of the camera trajectory extracted from (**a**,**c**) the ZED and (**b**,**d**) the applied Kalman filter against the ground truth (Boujou).

Table 1. Standard deviation in the comparison of the ground truth, raw trajectory data, and trajectory with the Kalman filter.

		Position X (%)	Position Y (%)	Position Z (%)
Indoor	Ground truth–raw trajectory	11.81081	1.875021	0.547256
	Ground truth–trajectory with Kalman filter	2.6780439	0.432794	0.112748
Outdoor	Ground truth–raw trajectory	1.740435	3.604414	0.147483
	Ground truth–trajectory with Kalman filter	1.537084	3.093616	0.077608

4.2. 3D Composition Using Pre-Visualization System

At the same time as the recording, a 360° video clip and the external parameters of the stereo vision were transmitted to the Unity 3D game engine to create a virtual camera for pre-visualization. Figure 9 shows the results of the pre-visualization of the indoor and outdoor scenes while recording the 360° video. The result displayed through the pre-visualization system was used to confirm the composition result. For the final video, further composition processes such as lighting, shadowing, and texturing in 3D software are needed.

(a)

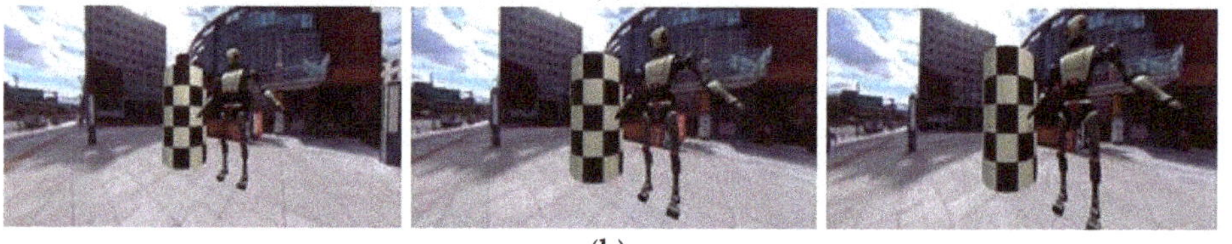

(b)

Figure 9. Results of pre-visualization: (**a**) indoor and (**b**) outdoor.

The final composition was conducted in 3ds Max 2018. When the recording was finished, the 3ds Max script, which included the trajectory information of the stereo vision, was immediately generated. It was used to create a virtual camera in the 3ds Max virtual space. Figure 10 shows the rendered images and the final 3D composition images. No difference can be seen in the camera trajectory because it uses the same trajectory data saved from a real-time pre-visualization system. As a result, it does not need an extra process for extracting the camera-tracking data, and thus our proposed system is more time efficient than the traditional method.

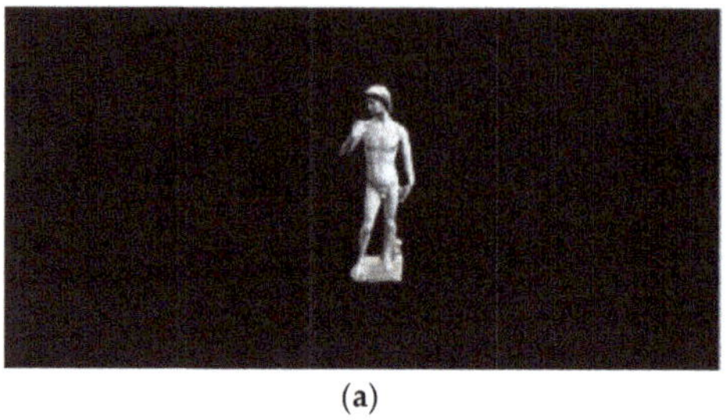

(a)

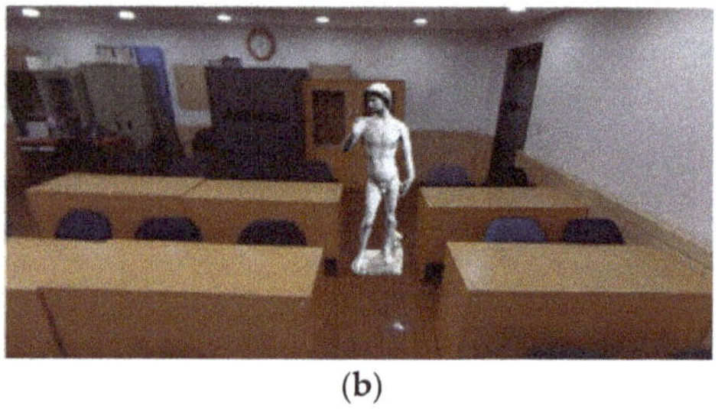

(b)

Figure 10. *Cont.*

(c)

(d)

Figure 10. Rendered images and final 3D compositing images: (**a**,**b**) indoor and (**c**,**d**) outdoor.

5. Conclusions

In this paper we proposed a real-time 3D composition method for 360° video production. The proposed system consists of two subsystems. Firstly, a stereo-vision ZED is used to obtain the parameters of the external cameras, which are mounted together to estimate the camera trajectory in real-time. Secondly, an efficient pre-visualization system is implemented to preview the results of the 3D composition during the recording.

In this study, we exploited a system that overcomes the limitations of the traditional method, which uses camera tracking after video recording. Our experimental results show that the 3D composition results of the proposed system are not significantly different than the results obtained using the traditional method. In addition, we implemented a stable trajectory by applying a Kalman filter to the raw data obtained from the ZED. The Kalman filter achieved better trajectory results than the raw data. Our system has an advantage over the traditional method because it does not need to extract feature points from the captured images. It can save the data of the external parameters during the recording process, and this was also verified in the composition results. However, as a limitation of the proposed system, it works using a USB port and not a network. In the future, the authors plan to implement a network communication system by installing a network device that will be able to send video and transformed data to a PC for further processing.

It can be predicted that, with the advancement of the virtual reality industry, interest in the 3D composition of 360° video will also increase, and therefore a more efficient system will be required. We expect that the system presented herein will be applicable for the effective 360° video production of 3D composition in low-budget production companies.

Author Contributions: Conceptualization, J.L.; methodology, J.L., P.G.; software, J.L.; validation, J.L., L.H., and S.K.; formal analysis, J.L. and S.H.; investigation, J.L.; resources, S.K., S.H. and S.L.; data curation, J.L., P.G. and L.H.; writing—original draft preparation, J.L. and L.H.; writing—review and editing, S.K., S.H. and S.L.; visualization, J.L., L.H. and P.G.; supervision, S.K. and S.H.; project administration, S.L., P.G.; funding acquisition, S.L., S.H., and S.K. All authors have read and agreed to the published version of the manuscript.

Funding: This research received no external funding.

Acknowledgments: This research was supported by the Ministry of Science and ICT (MSIT), Korea, under the Information Technology Research Center (ITRC) support program (IITP-2020-2015-0-00448) & (IITP-2020-01846) supervised by the Institute of Information & Communications, Technology, Planning, & Evaluation (IITP). This research was supported by Institute of Information & communications Technology Planning & Evaluation (IITP) grant funded by the Korea government (MSIT) (No.2020-0-00922, Development of holographic stereogram printing technology based on multi-view imaging).

Conflicts of Interest: The authors declare that they have no conflict of interest.

References

1. Munafo, J.; Diedrick, M. The virtual reality head-mounted display Oculus Rift induces motion sickness and is sexist in its effects. *Exp. Brain Res.* **2017**, *235*, 889–901. [CrossRef] [PubMed]
2. Jung, S.; Whangbo, T. Study on inspecting VR motion sickness inducing factors. In Proceedings of the 2017 4th International Conference on Computer Applications and Information Processing Technology (CAIPT), Kuta, Bali, 8–10 August 2017.
3. Erica, H. *The Art and Technique of Matchmoving: Solutions for the VFX Artist*, 1st ed.; Elsevier: New York, NY, USA, 2010; pp. 1–14, ISBN 9780080961132.
4. Dobbert, T. The Matchmoving Process. In *Matchmoving: The Invisible Art of Camera Tracking*, 1st ed.; Sybex: San Francisco, CA, USA, 2005; pp. 5–10, ISBN 0782144039.
5. Pollefey, M.; Van, G.L.; Vergauwen, M.; Verbiest, F.; Cornelis, K.; Tops, J. Visual modeling with a hand-held camera. *Int. J. Comput. Vis.* **2004**, *59*, 207–232. [CrossRef]
6. Davison, A.J. Real-Time Simultaneous Localisation and Mapping with a Single Camera. In Proceedings of the Ninth IEEE International Conference on Computer Vision, Nice, France, 13–16 October 2003.
7. Klein, G.; Murray, D. Parallel Tracking and Mapping for Small AR Workspaces. In Proceedings of the 2007 6th IEEE and ACM International Symposium on Mixed and Augmented Reality, Nara, Japan, 13–16 November 2007; pp. 225–234.
8. Davison, A.J. SLAM++: Simultaneous Localisation In addition, Mapping at the Level of Objects. In Proceedings of the IEEE Conference on Computer Vision and Pattern Recognition, Portland, OR, USA, 23–28 June 2013; pp. 1352–1359.
9. Bao, S.Y.; Savarese, S. Semantic Structure from Motion. In Proceedings of the IEEE CVPR 2011, Colorado Springs, CO, USA, 20–25 June 2011.
10. Hafeez, J.; Jeon, H.J.; Hamacher, A.; Kwon, S.C.; Lee, S.H. The effect of patterns on image-based modelling of texture-less objects. *Metrol. Meas. Syst.* **2018**, *25*, 755–767.
11. Lee, J.; Hafeez, J.; Kim, K.; Lee, S.; Kwon, S. A novel real-time match-moving method with HoloLens. *Appl. Sci.* **2019**, *9*, 2889. [CrossRef]
12. Huang, K.-C.; Chien, P.-Y.; Chien, C.-A.; Chang, H.-C.; Guo, J.-I. A 360-degree panoramic video system design. In *Proceedings of the Technical Papers of 2014 International Symposium on VLSI Design, Automation and Test*; Institute of Electrical and Electronics Engineers (IEEE): Piscataway, NJ, USA, 2014; pp. 1–4.
13. Michiels, N.; Jorissen, L.; Put, J.; Bekaert, P. Interactive Augmented Omnidirectional Video with Realistic Lighting. *Public Key Cryptogr. PKC 2018* **2014**, *8853*, 247–263. [CrossRef]
14. Huang, J.; Chen, Z.; Ceylan, D.; Jin, H. 6-DOF VR videos with a single 360-camera. In Proceedings of the 2017 IEEE Virtual Reality (VR), Los Angeles, CA, USA, 18–22 March 2017.
15. Cruz-Mota, J.; Bogdanova, I.; Paquier, B.; Bierlaire, M.; Thiran, J. Scale invariant feature transform on the sphere: Theory and applications. *Int. J. Comput. Vis.* **2012**, *98*, 217–241. [CrossRef]
16. Zhao, Q.; Feng, W.; Wan, L.; Zhang, J. SPHORB: A fast and robust binary feature on the sphere. *Int. J. Comput. Vis.* **2015**, *113*, 143–159. [CrossRef]
17. Leutenegger, S.; Lynen, S.; Bosse, M.; Siegwart, R.; Furgale, P. Keyframe-based visual–inertial odometry using nonlinear optimization. *Int. J. Robot. Res.* **2015**, *34*, 314–334. [CrossRef]
18. Sun, K.; Mohta, K.; Pfrommer, B.; Watterson, M.; Liu, S.; Mulgaonkar, Y.; Taylor, C.J.; Kumar, V. Robust stereo visual inertial odometry for fast autonomous flight. *IEEE Rob. Autom. Lett.* **2018**, *3*, 965–972. [CrossRef]
19. Vit, A.; Shani, G. Comparing RGB-D sensors for close range outdoor agricultural phenotyping. *Sensors* **2018**, *18*, 4413. [CrossRef] [PubMed]
20. ZED. Available online: https://www.stereolabs.com/zed/ (accessed on 10 October 2020).
21. Ibragimov, I.Z.; Afanasyev, I.M. Comparison of ROS-based Visual SLAM Methods in Homogeneous Indoor Environment. In Proceedings of the 2017 14th Workshop on Positioning, Navigation and Communications (WPNC), Bremen, Germany, 25–26 October 2017; pp. 1–6.
22. Alapetite, A.; Wang, Z.; Hansen, J.P.; Zajączkowski, M.; Patalan, M. Comparison of three off-the-shelf visual odometry systems. *Robotics* **2020**, *9*, 56. [CrossRef]
23. Iorns, T.; Rhee, T. Real-Time Image Based Lighting for 360-Degree Panoramic Video. In *Image and Video Technology—PSIVT 2015 Workshops*; Huang, F., Sugimoto, A., Eds.; PSIVT 2015, Lecture Notes in Computer Science; Springer: Cham, Switzerland, 2016; Volume 9555.

24. Rhee, T.; Petikam, L.; Allen, B.; Chalmers, A. MR360: Mixed reality rendering for 360° panoramic videos. *IEEE Trans. Vis. Comput. Graph.* **2017**, *23*, 1379–1388. [CrossRef] [PubMed]
25. Iorns, T.; Rhee, T.H. Real-Time Image Based Lighting for 360-Degree Panoramic Video. In Proceedings of the PSIVT Workshops, Auckland, New Zealand, 23–27 November 2015; pp. 139–151.
26. Kim, L.H.; Lee, J.H.; Kim, K.J.; Lee, S.H. A study on motion graphics virtual camera using real-time position tracking in post-production. *J. Mov. Image Technol. Assoc. Korea* **2019**, *1*, 133–149.
27. Z1. Available online: https://theta360.com/en/about/theta/z1.html (accessed on 10 October 2020).
28. Kalman, R.E. A new approach to linear filtering and prediction problem. *J. Basic Eng.* **1960**, *82*, 34–45. [CrossRef]
29. Welch, G.; Bishop, G. *An Introduction to the Kalman Filter*; Lecture; University North Carolina: Chapel Hill, NC, USA, 2001.
30. Prabhu, U.; Seshadri, K.; Savvides, M. Automatic Facial Landmark Tracking in Video Sequences using Kalman Filter Assisted Active Shape Models. In Proceedings of the European Conference on Computer Vision, Heraklion, Greece, 5–11 September 2010.
31. Chen, S.Y. Kalman filter for robot vision: A survey. *IEEE Trans. Ind. Electron.* **2012**, *59*, 4409–4420. [CrossRef]
32. Smeulders, A.W.M.; Chu, D.M.; Cucchiara, R.; Calderara, R.; Dehghan, A.; Shah, M. Visual tracking: An experimental survey. *IEEE Trans. Pattern Anal. Mach. Intell.* **2014**, *36*, 1442–1468. [CrossRef] [PubMed]

Publisher's Note: MDPI stays neutral with regard to jurisdictional claims in published maps and institutional affiliations.

applied sciences

MDPI

Article

Skeleton-Based Dynamic Hand Gesture Recognition Using an Enhanced Network with One-Shot Learning

Chunyong Ma [1,2,*], Shengsheng Zhang [1], Anni Wang [1], Yongyang Qi [1] and Ge Chen [1,2]

[1] College of Information Science and Engineering, Ocean University of China, Qingdao 266100, China; zhangshengsheng@stu.ouc.edu.cn (S.Z.); wanganni@stu.ouc.edu.cn (A.W.); qiyongyang@ouc.edu.cn (Y.Q.); gechen@ouc.edu.cn (G.C.)

[2] Laboratory for Regional Oceanography and Numerical Modeling, Qingdao National Laboratory for Marine Science and Technology, Qingdao 266200, China

* Correspondence: chunyongma@ouc.edu.cn

Received: 12 April 2020; Accepted: 20 May 2020; Published: 24 May 2020

Abstract: Dynamic hand gesture recognition based on one-shot learning requires full assimilation of the motion features from a few annotated data. However, how to effectively extract the spatio-temporal features of the hand gestures remains a challenging issue. This paper proposes a skeleton-based dynamic hand gesture recognition using an enhanced network (GREN) based on one-shot learning by improving the memory-augmented neural network, which can rapidly assimilate the motion features of dynamic hand gestures. Besides, the network effectively combines and stores the shared features between dissimilar classes, which lowers the prediction error caused by the unnecessary hyper-parameters updating, and improves the recognition accuracy with the increase of categories. In this paper, the public dynamic hand gesture database (DHGD) is used for the experimental comparison of the state-of-the-art performance of the GREN network, and although only 30% of the dataset was used for training, the accuracy of skeleton-based dynamic hand gesture recognition reached 82.29% based on one-shot learning. Experiments with the Microsoft Research Asia (MSRA) hand gesture dataset verified the robustness of the GREN network. The experimental results demonstrate that the GREN network is feasible for skeleton-based dynamic hand gesture recognition based on one-shot learning.

Keywords: one-shot learning; gesture recognition; GREN; skeleton-based

1. Introduction

With the rapid development of Kinect, Leap Motion, and other sensors in recent years, hand motion capture is getting much more efficient. By estimating the posture of the hand gesture, the position information of each joint can be detected from video or image sequences. Recent research [1–5] has tried various ways for dynamic hand gesture recognition based on 3D skeleton data characterized as strong correlations, temporal continuity, and co-occurrence relationships. Besides, the skeleton-based algorithm has fewer parameters, which is easier to calculate and more suitable for analyzing dynamic hand gestures. However, it is still challenging because hands are non-rigid objects, which can express a variety of different semantics [6]. With the gesture recognition technology being applied in more fields such as gaming and industry training, it is often necessary to make different customized annotation samples in large sizes. However, it is worth noting that the existing hand gesture database could not meet the needs of gesture interaction in various fields. The cost of large-scale gesture sample extraction artificially in each field is so high that it would limit the application of gesture recognition [7,8]. Meanwhile, the traditional gradient-based networks also require extensive iterative training to complete the model optimization. When encountering the new data, the models need to relearn their hyper-parameters to adequately incorporate the new information without catastrophic interference [9], which is inefficient. The existing networks fail to complete the optimization of the model with small size training samples, while one-shot learning could infer results as expected [10]. Therefore, the method of one-shot learning can be used to solve the problem that the model could not be optimized by the insufficient samples of skeleton-based dynamic hand gestures.

However, if the current algorithm of "one-shot learning" is directly applied to the hand gesture recognition, there will be three gradient-based optimization problems. Firstly, due to the small amount of data, many advanced and mature algorithms, such as Momentum [11] and Adagrad [12], cannot be optimized in limited iterations; especially when encountering non-convex problems, many hyper-parameters cannot achieve convergence. Secondly, for different tasks, the parameters of the network need to be initialized randomly. If the amount of data is too small, the final model cannot achieve convergence. This can be alleviated by conducting transferring learning methods, such as fine-tuning [13,14]. Finally, for the traditional neural network, its memory storage is limited. Additionally, the process of learning a new set of patterns will suddenly and completely erase a network's knowledge of what it had already learned, which is referred to as catastrophic interference [15]. Therefore, we need to find a memory module that can be used for large-scale storage and can also be accessed for relevant information. The large capacity enhanced memory neural networks, such as a neural Turing machine (NTM) [16], provides a feasible method for one-shot learning combined with hand gesture recognition. The NTM provides the capability to quickly encode and retrieve new information by limiting the changes in the output of the network before and after the network update [15,17]. In addition, it can also eliminate gradient-based optimization problems. On this basis, Santoro [9] introduced a new and pure content-based method for accessing an external memory, which is different from previous methods, additionally using a memory location-based focusing mechanism. The method can rapidly bind never-before-seen information to the external memory after a single presentation and combines the gradient descent to slowly learn an abstract method for obtaining useful representations of raw data. As a result, it can accurately identify the categories of data that have occurred only once.

This paper focuses on the architecture of enhanced neural networks based on skeleton-based algorithms and one-shot learning. Based on the memory-augmented neural network (MANN) [9], we propose skeleton-based dynamic hand gesture recognition using an enhanced network (GREN). The long short-term memory (LSTM) network is selected as the controller of the GREN network to enhance the recognition and memory ability of the network. Compared with the MANN network, which was originally applied to image recognition, the proposed GREN network classifies hand gestures by identifying skeletal sequences. Through the recognition of the GREN network, we conduct experiments on a dynamic hand gesture dataset (DHGD) [18] to show the effectiveness of our method. Then, we implement our method on the Microsoft Research Asia (MSRA) hand gesture dataset [19] to verify its contributions.

The rest of this paper is organized as follows:

- Section 2 details the related work of skeleton-based dynamic hand gesture recognition and one-shot learning.
- The GREN network is introduced in Section 3.
- The experiments of skeleton-based dynamic hand gesture recognition are explained in detail in Section 4.
- In Section 5, results and discussion are presented.
- The conclusions are given in Section 6.

2. Related Work

2.1. Skeleton-Based Dynamic Hand Gesture Recognition

Much research has been focused on skeleton-based dynamic hand gesture recognition [20–29]. Chen X. et al. [30] proposed a skeleton-based dynamic hand gesture recognition algorithm that has also been suggested to surpass depth-based methods in the aspect of performance. Chin-Shyurng et al. [31] created a skeleton-based model by capturing the palm position, and the dynamic time-warping algorithm was applied to the recognition of disparate conducting gestures at various conducting speeds, which achieves real-time dynamic musical conducting gesture recognition. Ding, Ing-Jr et al. [32] designed an adaptive hidden Markov model (HMM)-based gesture recognition method with user adaptation (UA) to simplify large-scale video processing to realize the natural user interface (NUI) of a humanoid robot device. Similarly, Kumar, Pradeep et al. [33] used the HMM to identify occluded gestures in line with a robust position invariant sign language recognition (SLR) framework.

Additionally, some studies have employed deep learning methods to conduct skeleton-based dynamic hand gesture recognition. Mazhar, Osama et al. [34] proposed that humans need neither to wear any specific clothing (motion capture clothes or inertial sensors) nor to carry a special remote control or learn complex teaching instructions in gesture recognition. As a result, they developed a real-time, robust, and background-independent gesture detection module in the light of convolutional neural network (CNN) transmission learning. Chen, XH et al. [29] exploited motion features of traits and global movements to augment features of recurrent neural networks (RNNs) for gesture recognition and improve the classification performance. Lin, C et al. [35] proposed a novel refined fused model in combination with the masked Res-C3D network and skeleton LSTM for abnormal gesture recognition in RGB-D videos, which learns discriminative representations of gesture sequences in particular abnormal gesture samples by fusing multiple characteristics from different models. Based on a combination of a CNN network and an LSTM network, Nunez, JC et al. [36] proposed a deep learning-based approach for temporal 3D pose recognition problems, and the proposed network architecture does not need to be adapted to the type of activity or the gesture to be recognized, as well as the geometry of the 3D sequence data as input. So far, there is no available deep learning network that can be directly used for skeleton-based dynamic hand gesture recognition based on small size samples.

2.2. One-Shot Learning

The implementations of one-shot learning can be divided into statistics-based, weight-based matching, and meta-learning. For the statistics-based, Lake [37] adopted the Bayesian framework realized one-shot learning of handwritten character pictures based on the statistical point of view and the way humans learn things, triggering the new wave of one-shot learning.

Besides the above statistics, there are also many methods on the basis of weighted matching for one-shot learning, which performs certain criteria modeling on known samples and then determines the class according to the distance of samples. The most typical method is the k-nearest neighbor (KNN), which is a nonparametric estimation method that can directly employ distance to determine the category without prior training. Another method is to learn an end-to-end nearest neighbor classifier, which can not only quickly learn new samples but also have a great generalization of known samples. Snell et al. [38] carried out classification by calculating the distance from prototype representations of each class, which turns into the nearest neighbor classification in the metric space. While Koch et al. [39] performed efficacious feature extraction on new samples by limiting input methods, then used supervised metric learning based on twin networks to train and finally reused features extracted by that network for small or no sample learning. Similarly, Oriol Vinyals et al. [40] also utilized metric learning based on deep neuro features, which uses external memory to enhance the neural network that maps a small labeled support set and an unlabeled example to its label, obviating the need for fine-tuning to adapt to new class types.

Meta-learning, also known as "learning to learn", aims to train a model on a variety of learning tasks, such that it can solve new learning tasks using only a small number of training samples [41]. A neural network with memory can implement meta-learning, but its memory storage is limited. A large number of new features may exceed the memory storage capacity so that the network cannot learn new tasks. The NTM network can solve this problem, as it is capable of both long-term storage via slow updates of its weights and short-term storage via its external memory module [16]. Based on the NTM network, Santoro et al. [9] introduced a memory access module that emphasizes accurate encoding of relevant (recent) information and pure content-based retrieval to implement meta-learning. Besides, Ravi et al. [42] proposed an LSTM-based meta-learner model, whose parameterization allows it to learn appropriate parameter updates specifically for the scenario where a set amount of updates will be made, while also learning a general initialization of another learner (classifier) network that allows for quick convergence of training.

In general, the current one-shot learning-based methods are in a booming period. However, there is still no appropriate method for one-shot learning with skeleton-based hand gesture recognition. Therefore, this paper will study the current advanced achievements and propose a suitable algorithm to realize hand gesture recognition in line with one-shot learning.

3. Dynamic Hand Gesture Recognition with the GREN Network

By improving a MANN network, this paper implements the GREN network based on one-shot learning, which is a variant of the NTM network from Santoro et al. [9]. Compared with the MANN network originally applied to image recognition, the proposed GREN network classifies hand gestures by recognizing skeletal sequences. The structure of the GREN network is shown in Figure 1.

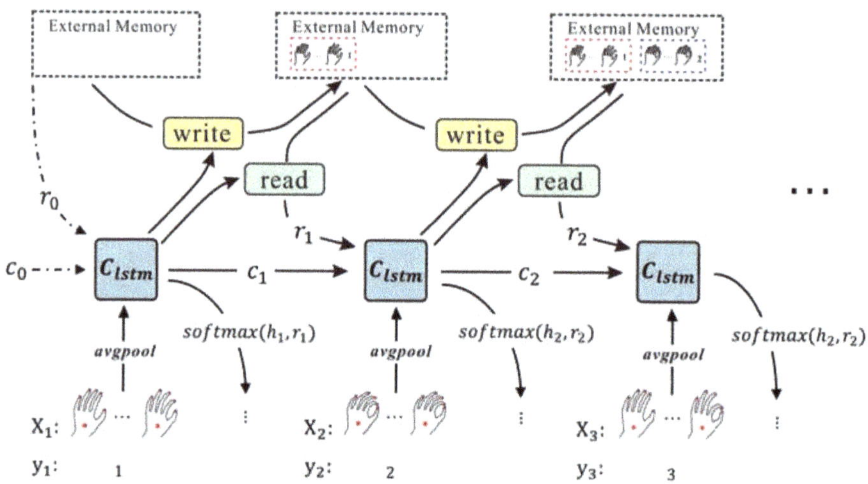

Figure 1. The structure of the GREN network. For the current time-step t, it takes the hand joint coordinate sequence X_t and the corresponding sample-class y_t as input and outputs the categorical distribution of prediction by a softmax layer. The controller, neuron C_{lstm}, generates h_t and c_t, which are the hidden state and the cell state of the LSTM used for the next time-step. A memory, r_t, is retrieved by the read heads from the external memory.

The GREN network consists of three components: a controller, read and write heads, and an external memory. The controller, neuron C_{lstm}, employed in our model is an LSTM network, which receives the current input and controls the read- and write-heads to interact with the external memory, respectively. Memory encoding and retrieval in an external memory are rapid, with vector representations being placed into or taken out of memory potentially every time-step [16], which makes it a perfect candidate for one-shot prediction. Additionally, it can be stored either for long-term storage by slowly updating the weights or for short-term storage by an external memory. Thus, when the model learns the type of representation of a gesture sequence, it will be placed into memory, and later these representations will be used to make predictions of data that it has only seen once. Besides, according to the difference of classification methods between the input of images and sequences, the average pooling layer (avgpool) is introduced to further focus on characteristics of sequence and improve the calculation efficiency in the network. For one-shot learning, the output distribution is categorical, which is implemented as a softmax function.

At the beginning, the initialized state of the GREN network is represented by *init_state*. The external memory is initialized, which does not store any data representations. Also, the memory r_0 retrieved from the external memory is empty. In addition, the cell state of the initialized controller, neuron C_{lstm}, is represented by c_0. Given the input sequence X_t, the controller receives the memory r_{t-1} and cell state c_{t-1} provided by the previous state *prev_state*, then produces a query key vector k_t used to retrieve a particular memory. When encountering sequences of the already-seen class, the particular memory vector row could be retrieved by read heads, which is addressed using the cosine similarity measure:

$$\mathrm{K}(k_t, M_t(i)) = \frac{k_t \cdot M_t(i)}{||k_t|| \cdot ||M_t(i)||} \tag{1}$$

where M_t is the memory matrix at time-step t and $M_t(i)$ is the ith row in this matrix. The row of $M_t(i)$ serve as memory "slots", with the row vectors themselves constituting individual memories.

After then, a read-weight vector w_t^r is produced by these similarity measures according to the softmax function:

$$w_t^r(i) \leftarrow \frac{\exp(\mathrm{K}(k_t, M_t(i)))}{\sum_j \exp(\mathrm{K}(k_t, M_t(j)))} \tag{2}$$

where the read heads can amplify or attenuate the precision of the focus by the read weights.

Those read weights w_t^r and corresponding memory $M_t(i)$ are used to retrieve the memory r_t:

$$r_t \leftarrow \sum_i w_t^r(i) \cdot M_t(i) \tag{3}$$

where the memory r_t is used by the controller as both an input to a classifier, namely, a softmax layer for class prediction and as an additional input for the next input sequence.

To achieve the combined learning in disparate classes and implement the one-shot learning, the least recently used access module (LRUA) proposed by Adam Santoro [9] is adopted, which is a pure content-based memory write head that writes memories to either the least used memory location or the most recently used one, and focusing on the accurate encoding of the most relevant information. In terms of a new sequence, it is written to a rarely-used location with the recently encoded information preserved or to the last used location, which can be used for updating with newer or possibly more relevant information:

$$w_t^u \leftarrow \gamma \cdot w_{t-1}^u + w_t^r + w_t^w \tag{4}$$

$$w_t^{lu}(i) = 1 \; if \; w_t^u \leq m\left(w_t^u, n\right) else \; 0 \tag{5}$$

$$w_t^w \leftarrow \sigma(\alpha) \cdot w_{t-1}^r + (1 - \sigma(\alpha)) \cdot w_{t-1}^{lu} \tag{6}$$

$$M_t(i) \leftarrow M_{t-1}(i) + w_t^w(i) \cdot k_t \cdot \forall i \tag{7}$$

where w_t^u is the usage weight updated at each time-step to keep track of locations most recently read from or written to; γ is the decay parameter; w_t^{lu} is the least-used weight computed using w_t^u for a given time-step; the notation $m(v, n)$ is introduced to denote the nth smallest element of the vector v; n is set to equal the number of the writer to memory; w_t^w is the written weight computed by the sigmoid function $\sigma(.)$, which combines the previous read weights w_{t-1}^r and previous least-used weights w_{t-1}^{lu}; α is a dynamic scalar gate parameter to interpolate between weights. Before writing to memory, the least used memory location is computed from w_{t-1}^u and set it to zero, then the memory M_t is written by the computed vector of written weights w_t^w. Thus, $M_t(i)$ can be written into the zeroed memory location or the previously used memory location; if it is the latter, then w_t^{lu} will simply get erased.

With the above analysis, we propose the following GREN algorithm, as shown in Algorithm 1.

Algorithm 1: GREN

Input: Given N samples $\{X_1, X_2, \ldots, X_N\}$ belonging to C classes with
Sample-classes $y_t \in Y = \{1, \ldots, C\}$, $for\ t = 1, \ldots, N$;
Output: A softmax layer for class prediction;
Initialization:
$prev\_state \leftarrow init\_state(N)\{$
    $c_0 \leftarrow C_{lstm}(N)$;
    $r_0 \leftarrow 0_{N\times(head\_num*memory\_size)}$;
    $w_0^r \leftarrow one\_hot\_weigh\_vector(N, head\_num, memory\_slots)$;
    $w_0^u \leftarrow one\_hot\_weigh\_vector(N, memory\_slots)$;
    $M_0 \leftarrow \varepsilon_{N\times memory\_slots\times memory\_size}$;
    $return\ \{c_0, r_0, w_0^r, w_0^u, M_0\}$;
$\}$;
$o = [\,]$;
for $t \leftarrow 1$ **to** N **do**
    $h_t, c_t \leftarrow C_{lstm}((X_t, y_t), prev\_state)$;
    for $i \leftarrow 0$ **to** $X_t.length$ **do**
        $output, curr\_state \leftarrow gren((X_t(i), x\_lable_t(i)), prev\_state)\{$
            Memory Retrieval:
                $K(k_t, M_t(i)) \leftarrow cosine\_similarity(k_t, M_t(i))$;
                $w_t^r(i) \leftarrow softmax(K(k_t, M_t(i)))$;
                $r_t += w_t^r(i) \cdot M_t(i)$;
            Memory Encoding (LRUA):
                $w_t^u \leftarrow \gamma \cdot w_{t-1}^u + w_t^r + w_t^w$;
                if $w_t^u \leq m\left(w_t^u, n\right)$ **then** $w_t^{lu}(i) = 1$ **else** $w_t^{lu}(i) = 0$;
                $w_t^w \leftarrow sigmoid(\alpha) \cdot w_{t-1}^r + (1 - sigmoid(\alpha)) \cdot w_{t-1}^{lu}$;
                $M_t(i) \leftarrow M_{t-1}(i) + w_t^w(i) \cdot k_t$;
            $return\ \{h_t, r_t\}, \{c_t, r_t, w_t^r, w_t^u, M_t\}$;
        $\}$;
        $prev\_state = curr\_state$;
        if $i == 0$ **then**
            $o2o\_w \leftarrow (output.length, M_{class}), rand\_unif\_init(minv, maxv)$;
            $o2o\_b \leftarrow (M_{class}), rand\_unif\_init(minv, maxv)$;
        end if;
        $output = output \cdot o2o\_w + o2o\_b$;
        $output = softmax(output)$;
        $o.append(output)$;
    end
    $learning\_loss = -cross\_entropy\_cost(y_t, o)$;
    $optimizer = AdamOptimizer(learning\_rate)$;
    $train\_op = optimizer.minimize(learning\_loss)$;
end

In the algorithm, the $one\_hot\_weigh\_vector(a, b, c)$ function generates a tensor of shape $a \times b \times c$ with $[:, :, 0]$ set to *one* (or $[:, 0]$, if the $one\_hot\_weigh\_vector(a, b)$ function generates a tensor of shape $a \times b$); $\{(a, b), rand\_unif\_init(minv, maxv)\}$ generates a tensor of shape $a \times b$ (or $\{(a), rand\_unif\_init(minv, maxv)\}$ generates a tensor of shape $a \times 1$) with a uniform distribution, and the value of all elements is set between *minv* and *maxv*.

In general, for the current time-step t, the sample data X_t and the corresponding sample-class y_t will be received by the controller C_{lstm}. The current state of the GREN network *curr_state* is used by the controller as an additional input for the next time-step. According to each sequence of the sample, the GREN algorithm randomly generates the class label x_label_t. If the sample date X_t comes from a never-before-seen class, it will be bound to the appropriate sample-class y_t and stored by the write heads in the external memory, which is presented in the subsequent time-step (see Figure 1). Later, once a sample from an already-seen class is presented, the controller will retrieve the bound sample-class information by the read heads from the external memory for class prediction. A softmax layer, $softmax(\cdot)$, is selected to output the standardized probability distribution of the model prediction, and combined with the cross-entropy cost function, *cross_entropy_cost*$(\cdot)$, to measure the loss between the predicted value and correct class label. Then, the adaptive moment estimation (Adam) [43], *AdamOptimizer*$(\cdot)$, is adopted to minimize the loss, and the back-propagated error signal from the current prediction updates those previous weights, which is followed by the updating of the external memory. Those processes would be repeated until the model converges.

4. Experiments

In this section, two hand gesture datasets named dynamic hand gesture database (DHGD) and MSRA are used for the experiments. Details about the experimental setup of the GREN network are introduced in the later part of this section.

4.1. Datasets

4.1.1. DHGD Hand Gesture Dataset

The public DHGD hand gesture dataset [18] contains sequences for 14 right-hand gestures performed in two ways: using one finger and the whole hand. Each class of gestures is performed 1 to 10 times by 28 participants in both of the above two ways, resulting in 2800 sequences, and the length of the gestures varies from 20 to 50 frames. Each frame contains the coordinates of the 22 joints in the 2D depth image space and 3D world space, and those joints are shown in Figure 2.

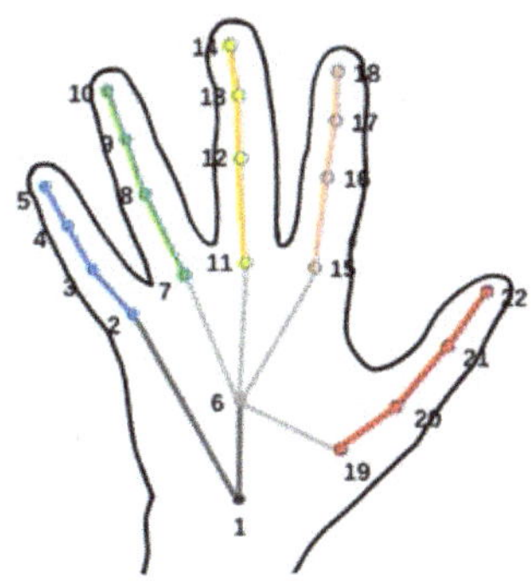

Figure 2. Twenty-two joints of a right-hand skeleton.

Some gestures (such as swipe and shake), which are defined by the movement of the hand, called the coarse gesture, while others are defined by the shape of the gesture, called the fine gesture. Table 1 shows the different classes of gestures in DHGD:

Table 1. List of 14 gestures in the dynamic hand gesture database (DHGD).

Name of the Gesture	Type of the Gesture	Type of the Gesture
1	Grab	Fine
2	Tap	Coarse
3	Expand	Fine
4	Pinch	Fine
5	Rotation Clockwise	Fine
6	Rotation Counter Clockwise	Fine
7	Swipe Right	Coarse
8	Swipe Left	Coarse
9	Swipe Up	Coarse
10	Swipe Down	Coarse
11	Swipe X	Coarse
12	Swipe +	Coarse
13	Swipe V	Coarse
14	Shake	Coarse

4.1.2. MSRA Hand Gesture Dataset

The public MSRA [19] hand gesture dataset, which contains skeleton-based sequence data of 17 right-hand gestures performed by 28 participants, is chosen to verify the robustness of the GREN network. The 17 right-hand gestures are manually chosen and are mostly from American Sign Language, to span the space of finger articulation as much as possible. Additionally, the length of each gesture varies from 490 to 500 frames. Each of these frames contains the coordinates of the 21 joints in the 2D depth image space and 3D world space, and those joints are shown in Figure 3.

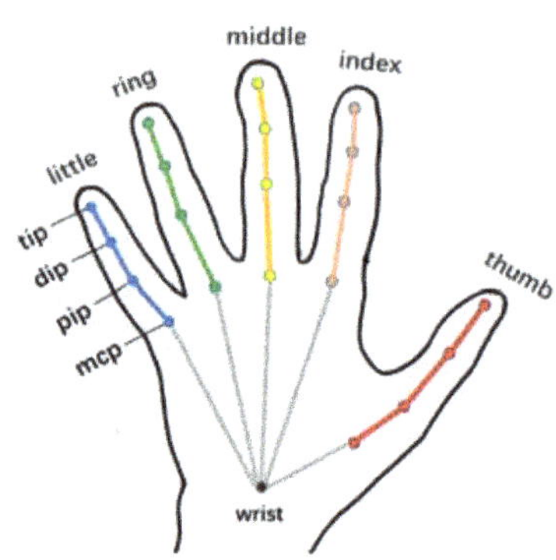

Figure 3. Twenty-one joints of a right-hand skeleton.

4.2. Experimental Setup

4.2.1. Data Pre-Process

The skeleton-based hand gesture datasets should be preprocessed as the input of our network. The whole framework of the data preprocessing is shown in Figure 4, in which the kth class gesture is processed by our method as an example.

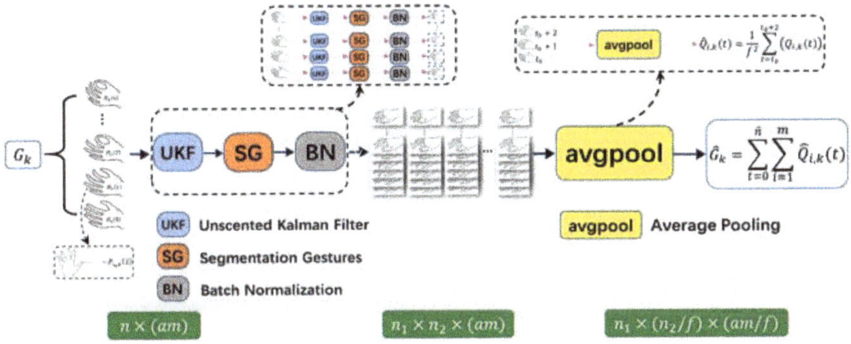

Figure 4. Framework of the data preprocessing.

First of all, the nested interval unscented Kalman filter (UKF) [44] is used to eliminate the possible noise in the hand gesture datasets. Moreover, due to some hand gesture datasets may contain unequal sequences from different participants, the short and long sequences should be changed into a standard sequence. The length of the standard sequence is set to a fixed value n based on both the average length of the sequence of each gesture. For short sequences, the length of them is increased by linear interpolation. For long sequences, we will eliminate the first few frames and the last few frames of the sequence because there are usually many pause actions at the beginning and the end, and they are not important to the whole gesture. The joint $P_{i,k}(t)$, a full hand skeleton $H_k(t)$ and the kth class gesture G_k are shown as follows:

$$P_{i,k}(t) = \left[x_{i,k}(t), y_{i,k}(t), z_{i,k}(t)\right] \tag{8}$$

$$H_k(t) = \sum_{i=1}^{m} P_{i,k}(t) \tag{9}$$

$$G_k = \sum_{t=0}^{n} H_k(t) \tag{10}$$

where n is the scale of the kth class gesture sequences; all of the joints i in one hand are combined into a full hand skeleton $H_k(t)$ when the time scale of the kth class gesture is at t; m represents the maximum number of joints in a full hand skeleton; the shape of the kth class gesture G_k is processed into $n \times (am)$; the feature scale is am, and a is the spatial scale.

The shape of the standard sequence is split into $n_1 \times n_2 \times (am)$ through the segmentation gestures (SG), where the kth class gesture forms n_1 sets of sequences and the time scale of each set is n_2.

Then, the skeleton-based hand gesture sequences can be mapped to the same specific interval by normalizing the changing hand joints, which is effective to improve the convergence rate of our network:

$$\mu_B \leftarrow \frac{1}{m} H_k(t) \tag{11}$$

$$\sigma_B^2 \leftarrow \frac{1}{m} \sum_{i=1}^{m} \left(P_{i,k}(t) - \mu_B\right)^2 \tag{12}$$

$$\hat{P}_{i,k}(t) \leftarrow \frac{P_{i,k}(t) - \mu_B}{\sqrt{\sigma_B^2 + \varepsilon}} \tag{13}$$

where μ_B is the mean of the sample and σ_B^2 is the sample variance; The linear transformation is added to these sequences and normalizes them to obtain $\hat{P}_{i,k}(t)$, which limits the distribution of them and makes the network more stable during training; ε is the role of the minimum number, which avoids zero in the denominator in the expression.

The network may lose its original feature representation capabilities by the normalization. A pair of learnable parameters γ and β are set for each normalization to eliminate hidden dangers, which is used to restore the original distribution to obtain $Q_{i,k}(t)$.

$$Q_{i,k}(t) \leftarrow \gamma \cdot \hat{P}_{i,k}(t) + \beta \equiv BN_{\gamma,\beta}\big(P_{i,k}(t)\big) \tag{14}$$

In the formula, $BN_{\gamma,\beta}\big(P_{i,k}(t)\big)$ is represented as a complete batch normalization (BN).

Additionally, the joint coordinates of the hand skeleton-based sequences are limited by the neighborhood, which increases the variance of the estimate and is not conducive to enhancing network learning. The average pooling layer (avgpool) can solve the above problems, which makes the structure of the skeleton-based sequence simpler and more stable, improves the calculation efficiency of the network, and avoids over-fitting during training. Here $Q_{i,k}(t)$ is introduced to represent the changes in the same joints of the adjacent multiple frames after the avgpool:

$$\hat{Q}_{i,k}(t) = \frac{1}{f^2} \sum_{t=t_0}^{t_0+2} \big(Q_{i,k}(t)\big) \tag{15}$$

$$\hat{G}_k = \sum_{t=0}^{\hat{n}} \sum_{i=1}^{m} \hat{Q}_{i,k}(t) \tag{16}$$

where f is the size of a filter of the average pooling layer; the size of $\hat{n}$ is set to the equal of $n_1 * (n_2/f)$; the shape of $\hat{G}_k$ is split into $n_1 \times (n_2/f) \times (am/f)$, which contains the features information of the kth class gesture, and as the input sequence of our network.

Finally, for one-shot learning, only a small part of the hand gesture datasets was taken as the training samples for subsequent experiments.

4.2.2. Implementation

The whole process of dynamic hand gesture recognition based on one-shot learning is shown in Figure 5.

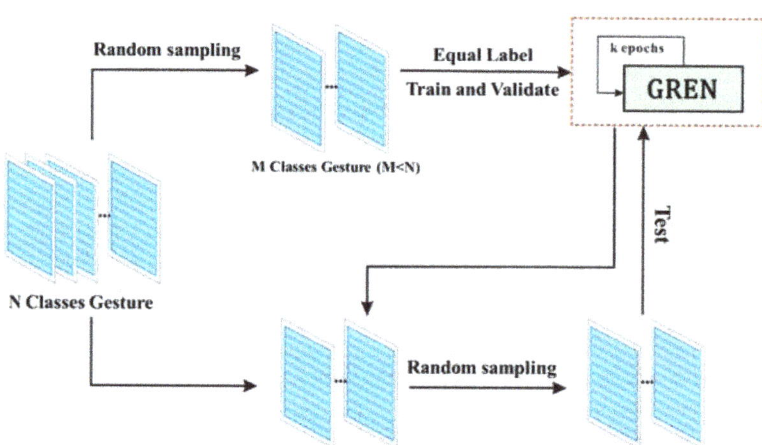

Figure 5. Flowchart of the implementation.

Firstly, the M different classes are randomly selected from the N classes already contained in the dataset, which prevents the network from simply mapping class labels to the output. From the episode to the next episode, those classes presented in the current episode with the associated labels and specific samples will be shuffled. Later, the sample sequences are equally singled out from each of the M classes, which are supposed to be of the same size. Each group from the randomly re-labeled M classes extracts 10 sets of sequences as the training data at random. Of course, it is not enough to take merely 10 sets of the sequence for each training. Additionally, the corresponding batch size is taken by random sampling as the input of training. Then, the model will be validated by the validation set every k epochs, and output the prediction accuracy with corresponding loss. Finally, the above processes are repeated until the model converges.

For the converged network model, the test set will be randomly selected to evaluate its generalization ability. After the test, the model's ability to recognize those new unrecognized sequences will be the criterion of model selection.

According to the public DHGD hand gesture dataset, the time scale is set to 60 so that the size of the gesture sequences will be at least 100 sets in each class. After the data preprocessing, the shape of $\hat{G}_k$ is split into $60 \times 20 \times 22$. For "one-shot learning", 60%, 20%, and 20% of the data are used for the training set, the validation set, and the test set, respectively.

The DHGD dataset contains two different ways of 14-classes gestures: one finger and the whole hand. N is set to 14 as the number of the unique class; M is set to 3 as the number of sample classes; k is set to 100 as the epoch-size in each training. For the 28-classes gestures encompassing the above two ways, N is set to 28 as the number of the unique class, while sizes of M and k remain unchanged in each training. A grid search [45] is performed over a number of hyper-parameters: controller size (200 hidden units for an LSTM), the learning rate (4e − 5), the number of read–write heads from memory (4), and training times (80,000). For the 14-classes, the batch size is taken as 8, while it is set to 16 in the case of 28-classes. The model presents the best results over those hyper-parameters configurations.

In this study, another comparison experiment has been conducted based on the MSRA dataset. The time scale is also set to 12. After the data preprocessing, the shape of $\hat{G}_k$ is segmented into $60 \times 5 \times 21$. Moreover, 50% of the data is used for the training set; 25% of the data utilized for the validation set; 25% of the data applied to the test set. For the MSRA dataset containing hand gestures of 17 classes, N is set to 17 as the number of the unique classes, and sizes of M and k remain unchanged in each training. Compared with the 14-classes and 28-classes, hyper-parameters for the 17-classes are shown: controller size (200 hidden units for an LSTM), the learning rate (4e − 5), the number of read–write heads from memory (4), batch size (16), and training times (70,000).

5. Results and Discussion

To visualize the process of the recognition accuracy measured on the validation set, we have separately analyzed two different ways of 14-classes: one finger and the whole hand, and the 28-classes encompassing both the above two ways. In addition, the accuracy curve is shown in Figure 6.

From Figure 6, the 14-classes, (1) represents right-hand gestures performed with one finger, and (2) represents gestures with the whole hand. The curve of the one-finger classified by our method is shown in blue, the curve of the whole-hand is shown with an orange line, and the curve of the 28-classes is shown with a grey line. It is observed that the recognition accuracy of the 14-classes (2) is superior to the 14-classes (1), and the 28-classes is between those two. Compared with the 14-classes (1), the 28-classes has better performance.

To assess the effectiveness of our algorithm for classifying the hand gestures of DHGD into 14-classes and 28-classes, we compare the standard LSTM network with regard to their DHGD recognition accuracy. Table 2 shows the comparison results of skeleton-based hand gesture recognition between LSTM and GREN networks.

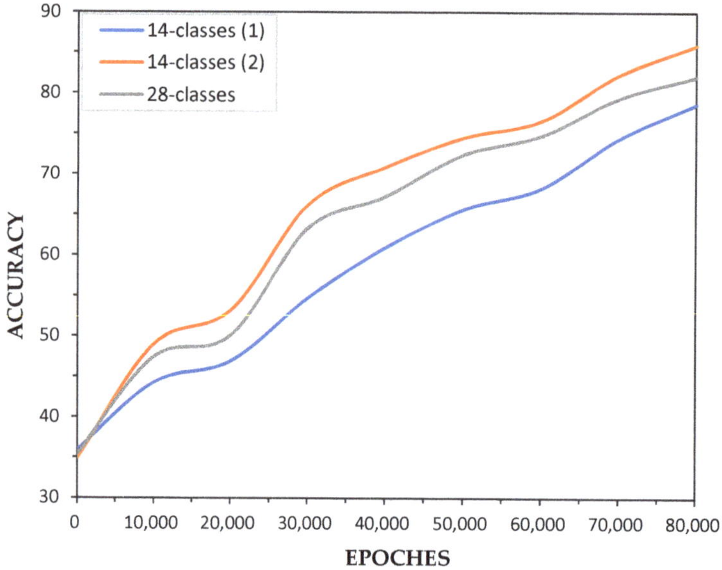

Figure 6. The accuracy curve of our method for 14-classes and 28-classes in the DHGD dataset.

Table 2. Comparison results between long short-term memory (LSTM) and gesture recognition using an enhanced network (GREN) networks based on the DHGD dataset.

Type		LSTM (%)	GREN (%)
14-classes	1	75.18	78.65
	2	79.82	85.90
28-classes	both	76.89	82.03

From Table 2, the final accuracy of our GREN network reaches 82.29% for the 14-classes classification that is the average of the two ways and 82.03% for the 28-classes classification. The proposed network indicates that recognition accuracy can reach 78.65% for the one-finger and 85.90% for the whole-hand. Thus, compared with the standard LSTM networks, the accuracy of the recognition increased by approximately 5.14%, the accuracy of the one-finger increased by approximately 3.47%, and the whole-hand accuracy increased by 6.08%, which show excellent performance of our method in one-shot learning.

We compare the GREN network with the state-of-the-art algorithm in DHGD, and the results are shown in Table 3.

For the different ways of learning, a mature scheme of one-shot learning combined with hand gesture recognition has not been proposed before. Those advanced methods of comparison adopt the way of recognizing large size samples for experiments. While our GREN network uses small size samples in the DHGD dataset and trains based on one-shot learning.

Compared with other advanced algorithms, our method also performs well. For the 14-classes classification, the final accuracy of our GREN network is 82.29%, which is higher than most other algorithms. Additionally, our GREN network presents a higher accuracy in the 28-classes recognition than does that of the other advanced algorithm. A comparison of other advanced algorithms shows that the accuracy of the GREN network will not reduce significantly with the increase of the classes of hand gestures in the 28-classes recognition. Experimental results suggest that the proposed GREN network is an efficient method for hand gesture recognition.

Table 3. Result of different method comparison for 14/28-classes gestures on dynamic hand gesture dataset using skeleton-based data.

Learning	Methods	Accuracy 14-Classes Gestures	Accuracy 28-Classes Gestures
Large-samples	HON4D: Histogram of Oriented 4D Normals for Activity Recognition from Depth Sequences [46]	75.53%	74.03%
	3-D Human Action Recognition by Shape Analysis Of Motion Trajectories on Riemannian Manifold [47]	79.61%	62.00%
	Joint Angles Similarities and HOG2 for Action Recognition [48]	80.85%	76.53%
	Key Frames with Convolutional Neural Network [18]	82.90%	71.90%
	Skeleton-Based Dynamic Hand Gesture Recognition [49]	83.07%	79.14%
	NIUKF-LSTM [44]	84.92%	80.44%
	SL-Fusion-Average [36]	85.46%	74.19%
	MFA-Net [29]	85.75%	81.04%
One-shot	GREN	**82.29%**	**82.03%**

Besides, to verify the robustness of the network, a similar experimental setup has also been performed on the MSRA hand gesture dataset. To more clearly demonstrate our network, we compared the experimental result with the LSTM network based on the MSRA dataset, which is shown in Table 4.

Table 4. Comparison results between LSTM and GREN networks based on the MSRA dataset.

Type	LSTM (%)	GREN (%)
17-classes	72.92	79.17

From Table 4, the final accuracy of our network is 79.17% for the 17-classes classification. Additionally, compared with the LSTM networks, the accuracy of the recognition increased by approximately 6.25%, which shows the better performance of the GREN network. The experiment verifies that this network could be replicated for other similar datasets, even if they are small sample size datasets.

6. Conclusions

This paper proposes the GREN network to recognize dynamic hand gestures based on a small number of skeleton-based sequence samples. According to the MANN network, the ability to store and update sequence data is further enhanced by introducing the average pooling layer (avgpool) and batch normalization (BN), so that we can combine the hand skeleton sequence with the GREN network to achieve dynamic hand gesture recognition based on one-shot learning. Experiments with the DHGD hand gesture dataset demonstrate the state-of-the-art performance of the GREN network for skeleton-based dynamic hand gesture recognition based on one-shot learning. Additionally, the MSRA hand gesture dataset verifies the robustness of our GREN network.

Author Contributions: Conceptualization, Y.Q.; methodology, A.W.; software, S.Z.; supervision, G.C.; validation, Y.Q.; writing—original draft, S.Z.; writing—review and editing, C.M. All authors have read and agreed to the published version of the manuscript.

Funding: This research was funded by Fundamental Research Funds for the Central Universities grant number 201762005, the National Natural Science Foundation of China grant number 41906155, and the Marine S&T Fund of Shandong Province for Pilot National Laboratory for Marine Science and Technology (Qingdao) grant number 2019GHZ023.

Acknowledgments: The authors gratefully acknowledge the support of the Fundamental Research Funds for the Central Universities (Grant No.: 201762005), National Natural Science Foundation of China (Grant No.: 41906155), and Marine S&T Fund of Shandong Province for Pilot National Laboratory for Marine Science and Technology, Qingdao (Grant No.: 2019GHZ023).

Conflicts of Interest: The authors declare no conflict of interest. The funders had no role in the design of the study; in the collection, analyses, or interpretation of data; in the writing of the manuscript, and in the decision to publish the results.

References

1. Si, C.; Chen, W.; Wang, W.; Wang, L.; Tan, T. An attention enhanced graph convolutional lstm network for skeleton-based action recognition. In Proceedings of the IEEE Conference on Computer Vision and Pattern Recognition, Long Beach, CA, USA, 16–20 June 2019; pp. 1227–1236.
2. Lv, Z.; Halawani, A.; Feng, S.; Ur Réhman, S.; Li, H. Touch-less interactive augmented reality game on vision-based wearable device. *Pers. Ubiquitous Comput.* **2015**, *19*, 551–567. [CrossRef]
3. Liu, J.; Wang, G.; Duan, L.; Abdiyeva, K.; Kot, A.C. Skeleton-based human action recognition with global context-aware attention lstm networks. *IEEE Trans. Image Process.* **2018**, *27*, 1586–1599. [CrossRef] [PubMed]
4. Nie, Q.; Wang, J.; Wang, X.; Liu, Y. View-invariant human action recognition based on a 3d bio-constrained skeleton model. *IEEE Trans. Image Process.* **2019**, *28*, 3959–3972. [CrossRef] [PubMed]
5. Lv, Z.; Halawani, A.; Feng, S.; Li, H.; Réhman, S.U. Multimodal hand and foot gesture interaction for handheld devices. *ACM Trans. Multimed. Comput. Commun. Appl.* **2014**, *11*, 10. [CrossRef]
6. Liu, X.; Su, Y. Tracking skeletal fusion feature for one shot learning gesture recognition. In Proceedings of the International Conference on Image, Vision and Computing, Chengdu, China, 2–4 June 2017; pp. 194–200. [CrossRef]
7. Zhu, W.; Lan, C.; Xing, J.; Zeng, W.; Li, Y.; Shen, L.; Xie, X. Co-occurrence feature learning for skeleton based action recognition using regularized deep LSTM networks. In Proceedings of the Thirtieth AAAI Conference on Artificial Intelligence, Phoenix, AZ, USA, 12–17 February 2016; pp. 3697–3703.
8. Liu, J.; Shahroudy, A.; Xu, D.; Wang, G. Spatio-temporal lstm with trust gates for 3d human action recognition. Proceedings of 14th European Conference on Computer Vision, Amsterdam, The Netherlands, 11–14 October 2016; pp. 816–833. [CrossRef]
9. Santoro, A.; Bartunov, S.; Botvinick, M.; Wierstra, D.; Lillicrap, T. Meta-learning with memory-augmented neural networks. In Proceeding of the International Conference on Machine Learning, New York, NY, USA, 19–24 June 2016; pp. 1842–1850.
10. Deng, L.; Yu, D. Deep learning: Methods and applications. *Found. Trends Signal Process.* **2014**, *7*, 197–387. [CrossRef]
11. Besak, D.; Bodeker, D. Hard thermal loops for soft or collinear external momenta. *J. High Energy Phys.* **2010**, *5*, 7. [CrossRef]
12. Duchi, J.; Hazan, E.; Singer, Y. Adaptive subgradient methods for online learning and stochastic optimization. *J. Mach. Learn. Res.* **2011**, *12*, 2121–2159.
13. Howard, J.; Ruder, S. Universal Language Model Fine-tuning for Text Classification. In Proceedings of the 56th Annual Meeting of the Association for Computational Linguistics (Long Papers), Melbourne, Australia, 15–20 July 2018; pp. 328–339. [CrossRef]
14. Bengio, Y. Deep learning of representations for unsupervised and transfer learning. In Proceedings of the ICML Workshop on Unsupervised and Transfer Learning, Edinburgh, UK, 26 June–1 July 2012; pp. 17–36.
15. Kirkpatrick, J.; Pascanu, R.; Rabinowitz, N.C. Overcoming catastrophic forgetting in neural networks. *Proc. Natl. Acad. Sci. USA* **2017**, *114*, 3521–3526. [CrossRef]
16. Greve, R.; Jacobsen, E.J.; Risi, S. Evolving neural turing machines for reward-based learning. In Proceedings of the Genetic and Evolutionary Computation Conference, Denver, CO, USA, 20–24 July 2016; pp. 117–124. [CrossRef]
17. Li, Z.; Hoiem, D. Learning without forgetting. *IEEE Trans. Pattern Anal. Mach. Intell.* **2017**, *40*, 2935–2947. [CrossRef]
18. De Smedt, Q.; Wannous, H.; Vandeborre, J.P.; Guerry, J.; LeSaux, B.; Filliat, D. 3D hand gesture recognition using a depth and skeletal dataset: SHREC'17 track. In Proceedings of the Workshop on 3D Object Retrieval. Eurographics Association, Lyon, France, 23–24 April 2017; pp. 33–38. [CrossRef]

19. Sun, X.; Wei, Y.; Liang, S.; Tang, X.; Sun, J. Cascaded hand pose regression. In Proceedings of the IEEE Conference on Computer Vision and Pattern Recognition, Boston, MA, USA, 7–12 June 2015; pp. 824–832. [CrossRef]
20. Tan, D.J.; Cashman, T.; Taylor, J.; Fitzgibbon, A.; Tarlow, D.; Khamis, S.; Shotton, J.; Izadi, S. Fits like a glove: Rapid and reliable hand shape personalization. In Proceedings of the IEEE Conference on Computer Vision and Pattern Recognition, Las Vegas, NV, USA, 27–30 June 2016; pp. 5610–5619. [CrossRef]
21. Supančič, J.S.; Rogez, G.; Yang, Y.; Shotton, J.; Ramanan, D. Depth-based hand pose estimation: Methods, data, and challenges. *Int. J. Comput. Vis.* **2018**, *126*, 1180–1198. [CrossRef]
22. Lv, Z. Wearable smartphone: Wearable hybrid framework for hand and foot gesture interaction on smartphone. In Proceedings of the IEEE International Conference on Computer Vision Workshops, Sydney, Australia, 1–8 December 2013; pp. 436–443. [CrossRef]
23. Oberweger, M.; Wohlhart, P.; Lepetit, V. Training a feedback loop for hand pose estimation. In Proceedings of the IEEE International Conference on Computer Vision, Santiago, Chile, 7–13 December 2015; pp. 3316–3324. [CrossRef]
24. Tang, D.; Taylor, J.; Kohli, P.; Keskin, C.; Kim, T.K.; Shotton, J. Opening the black box: Hierarchical sampling optimization for estimating human hand pose. In Proceedings of the IEEE International Conference on Computer Vision, Santiago, Chile, 7–13 December 2015; pp. 3325–3333. [CrossRef]
25. Ye, Q.; Yuan, S.; Kim, T.K. Spatial attention deep net with partial pso for hierarchical hybrid hand pose estimation. In Proceedings of the European Conference on Computer Vision, Amsterdam, The Netherlands, 11–14 October 2016; pp. 346–361. [CrossRef]
26. Guo, H.; Wang, G.; Chen, X.; Zhang, C.; Qiao, F.; Yang, H. Region ensemble network: Improving convolutional network for hand pose estimation. In Proceedings of the IEEE International Conference on Image Processing, Beijing, China, 17–20 September 2017; pp. 4512–4516. [CrossRef]
27. Chen, X.; Wang, G.; Guo, H.; Zhang, C. Pose guided structured region ensemble network for cascaded hand pose estimation. *Neurocomputing* **2019**, *395*, 138–149. [CrossRef]
28. Wang, G.; Chen, X.; Guo, H.; Zhang, C. Region ensemble network: Towards good practices for deep 3d hand pose estimation. *J. Visual Commun. Image Represent.* **2018**, *55*, 404–414. [CrossRef]
29. Chen, X.; Wang, G.; Guo, H.; Zhang, C.; Wang, H.; Zhang, L. MFA-Net: Motion feature augmented network for dynamic hand gesture recognition from skeletal data. *Sensors* **2019**, *19*, 239. [CrossRef]
30. Chen, X.; Guo, H.; Wang, G.; Zhang, L. Motion feature augmented recurrent neural network for skeleton-based dynamic hand gesture recognition. In Proceedings of the IEEE International Conference on Image Processing, Beijing, China, 17–20 September 2017; pp. 2881–2885. [CrossRef]
31. Chin-Shyurng, F.; Lee, S.E.; Wu, M.L. Real-time musical conducting gesture recognition based on a dynamic time warping classifier using a single-depth camera. *Appl. Sci.* **2019**, *9*, 528. [CrossRef]
32. Ding, J.; Chang, C.W. An adaptive hidden Markov model-based gesture recognition approach using Kinect to simplify large-scale video data processing for humanoid robot imitation. *Multimed. Tools Appl.* **2016**, *75*, 15537–15551. [CrossRef]
33. Kumar, P.; Saini, R.; Roy, P.P.; Dogra, D.P. A position and rotation invariant framework for sign language recognition (SLR) using Kinect. *Multimed. Tools Appl.* **2018**, *77*, 8823–8846. [CrossRef]
34. Mazhar, O.; Navarro, B.; Ramdani, S.; Passama, R.; Cherubini, A. A real-time human-robot interaction framework with robust background invariant hand gesture detection. *Robot. Comput. Integr. Manuf.* **2019**, *60*, 34–48. [CrossRef]
35. Lin, C.; Lin, X.; Xie, Y.; Liang, Y. Abnormal gesture recognition based on multi-model fusion strategy. *Mach. Vision Appl.* **2019**, *30*, 889–900. [CrossRef]
36. Nunez, J.C.; Cabido, R.; Pantrigo, J.J.; Montemayor, A.S.; Velez, J.F. Convolutional neural networks and long short-term memory for skeleton-based human activity and hand gesture recognition. *Pattern Recognit.* **2018**, *76*, 80–94. [CrossRef]
37. Lake, B.M.; Salakhutdinov, R.; Tenenbaum, J.B. Human-level concept learning through probabilistic program induction. *Science* **2015**, *350*, 1332–1338. [CrossRef] [PubMed]
38. Snell, J.; Swersky, K.; Zemel, R. Prototypical networks for few-shot learning. *Adv. Neural Inf. Process. Syst.* **2017**, 4077–4087.
39. Koch, G.; Zemel, R.; Salakhutdinov, R. Siamese neural networks for one-shot image recognition. In Proceedings of the ICML Deep Learning Workshop, Lille, France, 10–11 July 2015; Volume 2.

40. Cai, Q.; Pan, Y.; Yao, T.; Yan, C.; Mei, T. Memory matching networks for one-shot image recognition. In Proceedings of the 2018 IEEE/CVF Conference on Computer Vision and Pattern Recognition, Salt Lake City, UT, USA, 18–23 June 2018. [CrossRef]
41. Finn, C.; Abbeel, P.; Levine, S. Model-agnostic meta-learning for fast adaptation of deep networks. In Proceedings of the International Conference on Machine Learning, Sydney, Australia, 6–11 August 2017; pp. 1126–1135.
42. Ravi, S.; Larochelle, H. Optimization as a model for few-shot learning. In Proceedings of the International Conference on Learning Representations, Toulon, France, 24–26 April 2017.
43. Kingma, D.P.; Ba, J. Adam: A Method for Stochastic Optimization. In Proceedings of the International Conference on Learning Representations, San Diego, CA, USA, 7–9 May 2015.
44. Ma, C.; Wang, A.; Chen, G.; Xu, C. Hand joints-based gesture recognition for noisy dataset using nested interval unscented Kalman filter with LSTM network. *Visual Comput.* **2018**, *34*, 1053–1063. [CrossRef]
45. Pontes, F.J.; Amorim, G.F.; Balestrassi, P.P.; De Paiva, A.P.; Ferreira, J.R. Design of experiments and focused grid search for neural network parameter optimization. *Neurocomputing* **2016**, *186*, 22–34. [CrossRef]
46. Oreifej, O.; Liu, Z. Hon4d: Histogram of oriented 4d normals for activity recognition from depth sequences. In Proceedings of the IEEE Conference on Computer Vision and Pattern Recognition, Portland, OR, USA, 23–28 June 2013; pp. 716–723. [CrossRef]
47. Devanne, M.; Wannous, H.; Berretti, S.; Pala, P.; Daoudi, M.; Del Bimbo, A. 3-d human action recognition by shape analysis of motion trajectories on riemannian manifold. *IEEE Trans. Cybern.* **2014**, *45*, 1340–1352. [CrossRef] [PubMed]
48. Ohn-Bar, E.; Trivedi, M. Joint angles similarities and HOG2 for action recognition. In Proceedings of the IEEE Conference on Computer Vision and Pattern Recognition Workshops, Portland, OR, USA, 23–28 June 2013; pp. 465–470. [CrossRef]
49. De Smedt, Q.; Wannous, H.; Vandeborre, J.P. Skeleton-based dynamic hand gesture recognition. In Proceedings of the IEEE Conference on Computer Vision and Pattern Recognition Workshops, Las Vegas, NV, USA, 26 June–1 July 2016; pp. 1–9. [CrossRef]

applied sciences

MDPI

Article

Exploring Visual Perceptions of Spatial Information for Wayfinding in Virtual Reality Environments

Ju Yeon Kim [1] and Mi Jeong Kim [2,*]

1 School of Architecture, Soongsil University, Seoul 06978, Korea; kjy@ssu.ac.kr
2 School of Architecture, Hanyang University, Seoul 04763, Korea
* Correspondence: mijeongkim@hanyang.ac.kr; Tel.: +82-2-2220-1249

Received: 31 March 2020; Accepted: 15 May 2020; Published: 17 May 2020

Abstract: Human cognitive processes in wayfinding may differ depending on the time taken to accept visual information in environments. This study investigated users' wayfinding processes using eye-tracking experiments, simulating a complex cultural space to analyze human visual movements in the perception and the cognitive processes through visual perception responses. The experiment set-up consisted of several paths in COEX Mall, Seoul—from the entrance of the shopping mall Starfield to the Star Hall Library to the COEX Exhibition Hall—using visual stimuli created by virtual reality (four stimuli and a total of 60 seconds stimulation time). The participants in the environment were 24 undergraduate or graduate students, with an average age of 24.8 years. Participants' visual perception processes were analyzed in terms of the clarity and the recognition of spatial information and the activation of gaze fixation on spatial information. That is, the analysis of the visual perception process was performed by extracting "conscious gaze perspective" data comprising more than 50 consecutive 200 ms continuous gaze fixations; "visual understanding perspective" data were also extracted for more than 300 ms of continuous gaze fixation. The results show that the methods for analyzing the gaze data may vary in terms of processing, analysis, and scope of the data depending on the purpose of the virtual reality experiments. Further, they demonstrate the importance of what purpose statements are given to the subject during the experiment and the possibility of a technical approach being used for the interpretation of spatial information.

Keywords: virtual reality; area of interest; wayfinding; spatial information; perception

1. Introduction

Consumers have shifted from simply buying products to the consumption of improved quality in terms of culture and value [1–3]. At the same time, companies have adapted to locate consumers among diverse cultures, and large complex spaces combining commercial and cultural areas, such as library and music hall, have been created. However, as spaces have expanded to contain various content and scale, the distance and the complexity of visitor circulation in these indoor environments has increased. For this reason, consumers must recognize spatial information when traversing paths using signs and maps, even in a complex cultural space. Although designers have developed and provided various forms of signs for visitors to recognize spatial information, there are limitations in understanding how visitors perceive this information and use it for wayfinding.

Data for spatial cognitive processes may be collected through questionnaires and interviews. However, to collect the sensory physiological signal data of the direct visual perception process, it is necessary to use parallel scientific experimental methods as a supplement. Questionnaires and interviews are limited because they provide subjective opinions about the spatial experience, and survey participants can consciously change answers. The scientific measurement method using physiological signals can compensate for such limitations and directly extract any sensory signals generated unconsciously [4,5].

Human visual information can be measured by using eye-tracking equipment [6,7]. To collect information about the direct visual perception process, we experimented with a virtual reality (VR)-based head mounted display (HMD) system. VR-based eye-tracking equipment allows participants to experience spaces even if they are not in the real world [8]. The experience of the virtual space as if it is real using an HMD has the advantage of increasing the immersion in space, and it is possible to quantify the user's gaze data. There are various constraints in a real environment. There are many variables in extracting data because the real environment cannot be regulated constantly as a consequence of population congestion or external environmental factors. However, in the laboratory using a VR-based HMD, it is possible to control the variable parts while realizing the commitment to space.

Although the diffusion of VR and the expansion of the market are expected to continue, the design and the evaluation of the space through VR can be considered an initial stage. In this study, we contributed to the field by suggesting a research method that involves experiments to apply the VR environment to spatial design using visual perception. VR equipment was used to convert human visual information into the cognition of sensation, and visual searches were monitored using eye-tracking experiments with VR equipment. In addition, data analysis was proposed to track the unconscious spatial search according to the time taken for cognitive processing to occur. This is significant in proposing a method for utilizing the developing VR technology according to human sensory information in architectural space planning.

2. Literature Review

2.1. Theory of Visual Recognition Using the Eye-Tracking Mechanism

Many studies have attempted to explain human visual recognition, which occurs via the eyes in space [9–11]. Early research on eye tracking was technically limited to simple eye observation and eye endoscopy but has since evolved into interdisciplinary areas, including psychophysics, cognitive neuroscience, brain science, and computer science [12,13]. Such studies have tried to identify the nature of visual perception, attempting to define the gaze in terms of its focus with respect to identity, meaning, or expectation. Studying the gaze, researchers can identify what people's interests are; the gaze can be described as the act of focusing on something more clearly in the mind than several conceivable things considered simultaneously [14]. Julesz and Schumer noted that, when the gaze is grasped by sensation as an internally invisible mechanism manifesting as imagination, expectations, or thinking, the number of perceptions that can be immediately included in the present cognitive realm becomes small. That is, visual perception can be determined depending on what humans think [15]. Some concerns have been raised as to whether the essential mechanism of visual perception concerns the "where" of eye movement or visual gazes in relation to spatial location. In a study of physiological optics, a wandering exploration of the eye was identified and observed as the human gaze or visual attention. It was remarked that observing the field of view is the only way humans can see each individual part of a space as clearly as possible. Consciousness can voluntarily control the gaze, thus humans can pay attention to their surrounding objects without distraction [16]. The human eyes provide visual evidence of the world around them, and visual attention not only requires human cognitive function but is also related to preconceived factors. The features of visual attention have been developed into an important theory and inform the visual search of feature integration theory (FIT) [17,18], which maintains that the eye tracks things in space by encoding simple and useful properties, such as position and color, orientation, size, and stereo distance.

The modeling of computer applications for eye tracking in relation to visual attention is related to a bottom-up schema of dichotomous visual interest or a function-oriented description [19–21]. That is, when considering image stimulation, an interest in the cognitive stage considers a specific area of the image. This area can be recognized parafoveally in the sense of initially requesting detailed inspection through foveal vision. It should be emphasized that a complete model of visual attention requires a high level of visual and cognitive functioning. Human interest in space cannot be explained simply by considering visual features, but it is required to analyze the eye-tracking mechanism along with the cognitive factors of particular interest.

2.2. Visual Attention and Cognitive Processing Time

Visual perception responses can be measured by analyzing the reactions from the acquisition of visual information to the brain and the perception of visual stimuli. The cognitive stages are different depending on the time at which the subjects receive the visual information. Time associated with visual attention and the gaze toward objects and scenes can be interpreted as a cognitive process [22–26]. Related studies have ascertained the meaning of the relationship between cognitive processes and time, identified the temporal difference between cognition and exploration when reading, and examined the difference in gaze fixation for the perception process of vision in image scenes or in spatial information pertaining to complex objects or in-focus stimuli over time [27–31]. The time required for the human brain to perceive a visual stimulus and send commands to the body is 0.1 s [32–35]. In this time, it is not possible to judge the nature of the object by paying attention to it because this is the early stage of human perception. The minimum response time is 0.2 s before the acquired visual information is activated in the brain, and this visual gaze time applies to both objects and scenes [36]. The minimum response time from the visual stimulus to the reaction through the brain is 0.2 s, and the time it takes for the acquired visual information to move the mind is 0.3 s [37–39]. The visual understanding of an object is developed by the data obtained in a fixed view in about 0.3 s. The analysis of the data according to the time of eye movement is necessary for the gaze analysis because of differences in human perception and cognitive processing. The relationship between cognitive processing and time has been scientifically studied and verified by many scholars, as described above. Depending on the object and the purpose of each cognitive level, there may be a difference in the time analysis. In this study, based on the hypothesis of a difference in visual movement over 0.2 s and 0.3 s, which was verified in the previous study, these intervals were used as standards for the analysis of the experimental data.

2.3. The Potential of Virtual Reality for Eye-Tracking Experiments

The use of VR in cognitive psychology research has increased with the proliferation of VR tools [40]. The VR tool has many advantages over the existing experimental methodology, as it has the potential to create effective stimulus and response protocols by controlling the experimental environment [41]. Many studies have demonstrated the effectiveness of VR for experiments, and the potential scope of research can be expanded depending on the possibilities for immersion in and controllability of the VR environment [42]. In the VR environment, safe dynamic tests for brain-computer interfaces are possible, and a controlled experimental design is also conceivable in the form of an evoked environment. To obtain examples of cognitive experiments, the usefulness of the object situation or location in the experimental environment was investigated in a user-gaze comparative study, and the advantages of exploration in 3D were evaluated as increasing cognitive response speeds by 50% or more when compared to searching in a physical space, as VR can quickly induce an omnidirectional gaze during spatial exploration [43–46]. In addition, VR experiments create a useful paradigm because they allow for experimental control in cognitive psychology research. Psychological cognition experiments can discover new methodologies using the new research tools provided by VR and extending beyond traditional methods. As for the nature of cognitive psychological phenomena, approaches can vary and include specific perception, memory, problem solving, and mental image attention. However, the immersion and the visual fidelity of the VR experimental tool do not necessarily elicit realistic psychological responses from users [47–52]. It is necessary to supplement the experimental limitations by utilizing the advantages of immersive and controllable VR environments and data according to the search and the fixation of the user's cognitive process. In VR environmental experiments, researchers must supplement the cognitive process consciously for the experiment's participants through the steps of the questionnaire. By recognizing spatial information and applying the experimental method of intelligibility, the cognitive stage can be identified through VR environmental verification to provide guidelines not only for the spatial environment but also for future spaces [53–56]. In addition, the immersion and the visual fidelity in the virtual space were grasped through an analysis of the visual data, and the interview data of participants' cognitive processes were applied to verify the analysis of the visual data and the data analysis depending on the level of their recognition. This is our experimental configuration and data analysis method that verifies both existing technologies and psychological approaches to virtual reality as distinct from the approaches of related studies.

3. Materials and Methods

3.1. Visual Stimuli

The complex space connected to the subway station was constructed as an experimental setting. As visual stimuli in COEX Mall, Seoul, we included paths commencing at the entrance of Starfield and passing through the Star Hall Library to the COEX Exhibition Hall. The experiment stimuli were presented to participants as a way to recognize the paths in the immersive VR space. The experimental stages were as follows. First, to create the visual stimuli, a 360-degree camera was installed on the path from the entrance to the crossroads at the Star Hall Library. Several images were recorded, and visual stimuli were selected in consideration of the congestion, the location of the signs, and the visibility. During the preliminary investigation, it was found that the sign, the path, and the surrounding spaces were used to identify the destination at the crossroads of the experimental space. To obtain spatial information, the destination name of the sign and the direction of the arrow were found to be confusing factors in selecting the path of the space.

Second, through a pretest consisting of two participants, the field of view and its visibility to the participants were confirmed when the captured 360-degree images were applied to the VR-based HMD device (SMI-HTC vive). Third, by analyzing the precautions of the experimental steps presented to the participants of the pretest, a pre-explanation was added to improve the participants' understanding of the experiment. Fourth, in the experiment (with a total of 23 participants—ten males and 13 females), participants each wore an HMD headset and observed visual stimuli while responding to an interview conducted by the experimenter.

A total of four visual stimuli were presented in an order and under a time limit for the experiment. The perception processes of clarity and cognition according to the area of interest in this study were analyzed using one visual stimulus. In the visual stimulus shown in Figure 1a, the participant observed the environment for 30 s at the entrance of the experimental space. The experimenter explained to the participants the purpose of the experiment and the VR environment and asked the participant to recognize the visual information of the sign first. The next stimulus, shown in Figure 1b,c, was an image that allowed the first sign to be observed closely and could be seen as the participant moved toward the crossroads. Here, the participants could observe the guide signs at the front for 30 to 60 s, allowing them to understand and adapt to the next experimental stimulus. The next stimulus in Figure 1d was found at the first crossroad to the exhibition hall; participants were given the opportunity to select the next route to the exhibition hall after observing the spatial information on the signboard for one minute and recognizing the space. Finally, while observing the last visual stimulus in Figure 1e, the participants chose an appropriate path among the numbered crossroads and provided answers to the spatial perception items in the questionnaire interview. The data extracted from the last visual stimulus were analyzed as areas of interest (AOIs) for clarity and recognition.

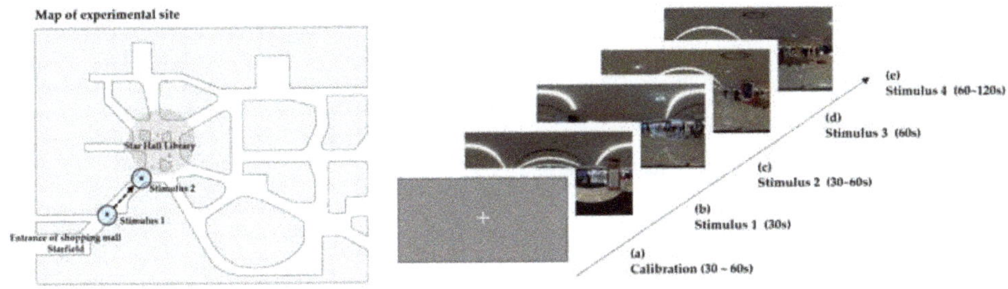

Figure 1. Visual stimuli for experimental procedures with a map of the experimental site. The stimulus images in the diagram are flattened, 360-degree images to indicate the experiment's setting. (**a**) As a focus adjustment step for extracting the gaze data of each subject, subjects performed the focus data test in a 360-degree space. The images (**b**) and (**c**) show the subjects' exploration of the surrounding environment, while the researcher explained the visual stimuli to the subjects, including signs and crossroads in the indoor space, via a "presentation of purpose statement". (**d**) As a step in extracting the gaze data, the participants viewed the signs and searched the space according to the purpose statement. (**e**) The researchers interviewed the participants about how they understood the space in the process of exploring it.

3.2. Experiment Procedure and Participants

The VR experiment was conducted over two days from 10 January 2019 to 11 January 2019, and the participants were 24 undergraduate or graduate students. All participants had a visual acuity or corrected visual acuity score of 0.5 or higher, with an average age of 24.8 years (standard deviation: ±1.89).

A preliminary experiment was conducted to verify the experimental setting. In the preliminary experiment, participants were asked to follow simple instructions after focus adjustment, which allowed them to adapt to the immersive VR environment before entering the main experiment (see Figure 2). The respondents observed the visual stimuli freely for 60 s, and as they had in the preliminary experiments, they explored the virtual space by freely turning their heads and bodies within the range allowed by the HMD equipment. The experimenter ended the preliminary experiment after monitoring the participants' behavior in real time. After the completion of the preliminary experiment, participants entered the main experiment, and the time required for each participant to complete it was different; the participants took a minimum of four minutes and a maximum of five minutes and 40 s. According to the validity rate of data collection, the gaze data of 11 final selected participants were selected and analyzed.

Figure 2. Experimental steps and visualization of the laboratory layout. (**a**) Layout of the virtual reality (VR) experiment. (**b**) The experimenter helps the participant wear the head mounted display (HMD) and adjusts its focus according to the nine points. (**c**) The experiment proceeds according to the visual stimulus stages. (**d**) After the VR experiment, the participant answer questions about the spatial information and their perceived consciousness.

The initial data for visual stimulus were reviewed to exclude participants whose validity rate was less than 85%. This was to increase the accuracy and the reliability of the experimental data by examining the validity rate in the process of recording the participants' gaze data in real time. Data from 11 out of 24 participants (45.83%) were valid and were selected. The percentage of valid initial data was 69.2%, but this rate was increased to 92.9% by the inclusion of 11 participants with validity rates of higher than 85%. The extracted data were analyzed according to the validation of the effective rate.

The visual stimuli used in the experiment were photographic images taken across 360 degrees. The participants performed in the experiment by observing a fixed virtual space image. AOI is an analysis method that can observe gaze fixation within the designated area to which the participants' gazes were directed. For the setting of AOIs and the extraction of data, the SMI BeGaze 3.6 program was used. The experimenter can designate a specific region of interest within the entire spatial image to extract gaze movements generated within that region. The change of gaze can be analyzed using the quantitative data of gaze movements, which is used to obtain information on wayfinding by mapping participants' cognitive processes of visual perception.

The area AOIs were construed as "path", "sign", and "background". "Sign" was used as a guide to obtain spatial information (see Figure 3), thus participants could receive information on the direction of the space. The AOIs used in the research analysis were identified using the data obtained by the participants looking for visual information and fixing their gazes accordingly in the VR images in adherence to the experiment's purpose statement. This allowed the researchers to analyze the research results to reveal characteristics of spatial perception according to the intelligibility and the recognition of the "sign" within the research site.

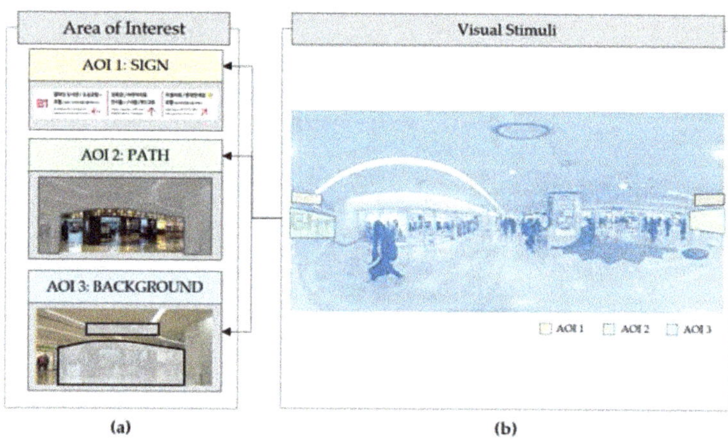

Figure 3. Areas of interest (AOIs) positioned in visual stimuli. (**a**) By setting the regions of interest in visual stimuli, the participants' gaze data that were created while searching for directions according to spatial information in the experiment were extracted. In the visual stimulus, the area excluding "sign" and "path" was set as the background. (**b**) A visual stimulus shown when a participant wore an HMD and explored the VR environment.

4. Results

4.1. Average Gaze Data for Understanding Spatial Information

Time-series analysis was performed to analyze the experimental data through the recognition of spatial information. To analyze the visual information process, the time was divided into six sequences of ten-second intervals. To extract the gaze information for the AOI set ("path", "sign", and "background") of visual stimuli, the number of gaze fixation data retrieved from each section was identified along with the AOIs that participants observed and the information they acquired over time.

In the interview conducted at the final stage of the experiment, participants answered questions about the clarity and the recognition of the crossroads selection process using a five-point scale ("very difficult", "difficult", "normal", "easy", and "very easy"). According to their answers, they were divided into two groups. Participants who selected "very difficult", "difficult", or "normal" were grouped as "difficult", and those who chose "easy" or "very easy" were classified as "easy". Dividing the group according to the level of spatial information recognition allowed the researchers to analyze the correlation between each group's changing characteristics in the cognitive understanding of spatial information and the movement of gaze information according to time-series analysis.

To analyze the spatial information, the movement of the gaze was studied over time to ascertain the average fixation number (see Figure 4) and the ratio of each AOI (see Figure 5). The data for the gaze increased and decreased over time in relation to the two groups ("easy" and "difficult"), which were divided according to their propensity for spatial understanding. Overall, using the averages and ratios extracted from the AOIs, the movement by which the gaze increases and decreases over time could be analyzed by applying the correlation between the understanding of space and the cognitive process.

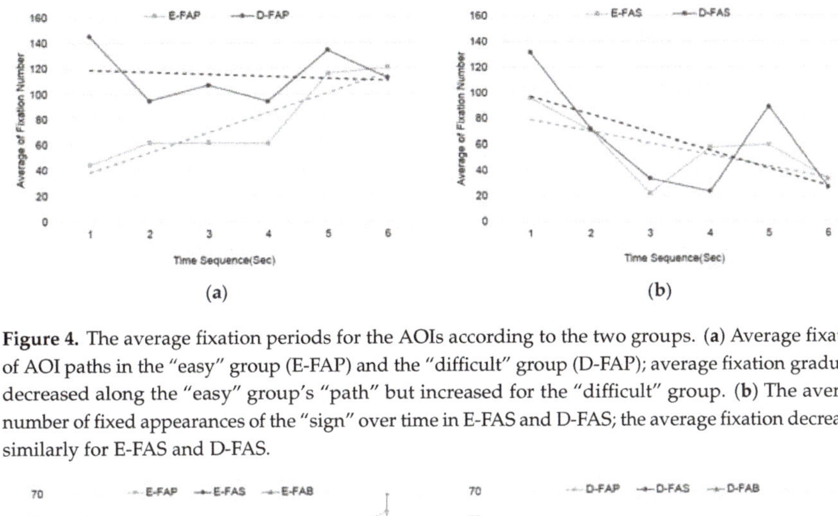

(a) (b)

Figure 4. The average fixation periods for the AOIs according to the two groups. (**a**) Average fixation of AOI paths in the "easy" group (E-FAP) and the "difficult" group (D-FAP); average fixation gradually decreased along the "easy" group's "path" but increased for the "difficult" group. (**b**) The average number of fixed appearances of the "sign" over time in E-FAS and D-FAS; the average fixation decreased similarly for E-FAS and D-FAS.

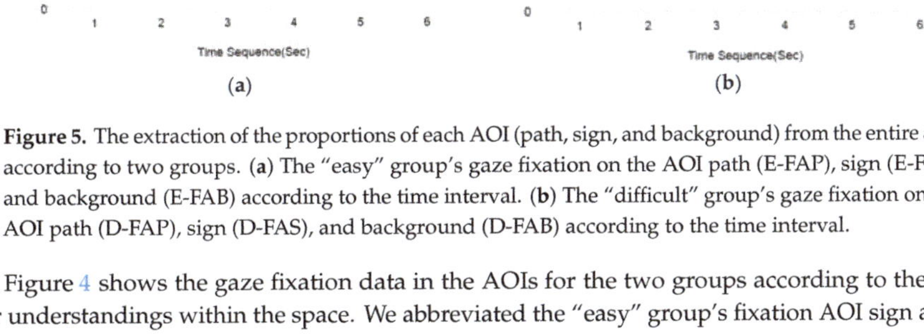

(a) (b)

Figure 5. The extraction of the proportions of each AOI (path, sign, and background) from the entire area according to two groups. (**a**) The "easy" group's gaze fixation on the AOI path (E-FAP), sign (E-FAS), and background (E-FAB) according to the time interval. (**b**) The "difficult" group's gaze fixation on the AOI path (D-FAP), sign (D-FAS), and background (D-FAB) according to the time interval.

Figure 4 shows the gaze fixation data in the AOIs for the two groups according to the scale of their understandings within the space. We abbreviated the "easy" group's fixation AOI sign as E-FAS, the "difficult" group's fixation AOI sign as D-FAS, the "easy" group's fixation AOI path as E-FAP, and the "difficult" group's fixation AOI path as D-FAP. After the start of the experiment, the gaze fixation of AOI for the "path" of the D-FAP group was 3.5 times higher than that of the E-FAS group in the experimental Time Sequence 1 (ten seconds). From the beginning of the gaze, the ratio of the D-FAP gaze gradually decreased throughout the entire experimental period, but the E-FAP gradually increased (see Figure 4a). The AOI gaze fixation for "sign" showed initially a high average gaze fixation rate in both the E-FAS and the D-FAS groups, which then gradually decreased (see Figure 4b).

Figure 5 shows the gaze fixation ratio for AOIs over the experimental period for the "easy" and the "difficult" groups. In the "easy" group, the gaze concentration ratio for the "sign" (55%) during the first section of the experiment (0 ~ 10 s) was approximately three times higher than that for the "path" (25%) and the "background" (19%). In the third section (20 ~ 30 s), which comprised the middle of the experiment, the "easy" group focused on the "path" rather than the "sign" in the order of "path" (52%), "background" (30%), and then "sign" (18%), unlike at the start of the experiment. After this, their attention was focused on the "path" rather than the other AOI areas, which shows that the gaze was fixed in accordance with the purpose of the experiment (see Figure 5a). During the first section of the experiment (0 ~ 10 s), the gaze fixation in the "difficult" group occurred in the order of "path" (40%), "sign" (36%), and "background" (23%), and the difference was less pronounced. In the second section of the experiment (10 ~ 20 s), the gaze fixation was higher for "path" (35%) and "background" (39%) than for "sign" (26%). In the middle of the experiment, a high gaze fixation ratio was shown for "path'" (50%). In the final section of the experiment, gaze fixation was also observed in the AOIs rather than the direction for wayfinding in the order of "path" (54%), "sign" (33%), and "background" (12%) (see Figure 5b).

In the interview during the final stage of the experiment, the groups were divided according to the difficulty experienced when perceiving space, and the extracted gaze fixation data were analyzed. As a result, differences in the capacity to understand the spatial information appeared according to the ratio of gaze fixation. The cognitive processing of information can be analyzed according to the time intervals of the experiment. In the experiment, the "easy" group's understanding of spatial information was obtained by perceiving the "sign", and their gaze was fixed on the wayfinding of the space for the purpose of the experiment. However, in the "difficult" group, the gaze fixation data appeared scattered in spaces other than the perception of the "sign" and the wayfinding.

4.2. Analysis of Conscious Gaze and the Visual Understanding of Data

To analyze the data of perceptual process, a time range of one visual movement was considered. The VR gaze tracking equipment recorded participant gaze data at 250 Hz per second, and the gaze time for each datum was 0.004 s—that is, one visual movement per 0.004 s was recorded and extracted at 250 Hz.

As seen in previous studies, the minimum time to consciously grasp an object in front of the eyes is 200 ms; it is understood that eyes perceive an object and consciously observe it when the gaze is fixed in one place. Therefore, to ensure a fixed gaze time of 0.2 s or more, at least 50 of the gaze data obtained in this experiment had to be continuous. However, less than 50 of the non-contiguous fixation data were also included in the gaze data recorded at 250 Hz. Using the above results, the raw gaze fixation data from the AOIs were analyzed by dividing the total experiment time (60 s) into six ten-second sections (see Figure 6). All fixation data were recorded regardless of dwelling time. The visual perception process was analyzed by extracting 50 or more 200 ms continuous gaze fixation data for "conscious attention" and 300 ms continuous gaze fixation data for "visual understanding" (see Figure 7).

The data for the participants' gaze movement that passed the validity rate test indicates the value of gaze fixation in the AOI areas. The AOIs were classified into FAP, FAS, and FAB. Each AOI region allowed for the extraction of gaze data from the FAS to understand the spatial information as well as gaze data from the FAP to understand the direction of wayfinding within a target space. FAB accounted for the remaining areas beyond FAP and FAS and was interpreted as searching gaze. In Time Sequence 1 (0 ~ 10 s) shown in Figure 6, the FAS value for gaze fixation was the highest, thus a conscious gaze fixation in the cognition of spatial information appeared. In Time Sequence 3 (20 ~ 30 s), the value of the FAS was the lowest, indicating that the search by the cognitive gaze for the "sign" was over at the beginning of the experiment. As the FAP value increased in Time Sequence 3, it continued until the end of the experiment. When analyzing the participants' perception of information about the space, it is understood participants used their gazes to search for 10 to 20 s at the beginning of the experiment, and then, after 30 s, with a greater understanding of the space, they proceeded to observe the direction of the destination. In the gaze movement analysis, there was a difference in the visual fixation data of the AOIs according to the time sequence (see Figure 6a).

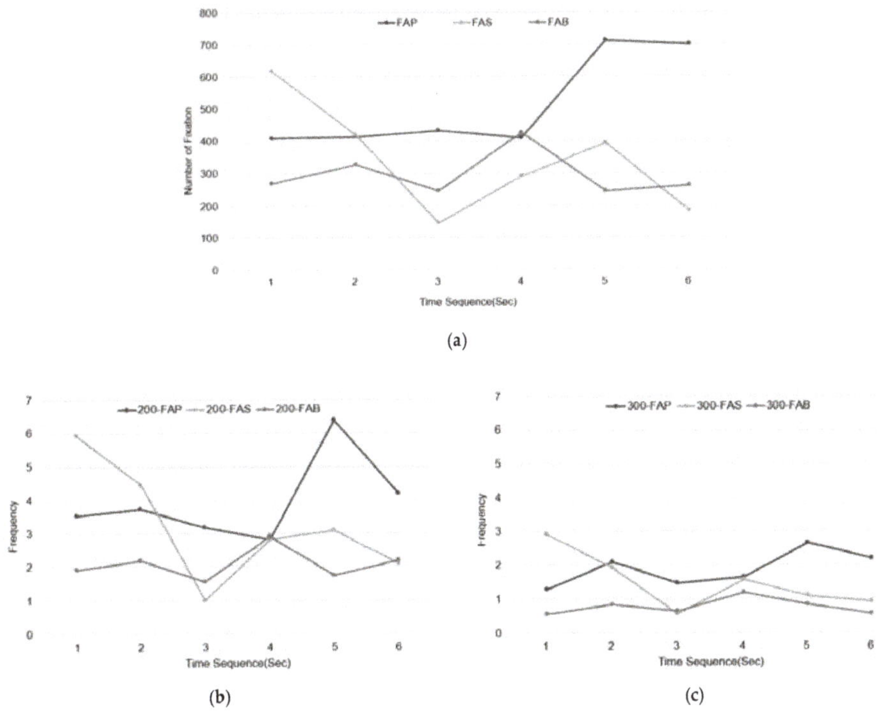

Figure 6. Fixation data analysis in the AOIs from the time series. (**a**) Participants' data from the AOIs. (**b**) 200 ms of visual fixation frequency; (**c**) 300 ms of visual fixation frequency.

A frequency of 200 ms when 50 or more gazes were fixed in succession was extracted from the data and analyzed as "conscious gaze" data. Figure 6c shows the analysis of more than 75 gaze fixation data at 300 ms as a "cognitive gaze" beyond consciousness (see Figure 6b). The 200 ms FAS (200-FAS) appears to have had a higher frequency than the other AOIs in Time Sequence 1 (0 ~ 10 s). Time Sequence 2 (10 ~ 20 s) revealed a decrease in raw data after the commencement of the experiment; however, it is understood that the "sign" area was perceived more consciously than the other AOI areas. The 300 ms FAP (300-FAP) increased in frequency from Time Sequence 2 (10 ~ 20 s) to Time Sequence 6 (50 ~ 60 s). It is conjectured that the spatial information was recognized in the "sign" during the initial Time Sequence 1.

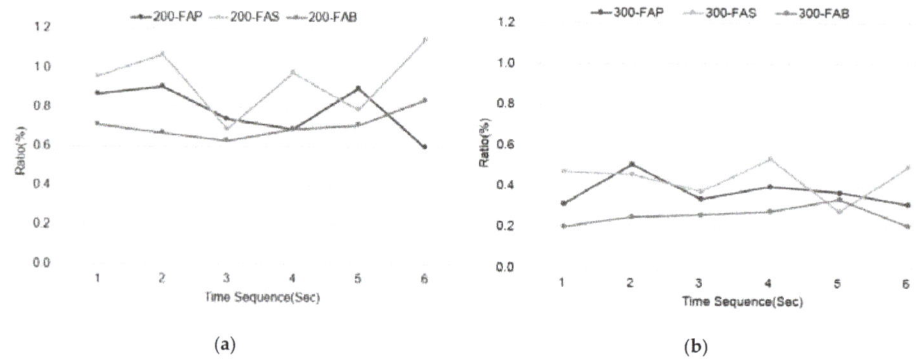

Figure 7. Relative values for eye-movement fixation in AOIs. (**a**) Conscious attention at 200 ms. (**b**) Visual understanding at 300 ms.

4.3. Relative Proportions of the Conscious Gaze and Visual Understanding

In the relative ratio analysis, each change in participant gaze movement was noted during the perception and cognitive processing of the virtual space. The raw data, the visual perception movement of the conscious gaze at 200 ms and the visual perception at 300 ms were analyzed using visual steps as the gaze movements according to the relative ratio values of each AOI (see Figure 7). Eye-movement data, for which the visual perception and cognitive stages were relative, were analyzed in terms of the stimulus and position of the visual perception in each time sequence. Analyzing the rate at which the highest level of perception was recorded for each time step shows conscious gaze for 200-FAS to be relatively high and increasing in contrast to that of the 200-FAP and the 200-FAB. Examining the detailed time sequence, the 200-FAS shows a lower rate than the 200-FAP in Steps 3 and 5, but the 300-FAS also exhibits a low rate in Steps 2 and 5. By the gaze fixation data analysis (see Figure 6), it was confirmed that the visual attention related to the stimuli increased with the FAP. By analyzing the ratio, it is possible to define the meaning of visual attention by considering where the gaze dwelled.

4.4. Analysis of the Perception and Understanding of Spatial Information According to Cognitive Differences

The data extracted from the visual exploration of the "easy" and the "difficult" groups were divided into the perception and the understanding of spatial information through the analysis of AOIs (see Figure 8). The stages of visual perception and exploration of space were investigated by analyzing the "easy" group's perceptions (see Figure 8a) and understandings (see Figure 8c) as well as the "difficult" group's perceptions (see Figure 8b) and understandings (see Figure 8d). The perception processes of the "easy" group were analyzed to be high for the "sign" (E-200-FAS), which was consistently searched for throughout the areas of the AOIs. However, the perception processes of the "difficult" group were analyzed to be either frequent or infrequent in terms of the gaze rate for the "sign" (D-200-FAS) relative to the areas of AOIs and depending on the time sequence. In the process of understanding the space at 300 ms, the E-FAS group continued to seek to comprehend the space, as the gaze was observed in the "sign" area at this time point. However, the "difficult" group was observed to stay focused on the "path", except for viewing the "sign" in the initial time step. Revealing the conscious difficulty response to spatial information, differences were witnessed between the E-FAS group and the D-FAP group in the visual scan path pattern throughout the experimental stages.

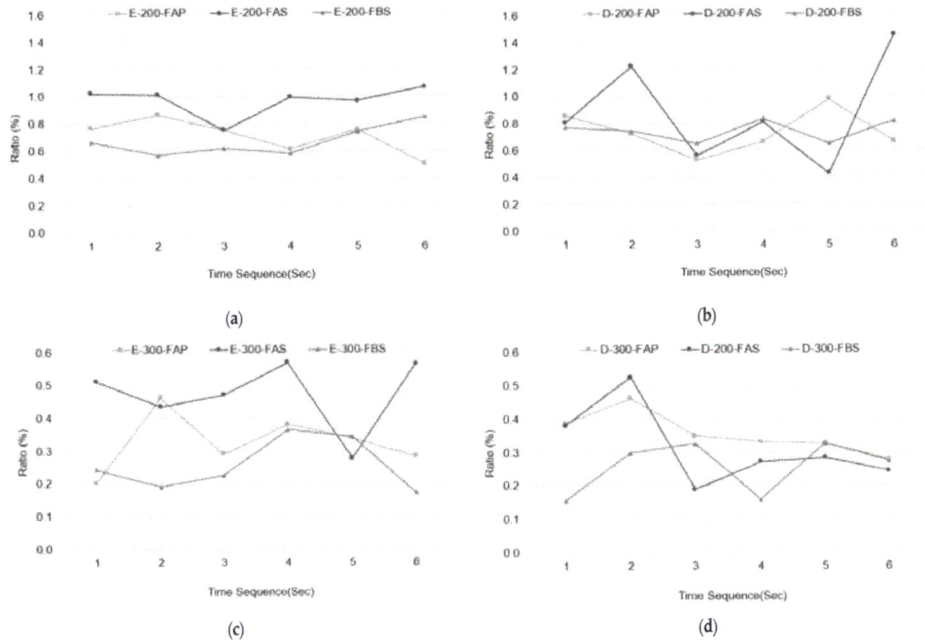

Figure 8. Analysis of visual perception and understanding in AOIs according to cognitive differences. (**a**) Visual data for 200 ms in the AOIs for the "easy" group. (**b**) Visual data at 200 ms in the AOIs for the "difficult" group. (**c**) Visual data at 300 ms in the AOIs for the "easy" group. (**d**) Visual data at 300 ms in the AOIs for the "difficult" group.

5. Discussion and Conclusion

Visual perception in the process of spatial experience was measured using various research methods. The visual perception process can be divided into conscious perception and unconscious perception, and the gaze movement can be classified according to James [14] "what you see" or Von Helmhotlz [16] "where you see". This research focused on "how" and "what" is seen as the gaze changes over time, focusing on the levels of perception and cognition throughout the VR experiment.

- **Extracting the meaning of gaze fixation data**

The processes of perception and cognition in information understanding were defined according to the time of the gaze fixation, and analysis was conducted accordingly. In the study of eye tracking involving statistical analysis according to the number of eye fixations and the differences in frequency, it cannot be established that people actually perceived the objects because they saw something. This study sought to ascertain whether the length of time at which a gaze is fixed can be regarded as the actual seeing of an object and to grasp the data that are meaningful to the process of visual perception. In the study of eye tracking, 0.1 s is interpreted as temporary eye fixation at a low level of spatial perception. Data above 0.3 s are analyzed as indicating visual understanding at the cognition stage. The interpretation of meaning may vary depending on how long the conscious and the unconscious gaze dwells rather than interpreting all data on gaze fixation as "seeing".

When comparing the average frequency of the gaze fixation data according to the ratio of the total AOIs, the "difficult" group exhibited greater gaze fixation than the "easy" group, and this trend continued over time (see Figure 4). However, examining the gaze fixation according to the ratio among the AOIs, the "easy" group focused on the "sign" at the beginning of the experiment, and the ratio of gaze fixation increased after the middle of the experiment in other areas (see Figure 5). That is, according to the degrees of perception indicated by the groups, it is possible to understand the process of directional gazing and the comprehension of spatial information through the ratio analysis of the gaze data, which remains unapparent in the frequency of the raw data for the gaze. Data interpretation and extraction methods according to the research purpose are emphasized to reveal information that is unknown within the number of gaze scan paths and fixation points.

- **Processing eye-tracking data according to time series**

A time-series analysis of data or a research method of data processing according to frequency and ratio is required to understand the stages at which the levels of consciousness and unconsciousness change in the processes of visual perception. In particular, in the study of spatial information, it is important to understand the flow of visual attention and information acquisition over time to identify the machinations of perception and cognition. When a cognitive process for wayfinding occurs as a result given information about a large and complex space, it is necessary to extract the data about when, what, and how to understand and make decisions. In this study, the gaze movement was tracked according to a time series, revealing the processes of perception and cognition using eye tracking in the VR environment.

This study sought to identify when and where to see and understand according to the purpose of wayfinding in response to spatial information. As a result, the characteristics concerning whether or not the spatial information was understood appeared in the classification of the participants, and the value of the gaze data according to the time flow was significantly different accordingly. In this study, signs affecting the processes of perception and cognition in the research of the nature of visual perception were validated, thus the thoughts and the conscious focus of participants appeared as gaze movements.

At 200 ms, defined as the recognition level of the gaze, the ratios of AOIs were evenly distributed over time for the "easy" group; however, a large deviation in the field of the "sign" was revealed for the "difficult" group (see Figure 8). It was interpreted that the "easy" group searched for "where to look" by determining the hierarchy of the gaze recognition area, whereas the "difficult" group was unable to locate this hierarchy of recognition. At 300 ms, defined as the understanding level of the gaze, it was shown that the "easy" group maintained intentional consciousness to continuously understand throughout the experiment, directing their gazes toward the "sign" for the entire experiment. However, the gazes of the "difficult" group were interpreted as searching the surrounding area rather than understanding the "sign" from the middle of the experiment. Accordingly, the tracking of eye movements is an invisible psychological mechanism for spatial information, which may be interpreted and verified.

- **Enhancing immersion using a VR eye-tracking experiment**

There are many limitations to such experimentation with the consumer's reaction to the spatial environment. However, an immersive experiment setting was created through the effectiveness of the VR technology, thus it was possible to quickly record the participants' reactions to the spatial environment. These strengths of the VR experiment, such as immersion and control, will not apply to all studies. In this study, research on the control of the VR experiment was conducted in advance and found a limit to the participants' immersion in the VR experiment over time. Thus, to enhance participants' immersion for the purposes of the experiment, visual stimuli were presented for exploring the space, and cognitive psychological phenomena were examined through interviews. In the resulting analysis, the participants were divided into groups according to the degree of their cognition responses, and the data were classified and analyzed accordingly.

In conclusion, it was identified that visual information correlates with human perception. The data of the gaze can be interpreted as indicating a cognitive stage of the exploration and understanding of visual perception and may be judged as a useful mechanism for psychological decision making. Because of the attributes of the gaze data, researcher interpretation is essential in extracting meaning and interpreting "what", "where", and "how much" a person looks and comprehends. Depending on what purpose statement is given over the course of the VR experiment, the extraction and the analysis of the gaze data obtained may vary. Even in an experimental environment implemented with an efficient VR technique, quantitative data extraction and interpretation must be conducted in combination with a qualitative research method. Most of the subjects in this study were in their early twenties, since the study conducted VR experiments as a verification of gaze perception and cognition. It would be desirable to conduct another experiment with a mixed aged group, as data collected according to age may be expected to reveal interesting results concerning sensory perception and cognition.

Author Contributions: Funding acquisition, J.Y.K.; Investigation, J.Y.K.; Methodology, M.J.K.; Supervision, M.J.K.; Writing—original draft, J.Y.K.; Writing—review & editing, M.J.K. All authors have read and agreed to the published version of the manuscript.

Funding: This work was supported by the National Research Foundation of Korea Grant funded by the Korean Government (Grant no.2017R1A2B2007276).

Conflicts of Interest: The authors declare no conflict of interest.

References

1. Sheth, J.N.; Mittal, B.; Newman, B.I. *Consumer Behavior and Beyond*; The Dryden Press: New York, NY, USA, 1999.
2. Sojka, J.; Tansuhaj, P.S. Cross-cultural consumer research: A twenty-year review. In *North American- Advances in Consumer Research 22*; UT: Association for Consumer Research: Minneapolis, MN, USA, 1995; pp. 461–474.
3. Venkatesh, A. Ethnoconsumerism: A new paradigm to study cultural and cross-cultural consumer behavior. In *Marketing in a Multicultural World*; SAGE Publications: Southend oaks, CA, USA, 1995; pp. 26–67.
4. Kim, J.; Kim, J. Method Extracting Observation Data by Spatial Factor for Analysis of Selective Attention of Vision. *Korean J. Sci. Emot. Sensib.* **2015**, *18*, 3–14. [CrossRef]
5. Kim, J. An Analyzed the Area of Interest based on the Visiting Intention and Existence of People in Cafe Space. *J. Korean Inst. Inter. Des.* **2016**, *25*, 130–139. [CrossRef]
6. Bozomitu, R.G.; Pasarica, A.; Tarniceriu, D.; Rotariu, C. Development of an Eye Tracking-Based Human-Computer Interface for Real-Time Applications. *Sensors* **2019**, *19*. [CrossRef] [PubMed]
7. Kim, J.; Kim, J. The Initial Value of Extraction for Visual Perception in VR(Virtual Reality) Eye-tracking Experiments. *J. Korean Inst. Inter. Des.* **2019**, *28*, 84–94. [CrossRef]
8. Bodden, V. *Virtual-Reality Headsets*; Checkerboard Library, an imprint of Abdo Publishing: Minneapolis, MN, USA, 2018; p. 32.
9. Madary, M. *Visual Phenomenology*; MIT Press: Cambridge, MA, USA, 2017.
10. Palmer, S.E. *Vision Science: Photons to Phenomenology*; MIT Press: Cambridge, MA, USA, 1999.
11. Siegel, S. How does visual phenomenology constrain object-seeing? *Australas. J. Philos.* **2006**, *84*, 429–441. [CrossRef]
12. Duchowski, A.T. Eye tracking methodology. *Theory Pract.* **2007**, *328*, 2–3.
13. Zeki, S. *A Vision of the Brain*; Blackwell scientific publications: Hoboken, NJ, USA, 1993.
14. James, W.; Burkhardt, F.; Bowers, F.; Skrupskelis, I.K. *The Principles of Psychology*; Macmillan London: London, UK, 1890; Volume 1.
15. James, W. *Principles of Psychology*; Harvard University Press (trad it. Principi di psicologia, Principato, Milano, 1983): Cambridge, MA, USA, 1890.
16. Von Helmholtz, H. *Treatise on Physiological Optics: Translated from the 3rd German Ed*; Optical Society of America: Washington, DC, USA, 1925.
17. Posner, M.I.; Snyder, C.R.; Davidson, B.J. Attention and the detection of signals. *J. Exp. Psychol. Gen.* **1980**, *109*, 160. [CrossRef]
18. Sparks, D.L.; Mays, L.E. Signal transformations required for the generation of saccadic eye movements. *Annu. Rev. Neurosci.* **1990**, *13*, 309–336. [CrossRef]
19. Bundesen, C. A theory of visual attention. *Psychol. Rev.* **1990**, *97*, 523. [CrossRef]

20. Connor, C.E.; Egeth, H.E.; Yantis, S. Visual attention: Bottom-up versus top-down. *Curr. Biol.* **2004**, *14*, R850–R852. [CrossRef]
21. Itti, L. *Models of Bottom-up and Top-down Visual Attention*; California Institute of Technology: Pasadena, CA, USA, 2000.
22. Bryant, J.S. Feature detection process in speech perception. *J. Exp. Psychol. Hum. Percept. Perform.* **1978**, *4*, 610–620. [CrossRef] [PubMed]
23. Fazio, R.H.; Herr, P.M.; Olney, T.J. Attitude accessibility following a self-perception process. *J. Personal. Soc. Psychol.* **1984**, *47*, 277–286. [CrossRef]
24. Schag, K.; Rauch-Schmidt, M.; Wernz, F.; Zipfel, S.; Batra, A.; Giel, K.E. Transdiagnostic Investigation of Impulsivity in Alcohol Use Disorder and Binge Eating Disorder with Eye-Tracking Methodology-A Pilot Study. *Front. Psychiatry* **2019**, *10*, 724. [CrossRef] [PubMed]
25. Allan, L.G. The perception of time. *Percept. Psychophys.* **1979**, *26*, 340–354. [CrossRef]
26. Aubry, F.; Guillaume, N.; Mogicato, G.; Bergeret, L.; Celsis, P. Stimulus complexity and prospective timing: Clues for a parallel process model of time perception. *Acta Psychol.* **2008**, *128*, 63–74. [CrossRef]
27. Gibson, W.C.; Herron, W.G. Psychotherapists' religious beliefs and their perception of the psychotherapy process. *Psychol. Rep.* **1990**, *66*, 3–9. [CrossRef]
28. Rayner, K. Eye movements and attention in reading, scene perception, and visual search. *Q. J. Exp. Psychol.* **2009**, *62*, 1457–1506. [CrossRef]
29. Rayner, K. Eye movements in reading and information processing: 20 years of research. *Psychol. Bull.* **1998**, *124*, 372. [CrossRef]
30. Chua, H.F.; Boland, J.E.; Nisbett, R.E. Cultural variation in eye movements during scene perception. *Proc. Natl. Acad. Sci. USA* **2005**, *102*, 12629–12633. [CrossRef]
31. Rayner, K.; Pollatsek, A. Eye movements and scene perception. *Can. J. Psychol. Rev. Can. Psychol.* **1992**, *46*, 342. [CrossRef]
32. Broadbent, D.E. *Perception and Communication*; Elsevier: Amsterdam, The Netherlands, 2013.
33. Blamey, P.J.; Cowan, R.S.; Alcantara, J.I.; Whitford, L.A.; Clark, G.M. Speech perception using combinations of auditory, visual, and tactile information. *J. Rehabil. Res. Dev.* **1989**, *26*, 15–24. [PubMed]
34. Lee, H.J.; Lee, J.; Kim, C.J.; Kim, G.J.; Kim, E.S.; Whang, M. Brain process for perception of the "out of the body" tactile illusion for virtual object interaction. *Sensors* **2015**, *15*, 7913–7932. [CrossRef] [PubMed]
35. Julesz, B.; Schumer, R.A. Early visual perception. *Annu. Rev. Psychol.* **1981**, *32*, 575–627. [CrossRef] [PubMed]
36. Abe, M.; Ueki, T.; Saeki, Y.; Tateoka, S.; Tomita, M. When and how of assistance in nursing. Report 1. Perception of the psychological process of the patient. *Kangogaku Zasshi* **1977**, *41*, 241–246.
37. Brown, J. On time perception in visual movement fields. *Psychol. Forsch.* **1931**, *14*, 233–248. [CrossRef]
38. Solso, R.L.; MacLin, M.K.; MacLin, O.H. *Cognitive Psychology*; Pearson Education New Zealand: Auckland, New Zealand, 2005.
39. Davidoff, J. *Differences in Visual Perception: The Individual Eye*; Elsevier: Amsterdam, The Netherlands, 2012.
40. Gaggioli, A. Using virtual reality in experimental psychology. In *Towards Cyberpsychology*; IOS Press: Amsterdam, The Netehrlands, 2001; pp. 157–174.
41. Wilson, C.J.; Soranzo, A. The use of virtual reality in psychology: A case study in visual perception. *Comput. Math. Methods Med.* **2015**, 1–5. [CrossRef]
42. Bayliss, J.D.; Ballard, D.H. Recognizing evoked potentials in a virtual environment. In *Advances in Neural Information Processing Systems*; MIT Press: Vancouver, BC, Canada, 2000; pp. 3–9.
43. Biocca, F.; Owen, C.; Tang, A.; Bohil, C. Attention issues in spatial information systems: Directing mobile users' visual attention using augmented reality. *J. Manag. Inf. Syst.* **2007**, *23*, 163–184. [CrossRef]
44. Checcucci, E.; Amparore, D.; Pecoraro, A.; Peretti, D.; Aimar, R.; De Cillis, S.; Piramide, F.; Volpi, G.; Piazzolla, P.; Manfrin, D.; et al. 3D mixed reality holograms for preoperative surgical planning of nephron-sparing surgery: Evaluation of surgeons' perception. *Minerva Urol. Nefrol.* **2019**. [CrossRef]
45. Giroux, M.; Barra, J.; Barraud, P.A.; Graff, C.; Guerraz, M. From Embodiment of a Point-Light Display in Virtual Reality to Perception of One's Own Movements. *Neuroscience* **2019**, *416*, 30–40. [CrossRef]
46. Sapkaroski, D.; Mundy, M.; Dimmock, M.R. Virtual reality versus conventional clinical role-play for radiographic positioning training: A students' perception study. *Radiography* **2020**, *26*, 57–62. [CrossRef]
47. Akizuki, H.; Uno, A.; Arai, K.; Morioka, S.; Ohyama, S.; Nishiike, S.; Tamura, K.; Takeda, N. Effects of immersion in virtual reality on postural control. *Neurosci. Lett.* **2005**, *379*, 23–26. [CrossRef] [PubMed]

48. Bowman, D.A.; McMahan, R.P. Virtual reality: How much immersion is enough? *Computer* **2007**, *40*, 36–43. [CrossRef]
49. McMahan, R.P. *Exploring the Effects of Higher-Fidelity Display and Interaction for Virtual Reality Games*; Virginia Tech: Blacksburg, VA, USA, 2011.
50. McMahan, R.P.; Bowman, D.A.; Zielinski, D.J.; Brady, R.B. Evaluating display fidelity and interaction fidelity in a virtual reality game. *Ieee Trans. Vis. Comput. Graph.* **2012**, *18*, 626–633. [CrossRef] [PubMed]
51. McMahan, R.P.; Lai, C.; Pal, S.K. Interaction fidelity: The uncanny valley of virtual reality interactions. In *International Conference on Virtual, Augmented and Mixed Reality*; Springer: Berlin/Heidelberg, Germany, 2016; pp. 59–70.
52. Plante, T.G.; Cage, C.; Clements, S.; Stover, A. Psychological benefits of exercise paired with virtual reality: Outdoor exercise energizes whereas indoor virtual exercise relaxes. *Int. J. Stress Manag.* **2006**, *13*, 108. [CrossRef]
53. Vatavu, R.D.; Pentiuc, Ş.G.; Chaillou, C.; Grisoni, L.; Degrande, S. Visual recognition of hand postures for interacting with virtual environments. *Adv. Electr. Comput. Eng.* **2006**, *6*, 55–58.
54. Wallet, G.; Sauzéon, H.; Pala, P.A.; Larrue, F.; Zheng, X.; N'Kaoua, B. Virtual/real transfer of spatial knowledge: Benefit from visual fidelity provided in a virtual environment and impact of active navigation. *Cyberpsychol. Behav. Soc. Netw.* **2011**, *14*, 417–423. [CrossRef]
55. Hahm, J.; Lee, K.; Lim, S.L.; Kim, S.Y.; Kim, H.T.; Lee, J.H. Effects of active navigation on object recognition in virtual environments. *Cyberpsychol. Behav.* **2006**, *10*, 305–308. [CrossRef]
56. Conroy-Dalton, R. Is spatial intelligibility critical to the design of largescale virtual environments? *Int. J. Des. Comput.* **2002**, *4-19*.

applied sciences

MDPI

Article

Semi-Immersive Virtual Reality as a Tool to Improve Cognitive and Social Abilities in Preschool Children

Maria Luisa Lorusso [1,*], Simona Travellini [1], Marisa Giorgetti [1], Paola Negrini [2], Gianluigi Reni [2] and Emilia Biffi [2]

[1] Scientific Institute IRCCS E. Medea, Unit of Neuropsychology of Developmental Disorders, Bosisio Parini, 23842 Lecco, Italy; simona.travellini@bp.lnf.it (S.T.); marisa.giorgetti@unicatt.it (M.G.)

[2] Scientific Institute IRCCS E. Medea, Bioengineering Lab, Bosisio Parini, 23842 Lecco, Italy; paola.negrini@gmail.com (P.N.); gianluigi.reni@bp.lnf.it (G.R.); emilia.biffi@lanostrafamiglia.it (E.B.)

* Correspondence: marialuisa.lorusso@bp.lnf.it; Tel.: +39-031-877592

Received: 31 March 2020; Accepted: 21 April 2020; Published: 24 April 2020

Featured Application: The system and the activities presented in this manuscript can be successfully employed for the empowerment of social abilities in pre-school children, promoting inclusion and preventing isolation. Potential applications are also the improvement of weak functions in the at-risk population (e.g., on children with neurodevelopmental disorders and children with language and communication disorders).

Abstract: Virtual reality (VR) creates computer-generated virtual environments where users can experience and interact in a similar way as they would do in real life. VR systems are increasingly being used for rehabilitation goals, mainly with adults, but also with children, extending their application to the educational field. This report concerns a study of the impact of a semi-immersive VR system in a group of 25 children in a kindergarten context. The children were involved in several different games and activity types, specifically developed with the aim of learning specific skills and foster team collaboration. Their reactions and behaviors were recorded by their teachers and by trained psychologists through observation grids addressing task comprehension, participation and enjoyment, interaction and cooperation, conflict, strategic behaviors, and adult-directed questions concerning the activity, the device or general help requests. The grids were compiled at the initial, intermediate and final timepoint during each session. The results show that the activities are easy to understand, enjoyable, and stimulate strategic behaviors, interaction and cooperation, while they do not elicit the need for many explanations. These results are discussed within a neuroconstructivist educational framework and the suitability of semi-immersive, virtual-reality-based activities for cognitive empowerment and rehabilitation purposes is discussed.

Keywords: semi-immersive virtual reality; children; cooperative games; interactive learning environments; empowerment; perception; motor planning; problem-solving

1. Introduction

Virtual reality (VR) has been defined as the "use of interactive simulations created with computer hardware and software to present users with opportunities to engage in environments that appear and feel similar to real-world objects and events," [1]. In computer science, definitions of VR emphasize the possibility and ability to combine software with hardware to create a fully immersive experience [2]. In health care, the term is used in a slightly different way; to describe both non-immersive and immersive experiences that create an alternative reality [3]. Its main applications seem to be in the field of motor learning, whereas only few randomized controlled studies address its effects on other cognitive functions [4]. Its advantages rest on the possibility to fine-control, personalize and hierarchize tasks. Furthermore, it gives the possibility of providing multimodal feedback in real time; indeed, the stimulation of multiple perceptual channels, implemented by the use of auditory and visual feedback, increases the patient's awareness of his performance and allows a sense of global wellbeing [5].

Indeed, VR represents a very promising technology for neurorehabilitation. Computers elaborate a simulation of the real world using real-time graphics, through which the subject can interact with the environment [6]. There are different types of VR, including (a) immersive virtual reality; (b) desktop virtual reality; (c) projection virtual reality (or semi-immersive virtual reality); and (d) simulation virtual reality [7]. VR ranges from non-immersive to fully immersive, according to the degree to which the user is isolated from the physical surroundings while interacting with the virtual environment [8]. Usually, immersive systems involve computer interface devices, such as head-mounted displays (HMD) or projection screens surrounding the subject (cave systems), fiber-optic wired gloves, position tracking devices, and audio systems providing 3D sound. Immersive virtual reality in particular provides an immediate, first-person experience and a deep "sense of presence" [9], i.e., the perception to be immersed in a different world created by the components of software and hardware [10]. The level of presence is a subjective feeling and depends on user experience [11]. Compared to immersive VR systems, semi-immersive devices do not provide the constant update of the visual information according to the participant's head movements. On the other hand, they are more immersive than typical 3D monitors, particularly in terms of the range of sensory modalities accommodated [12,13].

The use of VR in neurorehabilitation has grown considerably, and experimental evidence suggests that this technology could favor functional recovery in neuropsychological disorders [8,14]. With respect to other tools in neurorehabilitation, VR has a number of peculiarities. Among others, the possibility of creating tailor-made training programs, so that the rehabilitation process can be individualized and adapted to each patient's specific needs. Moreover, VR can foster active involvement, thanks to the possibility of creating new and appealing environments [14]. Very often, VR gives therapists the possibility to individualize treatment needs, as well as the opportunity for repeated learning trials, while gradually increasing task complexity and/or decreasing therapist support and feedback [1].

Immersive and semi-immersive VR systems provide the opportunity to practice cognitive and motor activities that cannot be practiced within the clinical or the educational environment, performing simulations of real-life scenarios and activities [5,15]. In many cases, clinicians need special software development tools for the design and coding of interactive simulated environments to achieve specific rehabilitation goals [1].

VR also offers the possibility to test the patient's progress within controlled, ecological, and secure testing environments, that reproduce the crucial characteristics of the real world and are selected ad-hoc for each patient and situation (e.g., [16,17]). In experimental settings, VR systems allow researchers to design dynamic and realistic environments (virtual environments or VE), while monitoring behavioral and physiological responses [18]. The high degree of control allowed during the investigation of the cognitive and behavioral components of a certain skill is an additional advantage.

It has been demonstrated [19] that the patients are able to transfer what they have learned from VE to real life. Spatial navigation skills have been the object of several studies involving VR (e.g., [20,21]. The results of such studies suggest that the mental representations of space in VE resemble those implicated in the navigation of the real world. Montana and colleagues' systematic review [14] has shown improvements in spatial memory after practicing with navigational tasks in VR. Most importantly, it has shown a transfer of the improvements to more general aspects of spatial cognition. Both immersive and non-immersive VR systems have been shown to improve navigation and orientation abilities. It has even been proposed that the improvement observed in visual-constructive abilities, attention and upper limb motricity could be due to the so-called shadow effect (i.e., the patient's shadow on the screen while performing VR training) of the immersive system [22]. At the neurophysiological level, the mechanisms through which VR works are only partially understood. It is hypothesized that VR entrains the same neural pathways that are involved in motor learning and motion-related cognitive processes [23]. VR training thus seems to promote brain plasticity through mechanisms related to the reactivation of brain neurotransmitters, and its results can be even better than those obtained by conventional treatment [24,25].

As to the applications of VR with children, most have addressed motor problems [26], for instance in cerebral palsy [27,28], or cognitive, social and emotional problems, especially in autism spectrum disorders [29–31]. However, VR applications are not restricted to neurorehabilitation and more and more experiences involving its use are found also in educational contexts [32].

The general psycho-educational framework is that of neuroconstructivism, where cognitive development occurs through the pro-activity of the child in exploring, manipulating, and interacting with his/her environment [33]. According to the theory of social constructivism [34], moreover, the learning environment should encourage the pupils to collaborate and participate actively, experiment, share and develop ideas, use language to reason, plan and reflect on one's actions.

In Richard et al. [35], it has been suggested that immersive virtual reality technology provides an alternative educational process by providing a knowledge-building experience. It is crucial that learning goals and solutions are established in collaboration with teachers [36]. The authors underscore that individual factors like age, gender, computer experience, psychological factors, cognitive and learning styles are likely to strongly affect learning outcomes, as well as technology-related factors such as immersivity, so that empirical studies are needed to determine which characteristics of virtual environments can really be pedagogically exploited. Less recent, but seminal studies, such as Winn's [37], suggest that among the main features contributing to learning, there are free navigation and first-person point of view, the manipulation of the relative size of objects in virtual worlds, the transduction of otherwise imperceptible sources of information, and the reification of abstract ideas.

The conceptual framework known as TEL—Technology Enhanced Learning—suggests that technology can help the construction of new rules through the interaction with data in the learning environment. Indeed, games encourage exploring, the exchange of ideas, communication and decision making [38]. Games offer an invaluable opportunity for learning. The players of a game have to interpret images, sounds and actions [39]. They need to understand and learn what Gee calls the "internal design grammar" [40] of the game, forming hypotheses, adjusting their behaviors according to those hypotheses and, based on feedback from the virtual world, accepting or revising the hypotheses. Thus, games, serious games, or edugames, represent a unique opportunity to pursue rehabilitative and educational goals, without necessarily involving the children in long and effortful therapy sessions, but rather exploiting their own interests.

Although it has been stated that motivation is a key factor in the success or failure of education [41] and that fun and passion are key ingredients of the learning process [42], there are only a few studies that show that game technology brings substantial benefits [43–45]. Various peripheral devices have been used in such projects, including head-mounted display gear, data gloves, or body suits, and employing different techniques from specially designed glass cubicles to wall projection. However, practical concerns and limitations, first of all related to the high costs of such devices, but also to teachers' and educators' difficulties in using them, restricted dissemination of this technology in K-12 and higher education settings [45]. In more recent years, however, costs have begun to decrease and technology has become more user-friendly, so that more studies are being conducted in educational settings also; yet, most studies involved older children or university students, and studies on preschoolers are very rare. Nonetheless, some studies using VR to stimulate collaborative behaviors [46], support art education [47] or vocabulary learning [48] showed positive results on both learning and motivation (the most frequently reported concern, usually from parents and teachers, is about side-effects or addiction in the use of digital devices).

The project in which the present study was included envisaged the creation of an intelligent space in which preschool children and caregivers could experiment with a wide range of activities (see [49,50]). The setting and the playing sessions were organized so as to stimulate particular functions and skills, and foster team collaboration, as well as group integration. The goal was playing together to learn while having fun. Through the direct observation of caregivers and psychologists, we also wanted to characterize the children's behaviors during the playing sessions, in order to describe the impact that the system had on their emotional states and on their interactions with the environment and with the people involved in the educational activity.

The educational principles that lead the choice and the design of the activities and games (see [32]) were:

- Define activities that can cover the whole range of neuropsychological functions, and can adapt to different children and different development profiles
- Choose activities that stimulate children's curiosity, motivation, creativity by promoting inclusion through collaboration (cooperative learning) and communication
- Define VR-based activities that can be seen as an empowerment rather than as a transposition of traditional activities (see [51]), emphasizing the aspects of active construction of knowledge and competence.

Given the requirements, Nirvana 1 system (BTS Bioengineering, Italy) was selected as a playing environment. Nirvana 1 is indeed a semi-immersive virtual reality device, where one or more subjects can interact with virtual activities projected on a wall or on the floor, without the use of markers or other sensors placed on the body. The system is also equipped with a platform for the design of activities, in terms of graphic effects and feedback. The system has already been used in some rehabilitation studies [5,15,30], mainly addressing adult patients and aimed at motor rehabilitation.

During the playing sessions, the children were led and helped by their teachers and by two trained psychologists, who filled an ad-hoc-constructed observation grid for each of three timepoints: the beginning, midpoint and the end of each session. This grid was used to record the children's reactions and behaviors during activity, and it was constructed so as to be able to highlight play and social behaviors as well as problem-solving attitudes and skills (in a similar way as other instruments created to describe preschoolers' behavior during play with toys [52]), and their changes through session time.

The aims of the study were multi-faceted:

The main aim was to observe the overall impact of the activities on both cognitive and social processes, including interaction, collaboration, participation, conflict management and strategic behavior, but also to give some insights on accessibility/usability of the system, as emerging from the children's observed behaviors (showing understanding of the functional principles of the games, asking questions about the proposed activities and requesting help from the adult observers).

Secondly, we wanted to describe the impact of the different kinds of activities and the different processes that could be stimulated and initiated by them. Indeed, the various activity types and the different games within each typology were expected to induce different reactions and different social dynamics, due to the varying degree of communication, turn taking and shared strategies required by the activities.

Finally, we wanted to describe how this impact is modified, by increasing familiarity with the games and activities from the beginning to the end of the session. Again, this aspect was investigated both at a general level, relating to the system as a whole, and at the level of single activity types. In general, it was expected that the comprehension of the game structure and function, as well as the strategic approach, could increase through time, and that also the ability to cooperate and positively interact would increase as a consequence of growing confidence and understanding. These changes, however, were expected to follow different trends for the easier and the more difficult activities, due to the different amount of time and effort needed for their comprehension and progressive mastering. Therefore, we expected that the interest elicited by the most difficult and complex games could decrease over time, at least for subgroups of children, and that intra-peer conflict could arise as a consequence of greater mastery of the activity and increased self-confidence in at least some of the children.

The present contribution describes observed effects at the group level as a first step, and as a pre-requisite in the validation of the instrument and preceding the investigation of its effects at the individual level. No attempt was made, at this point, to describe the effects of the single activities on the targeted functions and abilities. All these further steps are envisaged for successive studies in both clinical and educational settings.

2. Materials and Methods

2.1. The NIRVANA System

NIRVANA 1 (BTS Bioengineering, Italy) is a markerless system that allows the total immersion in a virtual environment, without limiting or altering the freedom of interaction and the human motion. It projects virtual environments on a wall or on the floor and one or more subjects can interact with these environments through simple movements. The activity is supported by a high sensorial, visual and auditory stimulation, that engages the user.

The system is equipped with a workstation, two optoelectronic infrared cameras and two projectors 4000 ANSI lumens (one for the wall and one for the floor projection), a webcam and a Dolby surround system. The configuration of the system used in this work is shown in Figure 1. A1 and B1, the optoelectronic camera and the projector, respectively, for the floor activities, were installed on the ceiling in the middle of the active area; A2 and B2, the optoelectronic camera and the projector for the wall activities, were placed 2.5 m high. The required total electrical power was about 3500 W.

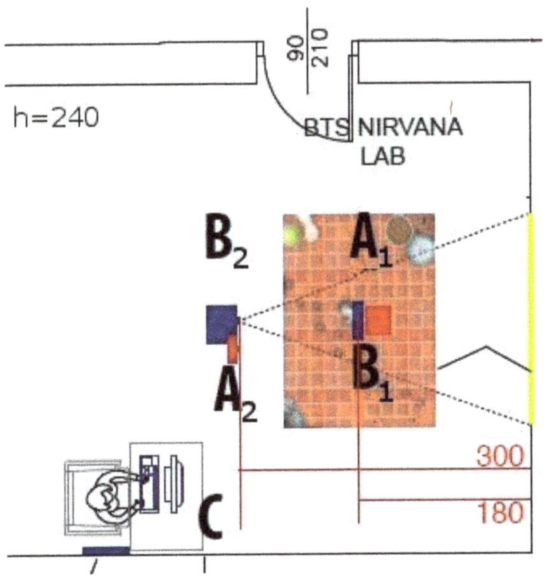

Figure 1. Nirvana Lab. A1 and A2 are the two optoelectronic infrared cameras (for the floor and the wall projection, respectively); B1 and B2 are the two projectors. C represents the workstation position.

The interaction with NIRVANA is straightforward: whenever infrared rays emitted by the optoelectronic infrared camera intercepts a body or a part of it, or even an object, an event occurs. This means that multi-touch and multi-subject approaches are possible.

The NIRVANA system is supplied with a graphic environment for the design of exercises; the software section is named Contents and it allows one to define new exercises according to six different typologies, with the following features:

Sprites: in this kind of exercise, it is possible to define a landscape and a variable number of objects, placed in user-selected locations (the user can specify X and Y coordinates with reference to the background). For each element, it is possible to define a size. These objects have two states and switch from one state to the other (another image or an animation) when touched. The user can define the timeout between different statuses of the element.

Particles: in this exergame, body movement removes the elements that cover the underlying background image. When the movement stops, elements cover the background again.

Reveal: in this kind of exercise there is a landscape with four states (images or animations with or without sounds). Changes in these graphic layers are allowed. The transition from one state to the following is defined by the adjustable percentage of area covered with the movement. If the movement stops, the sequences go back to the previous state. The default effect is water movement.

Move to: in this exercise it is possible to set a landscape and user-defined elements (e.g., dots, flowers, leafs). When an area is touched, elements gather in that area. A default activity of the system with these features is named "Dog" that can be customized by changing wallpapers.

Move away: this kind of exercise is similar to the previous one, but elements escape from the touched area (this typology has not been used for the present study).

Follow me: moving items should be tracked and an effect occurs when you reach the target. A default activity with these features is "Whack-a-mole".

2.2. Design of Activities

A series of activities have been defined, either developing completely new activities and games, by adding images, videos and audio-files to the pre-existing activities and defining new tasks and goals, or (in the few cases where customization options were limited) using the activities and games already provided by the system, but re-organized and structured according to a precise educational and cognitive framework, so as to stimulate the targeted functions in the most effective way. Adjustable parameters have been set so as to redefine the tasks and model the activity according to very specific cognitive goals. In many cases, structural or functional adaptation was aimed at extending individual activities to group activities, where each individual has a specific task or needs to take turns and/or monitor/interpret the other individuals' actions and intentions to modulate his/her own activity.

The creation of an immersive environment with wall and floor projection offers the possibility to carry out activities that can involve a medium-large group (5–10 children), with a variable need of mediation on the caregiver's side, according to the planned activities. Access to the activities is not determined by the age of the children, but it is rather tied to the contents proposed by the caregivers and to the organization they decide to impose on the activity itself.

From the cognitive point of view, the most stimulated abilities were gross motor-praxic and fine motor coordination skills, perceptual, attentional and memory functions, along with problem-solving skills. Both wall projection and floor projection were used in the design of the activities, trying to exploit their specific characteristics and potentials, in order to have a greater impact on both motor demands and collaboration/strategic planning requirements at the group level. Indeed, floor projection allows for more complex motor tasks and for more structured group arrangements and coordinated motor activities. However, wall projection allows for the greater stimulation of fine motor coordination of movements performed with arms and hands.

Table 1 provides a list of the different task typologies that are provided by the system, along with a description of the specific tasks that were developed for the project.

Table 1. Exercise Typologies, implemented activities, type of system projection and neuropsychological functions involved in each activity.

Exercise Typology	Activity	Projection	Neuropsychological Function
Sprites	Eggs	floor	Visual discrimination Motor coordination
Sprites	Sound environment	floor	Auditory discrimination Auditory-visual-matching
Sprites	Old MacDonald Had A Farm	floor	Auditory discrimination Auditory sustained attention Executive functions (Planning and Inhibition) Motor coordination
Sprites	Musical canon	wall	Auditory working memory Auditory discrimination Auditory sustained attention Auditory divided attention
Sprites	Musical puzzle	wall	Auditory discrimination Auditory working memory Executive functions (Planning)
Sprites	Jam Session	wall	Auditory discrimination Auditory sustained attention Auditory divided attention Executive functions (Planning) Auditory working memory
Reveal	Water Lily Pond	floor	Motor coordination Proprioceptive awareness Executive functions (Inhibition)
Particles	What's hiding?	floor/wall	Visuospatial attention visual-motor integration Lexical access
Follow Me	"Whack-a-mole"	floor	Visuospatial attention visual-motor integration
Move to	Dog	floor	Motor coordination Proprioceptive awareness Executive functions (Planning)

2.2.1. Typology "Sprites"

The "Eggs" activity was designed by choosing different egg types and different arrangements (Figure 2A). Moreover, a full educational framework was defined, in order to stimulate visual-motor and planning functions in a systematic and meaningful way. In this framework, children play according to defined rules and times, within an educational framework. The goal is to follow increasingly complex tracks with the steps. A starting level is provided to familiarize them with the task. Subsequent presentations include different egg tracks, distinguished by color and sound feedback (Table 2). The presentation of the tracks on the floor creates the condition for the involvement of two teams or, alternatively, the instruction to alternate the sequence of eggs to break by stepping on them.

Figure 2. Examples of scenes designed for the activities: (**A**) Eggs—find it 1 (see Table 2); (**B**) the "Old MacDonald Had A Farm" background; (**C**) Musical canon with the "Fra' Martino" (Brother John) wallpaper; (**D**) What's hiding? activity with the set "What game is it?"—a basket hidden by small balls (see Table 5); (**E**) What's hiding? activity with the set "What animal is it?"—a frog hiding under the clouds (see Table 5); (**F**) "Dog" activity, with an S-shaped path.

Table 2. "Eggs" activity arrangement. The sound effects "Broken shell 1" and "Broken shell 2" represent two different sound effects.

Name	Number of Eggs	Arrangement	Visual Effect	Sound Effect	Background
Eggs 1	10	a row in a straight line	Bullseye	Broken shell 1	Green lawn
Eggs 2	20	two parallel zigzag rows two types of eggs, different in color	Bullseye	Broken shell 1 Broken shell 2	Green lawn
Eggs 3	20	two irregular rows two types of eggs, different in color	Bullseye	Broken shell 1 Broken shell 2	Green lawn
Eggs 4	20	random arrangement two types of eggs, different in color	Bullseye	Broken shell 1 Broken shell 2	Green lawn
Eggs—Find it 1	20	The eggs are scattered on the lawn. They camouflage with the background	Bullseye	Broken shell 1	Pebbles floor
Eggs—Find it 2	10	Smaller and better hidden eggs are scattered on the lawn. They blend in the background.	Bullseye	Broken shell 1	Pebbles floor

The newly designed "Sound environment" activity is structured to facilitate recognition and association between a sound environment presented in acoustic mode with the corresponding photographic representation, located in the central area of the wall or floor. Around the image, there are four different colored trumpet-shaped buttons, which, once pressed, emit four different ambient sounds. The activity requires one to correctly match the auditory stimulus to the sound environment represented at the center of the layout. The location of the target-sound and the other sounds that act as distractors vary for each game set, to prevent a learning effect (Table 3).

Table 3. "Sound environment" activity: stimuli and matching.

Game	Central Figure (Target)	Sound Effect (Target)	Sound Effect 2	Sound Effect 3	Sound Effect 4
Sound environment 1	Asian megalopolis	Traffic noise	Kids playing voices	Jungle with animal noises	Symphonic orchestra
Sound environment 2	Animal farm	Animal farm noises	Birds	Symphonic orchestra	Traffic noise
Sound environment 3	Jungle	Jungle with animal noises	Animal farm noises	Traffic noise	Thunder
Sound environment 4	Mexican Mariachi	Music from Mariachi band	Rock concert	Kids playing voices	Flute solo
Sound environment 5	Thunderstorm with lightning	Thunder	River water	Traffic noise	Murmur of the sea
Sound environment 6	Circus tent	Circus theme	Music from Mariachi band	Symphonic orchestra	Kids playing voices
Sound environment 7	Stadium	Supporters choir at the stadium	Kids playing voices	Traffic noise	Rock concert
Sound environment 8	Symphonic orchestra	Symphonic orchestra	Rock concert	Flute solo	Music from Mariachi band

Among the proposals of the "Sound environments", the original activity "Old MacDonald Had A Farm" was designed. In this variant, the different characters in the well-known song replace the colored buttons to be pressed (Figure 2B). The presentation can be used both on the wall and on the floor. In the layout, there are drawings of the farmer and farm animals. The motor interaction with the character "Farmer" is coupled with the musical base of the song "Old MacDonald Had A Farm". To make the activation of the song accessible only to the educator, the farmer's character was positioned high and away from the other animals, near where the children were positioned. The track has been modified by adding silent pauses, in correspondence with the noises of each named animal. Five farm animals were identified in the song and for each one, a button with the animal's drawing was inserted, which was associated with an audio file of the noise. The activation of each character can be viewed, thanks to the colored borders that remain lit as long as the associated sound is produced. Children are required to activate their character at the exact moment when the song names their animal, producing the associated voice. The activity stimulates auditory attention functions, auditory work memory, executive functions of planning and inhibition; it also promotes the enhancement of shift management skills and collaboration, offering correctness feedback that can be easily recognized by children without adult mediation.

In the ad-hoc conceived "Musical canon" activity, the goal is the creation of a canon composition, starting from four different guitar strings presented on the wall. Each string of the virtual guitar is associated with a complete verse of a song, which lasts about 15″–20″. Children are required to select the strings in sequence, with progressive timed insertions, creating the canon effect (Figure 2C).

Additionally, the "Musical Puzzle" activity has been designed by the authors as a task requiring one to reconstruct a song from musical fragments associated with four different trumpets presented on the wall. Each audio track has a duration of 5″–10″. The trumpets must be activated in the correct order to obtain the complete song.

In the "Jam Session" activity, the goal is the composition of creative music sessions. Ad-hoc selected audio tracks of the distinct musical instruments that make up a song are presented. Each of the four strings of the virtual guitar projected on the wall is associated with an audio file with an isolated partition of an instrument. The tracks each last 10″–15″. The strings can be played sequentially or simultaneously to render the effect of the entire song.

Table 4 lists the content of musical activities.

Table 4. Contents of musical activities.

Activity	Song or Music Track	Procedures	Projection
Musical Canon	"Fra' Martino" ("Brother John", traditional song)	Mediation and instructions required	wall
	"Capra Capretta" (nursery rhyme)	Mediation and instructions required	
	"Stella Stellina" (nursery rhyme)	Mediation and instructions required	
	"La canzone del cuculo" (nursery rhyme)	Mediation and instructions required	
Musical Puzzle	"La casa" (nursery rhyme)	Whole audiofile to be presented first	wall
	"Il leone si è addormentato" —Italian version of "The Lion Sleeps Tonight" (children's song)	Whole audiofile to be presented first	
	"Ci vuole un fiore" (children's song)	Whole audiofile to be presented first	
Jam Session	"House of the rising sun" (traditional folk song)	Free access or mediated by educator	wall
	"La Bamba" (Mexican folk song)	Free access or mediated by educator	
	Reggae melody (instrumental)	Free access or mediated by educator	

2.2.2. Typology "Reveal"

In the "Water Lily Pond" activity, motor responses such as inhibition and movement and body control are requested. The goal of the ad-hoc designed tasks is to remain as still as possible on some lotus leaves, assuming the positions assigned by the educator (for example, standing still on one foot, the position of the frog, two children on the same leaf, etc.) Visual feedback is provided by the water surface of the pond, sensitive to movement. A motor variant of the task is proposed, requiring children to jump from one leaf to another. In this version, the goal is to be very accurate in positioning itself on the leaf, minimizing water movements.

2.2.3. Typology "Particles"

In the "What's hiding?" activity, children are invited to organize themselves to find out what (object or short sequence) is hidden under a layer of elements ("particles"), that move due to the interaction of the body on the floor (Table 5). The particles become rare with the movement of the body and re-thicken in a short time (Figure 2D,E). This activity was largely manipulated so as to address several cognitive functions. The images to be unveiled were selected following the general criterion of being difficult to either spot or recognize, thus involving visuospatial attention, visual discrimination, visual representation/integration and, on the other hand, the need for cooperation between many several children, to achieve the objective of identifying the target. Three types of visual combinations were envisaged:

(1) Very small target-element with slow-moving particles
(2) Very large target-element with fast-moving particles
(3) Chronological sequences (with several elements in a sequence) with fast-moving particles

Table 5. "What's hiding?" activity arrangement.

Name of the Set	Content Elements	Arrangement
In the grass	Ladybug; Mushrooms; Apple; Shoe; Ball; Can; Watch	Small red elements spread in the grass
What fruit is it?	Watermelon; Strawberry; Kiwi; Melon; Pineapple; Orange; Apple; Chestnut	Enlarged significant details of each element are shown on the screen
What animal is it?	Tiger face (very close); Cat face (very close); Starfish (small); Seashell (small); Crab (small); Frog (camouflaged); Butterfly (camouflaged)	Mixed enlarged, small and camouflaged elements are shown on the screen
What object is it?	Wall clock; Corkscrew; Sharpener; Scissors; Nutcracker; Dominoes; Pen; Key; Brush; Clew	Enlarged significant details of each element are shown on the screen
What game is it?	Football; Tennis; Ping-pong; Basketball; Table football; Golf	Enlarged significant details of each element are shown on the screen
Video 1	A butterfly comes out of the cocoon; Lightning pierces the night sky; Cookies baking in the oven; Birds fly in the sunset sky; A pigeon walks on an urban background; A monkey crosses a road	Short movie clips are shown in succession on the screen
Video 2	An airplane flies in the sunset sky; A car drives on a road; A man paddles on the sea while a shooting star appears in the sky; A pedestrian walks in an urban background; A flower blooms; Movements of the sponges on the coral reef	Short movie clips are shown in succession on the screen
Video 3	A man rowing; A squirrel peeks out; A car crosses an urban background; Some elephants cross a road; A seagull flies in the sunset sky; A big monkey crosses the road	Short movie clips are shown in succession on the screen

2.2.4. Typology "Follow Me"

The "Whack-a-mole!" activity was used as provided by the system. Selective and diffused visual attention functions are involved and the game requires a gross-motor response while performing the activity on the floor. The goal of the game is to tap the moles as soon as they emerge from the ground. The speed of presentation of the stimuli is variable and the players receive visual and sound feedback when they "catch" the mole.

2.2.5. Typology "Move to"

The "Dog" activity was customized by adding special floor backgrounds, on which different paths are traced through the use of images of footprints or path signs (circles, start and finish lines, platforms, etc.) (Figure 2F). Children were invited to divide themselves into small groups (2–3 children), according to different rules given by the educator (hand in hand, back to back, etc.). The aim of the activity is to coordinate with each other to follow the paths on the floor, without being joined by the dog that chases them. The system reports how many times the dog "touches" a child on the floor projection.

2.3. The Game Sessions

A subgroup of designed activities was devoted to the stimulation of visual-perceptual and visual-attentional functions combined with motor coordination skills, especially supported by floor projection. Children were requested to walk and reach objects or targets disposed according to configurations of varying complexity and to move within predefined spaces maintaining fixed configurations with other children, while the outer context may be varying, or to organize group strategies to reach a given goal.

In another set of activities, auditory discrimination and auditory attention were targeted, as well as the abilities to form visual-auditory associations or to train auditory working memory, executive functions and creative expression. At the same time, children were offered an opportunity to refine their musical sensitivity and their perception of rhythm and melody. Various types of auditory-based activities were proposed to the children subdivided into small-medium groups (3–5 participants). The activities stimulated the ability to recognize and associate certain sounds typical of a natural or human environment, or they included musical tracks to be rearranged or to be combined logically or creatively. Most of the proposals were designed for wall projection, to decrease the motor activation of the body and to promote cooperative interaction between children. Cooperation was stimulated both in terms of sharing information and in making decisions.

The psychologists who took care of conducting the activities with the children built a narrative frame, on the theme of the "Journey" as a metaphor for a journey through various areas of expertise. A fictitious character named "Auntie Simalù" was invented and presented through drawings and animations, playing the role of a guide who leads the journey and acting as a glue and connecting element between the various proposals.

2.4. Participants

The games were proposed to 25 children (12 males), aged 4–5 years, attending a kindergarten in Lecco, Italy, at the beginning of the new school year. All the children whose parents gave informed consent to participation were included in the study. All children had been invited to participate by the team of school teachers, provided they were at least 4 years of age and there were no organizational factors that would make participation discontinuous (e.g., children who could not be at school in the first observing session due to late morning arrival were not asked to join the project). Children from different sections (four classes) were mixed into different groups, so as to obtain comparable groups with respect to age, level of mutual knowledge (which was generally low since the school year had just started) and multicultural features. There were 3 children of non-native origin in the group. There were no known disabled children, except for a child with mild mobility difficulties.

All of the children had had previous experience with electronic devices, and at least a PC or a tablet was present in each of the families, as reported by the parents in a questionnaire. The social background of the families was that of a small city, and the school was located in a part of the town where no particular social problems (crime or deviant behaviors) are usually reported.

The children were divided into three different groups of 8–9 children each and were guided through the activities by the psychologists and class teachers. The choice to divide the children into three groups allowed one to plan game sessions of one hour each, which was deemed to be a reasonable time, considering the attention times and play skills of children of this age group. Moreover, groups of such size allow rich and dynamic group interactions to be observed, and at the same time are easily managed by adult observers. All sessions were video recorded.

The children's parents signed informed consent forms, as well as permission and release forms for images, videos and sound recordings, in accordance with the Declaration of Helsinki.

2.5. Observation Procedures

A structured observation grid about the observed behaviors and interactions was filled by the adult participants (teachers and psychologists) at the end of each session. A copy of the grid can be found in the Supplementary Materials.

In the observation form, 10 descriptive variables have been provided, assessed independently by each observer involved, for each activity proposed in the different days of experimentation in the school context. The grid required evaluation on a Likert 5-point scale (from 1 = "not at all" to 5 = "very much") of a series of variables concerning: "Game" (understanding of the game; strategic behaviors; participation; enjoyment), "Interactions with peers" (interaction; conflict; cooperation) and "Clarification requests" (questions about the device = Q-device; questions about the activities = Q-activity; active help requests = Q-active help). Each variable was rated at three different moments of the playing session: initial (T1), intermediate (T2) and final (T3). Since the emphasis was on the groups and not on the single individuals, and since variables such as interaction and cooperation are most meaningful if evaluated at the group level, the raters were asked to fill a rating form for each of the observed groups.

2.6. Data Analysis and Statistics

Data from the ratings collected on the groups performing the same kind of activity were collapsed by averaging within each rater, since there was no hypothesis of any difference between groups. Then, a reliability analysis was performed among the raters, to assess their agreement. Specifically, intraclass correlation coefficients (ICC) were computed for each activity, considering a two-way random model and looking for the absolute agreement. When the agreement was considered optimal, data from different observers were averaged. Subsequently, the scores of all the different activities were analyzed together, in order to evaluate the system as a whole. First of all, a non-parametric paired analysis was performed to check for differences among timepoints for each variable: Friedman test was run among T1, T2 and T3 and, when statistically significant, a post hoc Wilcoxon test was performed between couples of timepoints. The level of significance was set at 0.05.

Finally, a non-parametric correlation analysis was run between all the variables; Bonferroni correction was applied to account for multiple comparisons. Due to the high level of similarity of two variable pairs (Participation and Enjoyment on one hand, Interaction and Cooperation on the other hand), as reported by the raters and as confirmed by pairwise correlations (rho = 0.86 and 0.91, respectively), it was decided to consider such pairs as expressions of a same underlying construct and therefore, to have each pair count as 1 for the Bonferroni correction (thus, correction was applied for a total of 7 variables instead of 9, with alpha set at 0.05/28 = 0.002).

Analyses were performed in SPSS 21.

3. Results

The test in the kindergarten setting allowed evaluation of the app functionality.

The activities were highly appreciated by the children, who experimented with the device with curiosity and enthusiasm. A video illustrating various moments of the study, with the children performing different activities, is provided in the Supplementary Materials.

Figure 3 depicts specific moments during activities, with virtual environments projected on the floor (Figure 3A,B) or on the wall (Figure 3C,D).

A total of 58 filled grids were collected on a set of 31 total game sessions (each game session was devoted to one activity). The forms were filled by the different observers who participated in the activities (educators and psychologists).

The number of observations for the various activities (9 different activities) ranged from 12 (What's hiding) to 2 (musical canon), with a mean of 6.44. The different numbers depended on the teachers' time schedules that allowed for participation in the activities, only at certain times of the day. At least one of the two psychologists was always present.

First of all, ICC values computed among the raters for each activity were larger than 0.9, but for Musical Puzzle (ICC = 0.5). Therefore, the data from different observers, except for Musical Puzzle (for which ICC was deemed to low) and Canon (for which too few observations were collected to be considered reliable), were collapsed by averaging.

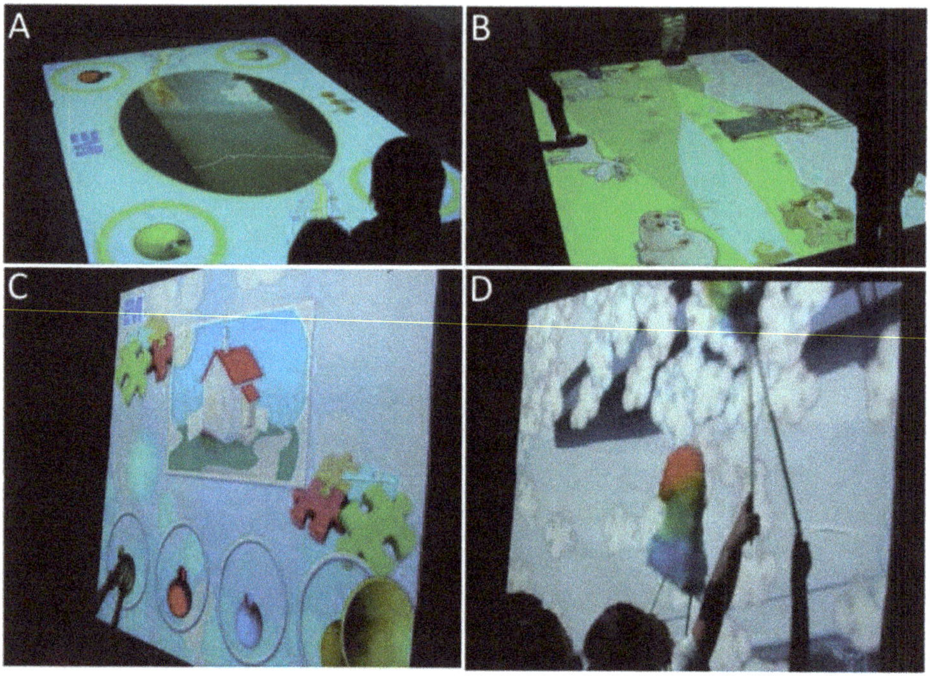

Figure 3. Moments of playful activities. (**A**) Sound environments: children have to find the trumpet with the sound corresponding to the central image (a storm). (**B**) the "Old MacDonald Had A Farm": children have to jump on the animal in the correct instant during the song. (**C**) Musical Puzzle: children have to find the correct order of the trumpets to play the song ("La casa"—the house) correctly. (**D**) What's hiding activity: children use feather dusters to discover what clouds are hiding (a ballpoint pen detail).

Figure 4 shows the results about the variables rated at the different timepoints, T1 (Initial), T2 (Intermediate) and T3 (Final), considering the system as a whole (for this analysis, Musical Puzzle and Canon were not considered). As supported by the statistical analysis (Table 6), a significant increase in comprehension, strategy, enjoyment, participation, interaction and cooperation along timepoints was highlighted. A post hoc analysis stated that significant differences were mainly between T1–T2 and T1–T3, while they stabilized at T3, but for comprehension, that significantly increased also at T3.

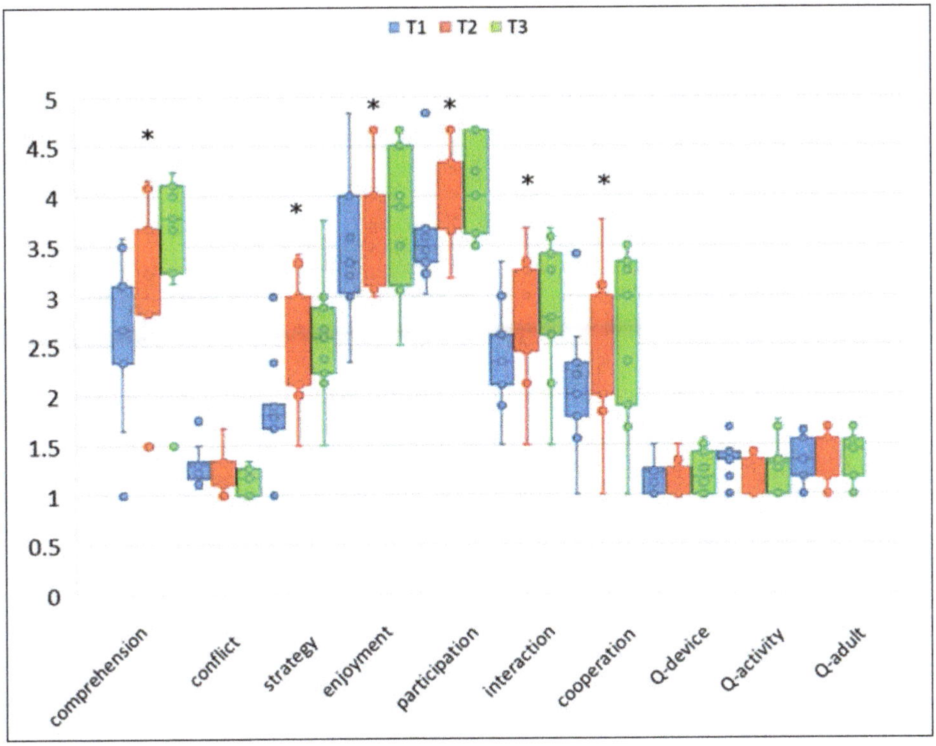

Figure 4. Box-and-whisker plots of each variable at T1, T2 and T3. Statistically significant differences among timepoints, as defined by Friedman test, are marked with a star.

Table 6. Comparison among timepoints for each variable. Data in T1, T2 and T3 are reported as mean ranks. Statistically significant correlations ($p < 0.05$) are shown in bold. When the Friedman test was statistically significant, p-values of post hoc analysis (Wilcoxon test) are reported. Legend: COMP—Comprehension; CONF—Conflict; STRAT—Strategy; ENJ—enjoyment; PART—Participation; INT—Interaction; COOP—Cooperation; Q-DEV—Q-device; Q-ACT—Q-activity; Q-ADU—Q-adults.

	Friedman Test					*Post Hoc*—Wilcoxon Test		
	T1	**T2**	**T3**	**Chi-Squared**	***p*-Value**	**T1 vs. T2**	**T1 vs. T3**	**T2 vs. T3**
COMP	1.00	2.13	2.88	15.20	**0.001**	**0.012**	**0.012**	**0.028**
CONF	2.19	2.38	1.44	5.25	0.072			
STRAT	1.00	2.38	2.63	13.07	**0.001**	**0.012**	**0.012**	0.674
ENJ	1.31	2.19	2.50	6.69	**0.035**	**0.043**	**0.035**	0.173
PART	1.25	2.25	2.50	8.00	**0.018**	**0.021**	**0.017**	0.144
INT	1.19	2.13	2.69	10.14	**0.006**	**0.017**	**0.018**	0.092
COOP	1.13	2.31	2.56	9.74	**0.008**	**0.017**	**0.012**	0.612
Q-DEV	1.94	1.88	2.19	1.40	0.497			
Q-ACT	2.44	1.63	1.94	3.91	0.142			
Q-ADU	2.13	1.88	2.00	0.50	0.779			

Figures 5–7 show the impact of activities on each variable; for this analysis, we decided to take all of the activities into consideration (including Musical Puzzle and Canon), for the sake of completeness.

As can be seen from Figure 5, the understanding of the activity improves from the beginning of the session to the end, and achieves good absolute ratings in general. The exceptions are some of the music-based activities, Musical Puzzle and Canon, which remain below 2 at the end.

Strategic behaviors in the groups of children tend to increase over time, with both increasing practice and familiarity with the task, especially from T1 to T2, reaching high ratings in "Whack-a-mole" and Eggs activities.

Participation is generally high in all of the activities; it increases from T1 to T2 and remains stable over T3. Enjoyment is high in all the activities and increases over time, with the exception of music-based activities: in Musical Puzzle, Canon and Jam session, enjoyment at T3 is lower than at T2.

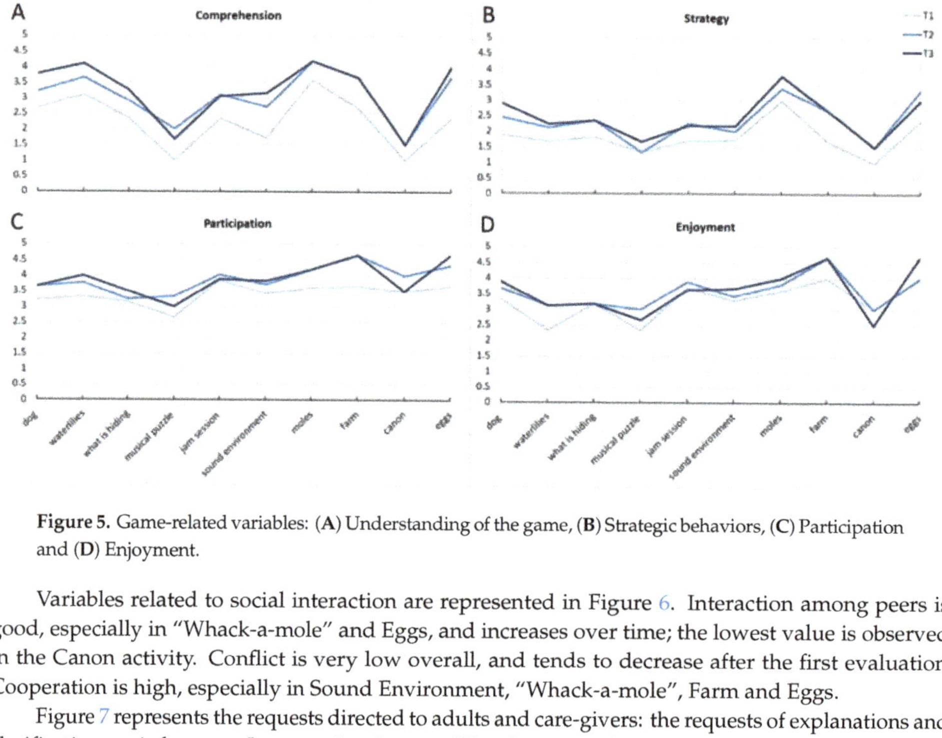

Figure 5. Game-related variables: (**A**) Understanding of the game, (**B**) Strategic behaviors, (**C**) Participation and (**D**) Enjoyment.

Variables related to social interaction are represented in Figure 6. Interaction among peers is good, especially in "Whack-a-mole" and Eggs, and increases over time; the lowest value is observed in the Canon activity. Conflict is very low overall, and tends to decrease after the first evaluation. Cooperation is high, especially in Sound Environment, "Whack-a-mole", Farm and Eggs.

Figure 7 represents the requests directed to adults and care-givers: the requests of explanations and clarification are infrequent (between 1 = "not at all" and 2 = "rarely") concerning both the device and the activity, and direct requests of help are quite uncommon, except for Musical Puzzle, where requests increase during the session.

Finally, results of the correlation analysis are reported in Table 7. The understanding of the activity strongly correlates (Spearman's rho > 0.7) with strategic behaviors and participation, and has a moderate correlation with interaction. Furthermore, strategy, enjoyment, and participation moderately correlate with interaction. Enjoyment has a very strong correlation with participation and a moderate correlation with cooperation. Finally, a very strong correlation was found between cooperation and interaction.

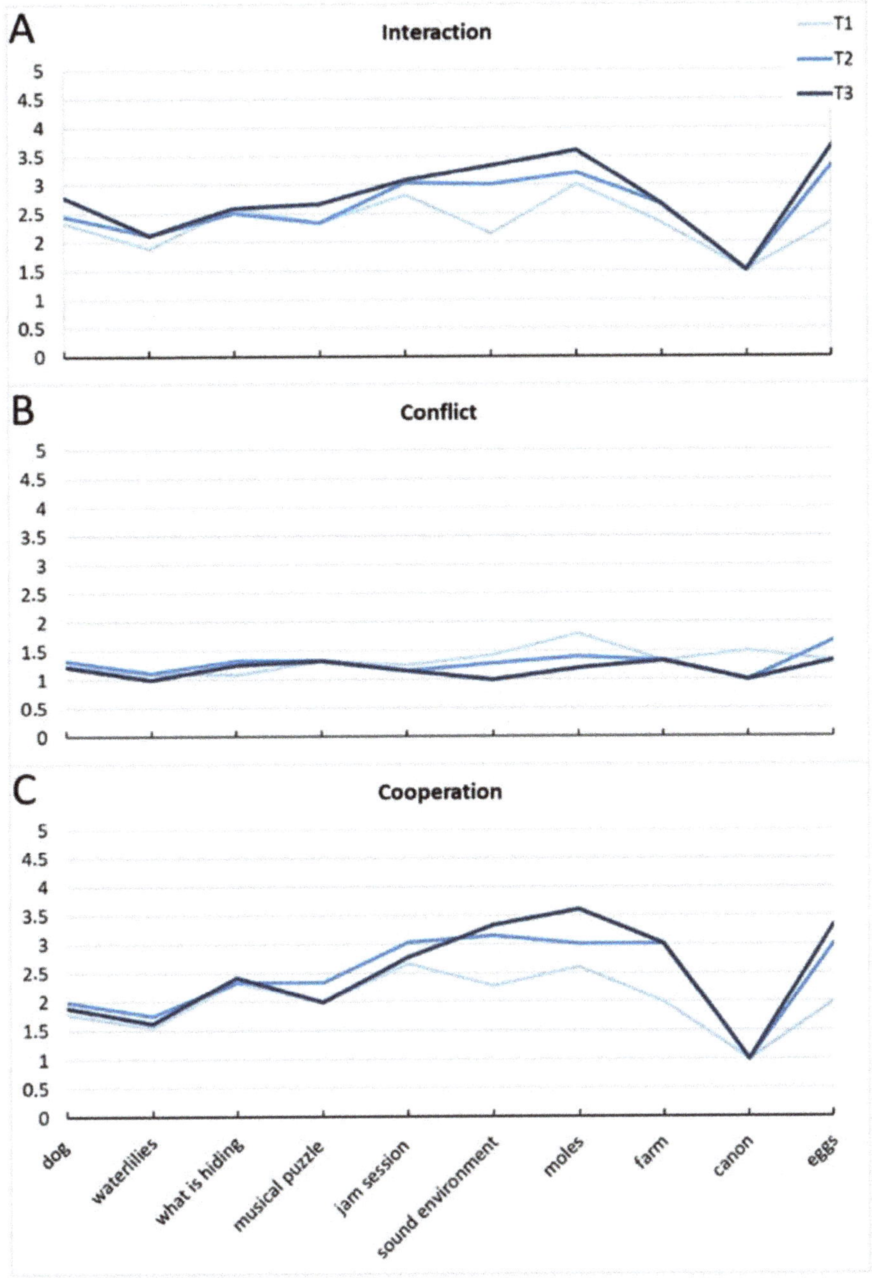

Figure 6. Interaction with peers: (**A**) Interaction, (**B**) Conflict and (**C**) Cooperation.

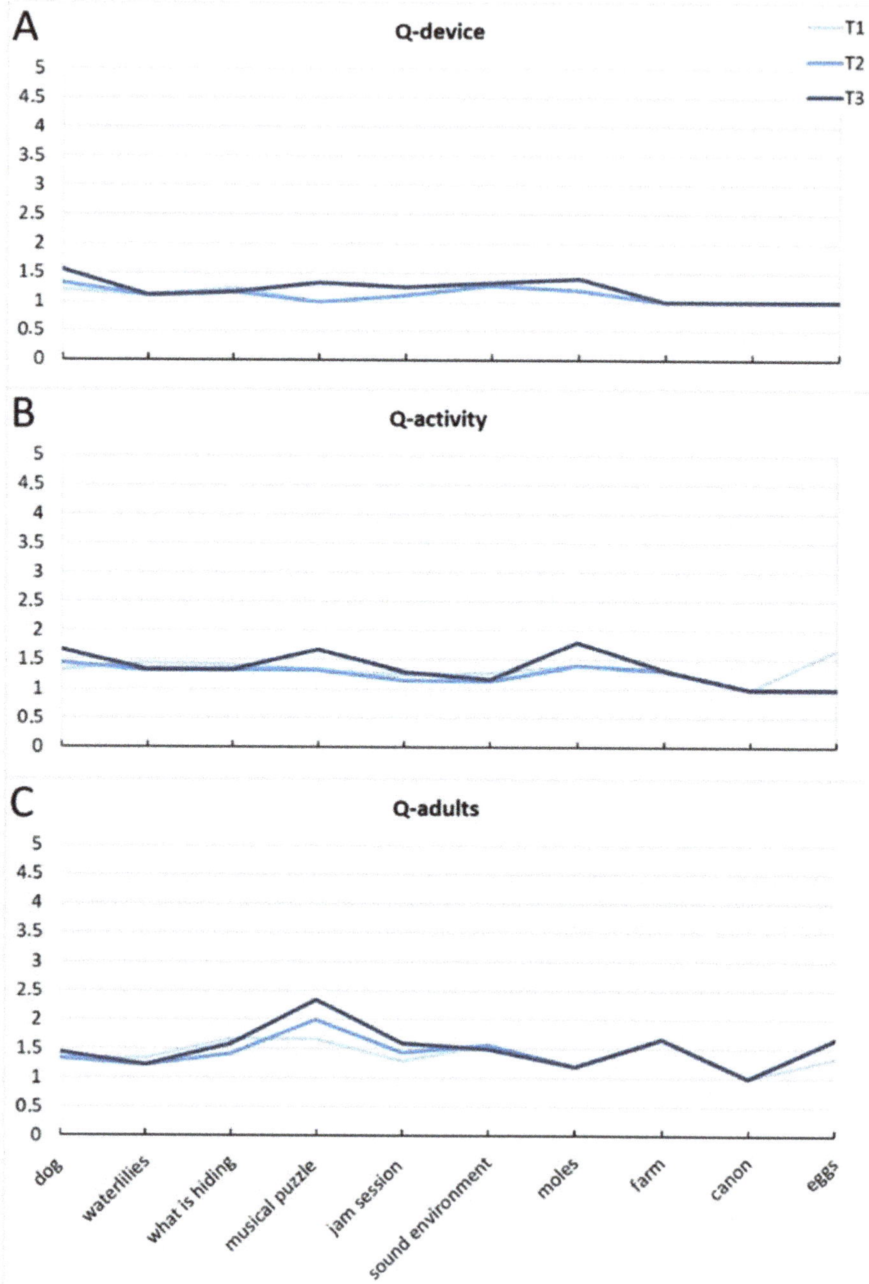

Figure 7. Clarification requests: Questions concerning the device, activity and help requests.

Table 7. Correlations among variables. Data are reported as Spearman's rho (p-value). Statistically significant correlations ($p \leq 0.002$ to correct for multiple testing) are shown in **bold**. Legend: COMP—Comprehension; CONF—Conflict; STRAT—Strategy; ENJ—enjoyment; PART—Participation; INT—Interaction; COOP—Cooperation; Q-DEV—Q-device; Q-ACT—Q-activity; Q-ADU—Q-adults.

	CONF	STRAT	ENJ	PART	INT	COOP	Q-DEV	Q-ACT	Q-ADU
COMP	−0.15 (0.477)	**0.74 (<0.001)**	0.46 (0.024)	**0.73 (<0.001)**	**0.60 (0.002)**	0.44 (0.030)	−0.05 (0.819)	−.19 (0.375)	−0.40 (0.055)
CONF		0.35 (0.098)	0.26 (0.213)	.06 (0.800)	0.01 (0.958)	.08 (0.714)	−0.41 (0.049)	.19 (0.368)	0.43 (0.035)
STRAT			0.50 (0.014)	0.53 (0.008)	**0.69 (<0.001)**	0.51 (0.010)	−0.03 (0.892)	−0.03 (0.885)	−0.16 (0.447)
ENJ				**0.86 (<0.001)**	**0.65 (0.001)**	**0.63 (0.001)**	−0.23 (0.287)	−0.12 (0.166)	−0.03 (0.892)
PART					**0.60 (0.002)**	0.58 (0.003)	−0.30 (0.161)	−0.37 (0.076)	−0.16 (0.449)
INT						**0.91 (<0.001)**	.19 (0.379)	−0.46 (0.024)	−0.17 (0.419)
COOP							0.18 (0.390)	−0.52 (0.009)	−.13 (0.537)
Q-DEV								0.20 (0.340)	−0.34 (0.110)
Q-ACT									−0.11 (0.595)

4. Discussion

The present study describes a system intended to stimulate motor skills, perceptual functions, executive functions, and social skills.

The system was tested with three groups of normally developing children during kindergarten activities. The number of different observations and the varying composition of the observed groups does not allow one to statistically compare initial, intermediate, and final observations for the single activities. Therefore, the general trends concerning changes over time in the observed behavioral variables have been analyzed based on the averages of the various activities. Nonetheless, a general evaluation of the graphs representing the mean levels of the observed variables for the single activities sometimes show very clear patterns and trends that can be considered significant at a qualitative level and will be discussed in the next paragraphs.

Overall, the results of structured observation reveal that the system is easy to understand, enjoyable, elicits high levels of participation and little conflict, moderately favors strategic behaviors, cooperation and social interaction and can be used by the children with a limited need of instruction or support from an adult. Notably, the children's engagement and their cooperation during the activities improve with the familiarization and use of the system. More precisely, there is a clear improvement in these aspects from the beginning to the midpoint of the activity, and a general stabilization from the latter to the end of the activity. Conflict, by contrast, remains stable (and low) across the whole session, and so do questions about the activity and help requests directed to adults.

Comprehension of the tasks generally improves during the session, and achieves a very good level in the end. An exception is represented by two of the music-based activities, Musical puzzle and Canon, which start with rather low levels of comprehension and improve only minimally across the session. This probably depends on the nature of these activities, which require a high level of coordination among several members of the team, and very rigid turn-taking, based on the accurate reconstruction of song texts in one case, and the ability to sing out-of-phase in the other case. These are highly complex skills that are probably developed later on. The easiest activities are the motor-based activities, for which both initial comprehension and progressive increase in comprehension reach higher levels.

Additionally, strategic behaviors tend to increase with time. This is particularly true for motor games, such as Dog, "Whack-a-mole" and Eggs, where organization and coordination between participants in the group make their action much more effective. This advantage has probably been noticed and exploited by the children, and it is also reflected in the significant, positive correlation between strategic behaviors and both task comprehension and interaction (the correlations with participation and cooperation, even if non-significant after Bonferroni's correction, also represent moderate associations between variables). In other activities where the need for structured, organized action of the groups is even more explicit, such as the music-based activities, a clear increase in strategic behaviors is also observed, and it is present even for the most difficult activities, such as the Musical Puzzle (where an improvement is reached at the end of the session only) and Canon. These results suggest that accessibility and usability of the proposed activities are adequate and satisfactory.

Participation is generally high, with an initial increase and then stabilization. The only exceptions are the two most difficult activities, Musical puzzle and Canon, for which there is an initial increase in participation, followed by a decrease towards the initial levels. A similar trend is observed for both Interaction and Cooperation among peers, which tend to increase over time (also from midpoint to end of the session) for all activities but Canon, which remains at rather low levels for both parameters. This result is probably to be interpreted with regard to the difficulty in understanding the game, even after some time has passed and some experience has been gained. Indeed, there are high and significant correlations among comprehension, participation and interaction scores, supporting this interpretation. Increases in interaction and cooperation are particularly evident for activities like Sound Environments and "Whack-a-Mole", which have a clear group structure but are easier to understand.

Conflict is generally very low, and tends to be stable, or even decrease for some activities such as Sound environment, "Whack-a-mole" and Canon, i.e., the most challenging activities, where the advantages experienced through strategic interaction and cooperation may have encouraged group cohesion, rather than competitive behaviors.

Enjoyment of the activities is high and shows an increase from the first to the second time measure. This is especially true for motor activities requiring structured, organized group action, such as Waterlilies and Farm. The most difficult activities (Musical puzzle and Canon) even show a first increase followed by a decrease in enjoyment, suggesting that such activities could be particularly tiring and possibly boring for children as young as preschoolers. Correlations with enjoyment confirm that this variable is strictly linked to participation, which can be seen as conceptually related to it (both express the capacity of the activity to attract and involve the child), but also to interaction and cooperation, which suggests that sharing the activity with peers increases the pleasantness of the activity itself.

As to the questions directed to the adults, including requests of explanation, clarification (concerning both the device and the activity) and help, these are very infrequent and confirm the good levels of accessibility and usability of the system. The absence of correlations with other variables, however, may indicate that all types of questions were rather an expression of participants' needs, rather than reflecting the effects of other factors.

Such results suggest that VR and digital applications have the potential to become important instruments in promoting children's cognitive and social development, and in improving routine educational work in kindergarten settings. The system turned out to be flexible and able to adapt to the various goals of the educators and psychologists: even starting from the same typology of activity, it was possible to address very different skills and functions, with a moderate amount of programming work. The possibility to vary both images and sounds and combine them in various ways offered many different pathways to stimulation of a certain capacity, so that even a single, very specific rehabilitation goal could be pursued without having the child performing repetitive and boring activities. Another valuable characteristic of the activities is the ability to keep high levels of attention and motivation during the whole session, even increasing them from the beginning to more advanced phases. In other terms, the "novelty effect", which is often reported in studies on the use of VR in educational contexts [32,51], was not observed here. Last but not least, the system (which has mainly been developed for individual rehabilitation) revealed to be very suitable for group activities, provided that educators or trainers have good mastery of the programming part and can exploit the various options in changing, creative ways.

Floor projection turned out to be very suitable to motor games, not only in standing position (jumping, running, balancing etc.), but also in lying position, where the children could roll and crawl, and use their legs, hands and arms to produce effects on the underlying images. This was very entertaining and motivating for the children, and also evoked many different ways to interact with peers and organize various motor strategies to improve effectiveness. Wall projection, by contrast, was used mainly for music-based activities ("pulling" the strings to produce notes or melodies or song parts). This was performed with hands and fingers, but it was less immediate for the children to perceive the direct link between finger/hand movements and their effects, due to the characteristics of the projection/detection system. The feedback that the system gives is indeed driven not only by the direct contact of hands and fingers with the wall, but also by the movement in the space between the optoelectronic camera emitting infrared rays and the wall. Another application of wall projection was for the "what's hiding" activity, where the children had to reveal, through their movement, objects and object parts hiding under floating elements like balls or clouds. In this case, the effect of hands and arm movement was amplified by using colored feather dusters, which the children greatly enjoyed, and which produced a greater feeling of directly acting on the wall surface.

The single activities appear to have some unique characteristics that should be kept in mind when planning an intervention for cognitive empowerment or rehabilitation. In particular, motor games and, more generally, activities involving gross body movement, such as following paths, jumping, chasing, appear more motivating and yet, are able to stimulate strategic behaviors. Music-based activities can be difficult in this age-range and should be kept as simple as possible, allowing the free expression of creativity (like in the Jam session activities or in the Sound environments tasks). They appear to stimulate cooperation and interaction more than strategic behaviors. More complex activities may suit older children.

The system used, based on the principles of virtual reality, was demonstrated to be suitable for young pre-school children, as it was easy to understand and to interact with. The multiplayer feature of the system is a strength in this playful context, with respect to other systems, used by a single player in a rehabilitation session [53,54]. However, the main disadvantage of the system, at least in the wall configuration, is the lessened realism of interaction, due to the method of capturing movements with the optoelectronic system.

On the whole, a positive impact of the activities at both the cognitive and the social level is observed, which suggests that the system can be effectively used in a kindergarten setting to empower motor planning, strategic behavior, and cooperative skills. The system seems to be suitable also for rehabilitation applications in children with cognitive or motor disorders. Most of the principles described by Gee [40,55,56] are implemented in the activities. Among others, the "Active, Critical Learning Principle": the learning environments encourage active and critical learning; the "Multiple Routes Principle": there are multiple ways to move ahead, and this allows learners to make choices, exploit their own strengths and styles of learning, explore alternative styles; "Skills as Strategies": the learner isn't simply practicing a skill for its own sake, but with the goal of solving a problem; the "Multimodal Principle": meaning and knowledge are built up through various modalities (images, sounds, words, symbols, interactions, etc.); "System Thinking": games make players think in a bigger picture, helping them see how the pieces can fit together; and the "Intuitive Knowledge

Principle": intuitive knowledge is built up in repeated practice and experience, and this occurs in association with an affinity group.

A limitation of the present study is the lack of direct indexes of the children's performance during the activities. Future studies will be needed to collect objective data on the effectiveness of the system and activities in improving the cognitive and neuropsychological functions targeted in the various games and observing the impact on social attitudes and behaviors at the individual and group level in a more systematic way and over longer periods of time. In order to do this, it would be very useful if the system could offer more possibilities to record data about accuracy and qualitative information during the activities (e.g., number of different children actively participating in a certain activity at a certain moment, number of steps taken to reach a certain goal, etc.) Generally speaking, it would be desirable that this type of semi-immersive system, essentially designed and used in rehabilitation settings, could be adapted (in terms of design, user-friendliness and cost) for use in educational settings. In this perspective, technological solutions (not conceived as mere transpositions of traditional activities, but exploiting all the unique possibilities of virtual environments) could, at the same time, extend the range and breadth of situations to be experienced and skills to be learned, and also constitute an introduction to the more and more pervasive ICT reality.

A last note should be devoted to the multidisciplinary teamwork that led the programming of the activities, the games, and the organization of the game sessions, involving psychologists, educators, teachers, engineers, and technicians. This allowed for pedagogical and cognitive, as well as functional, and usability-related issues to be implemented and harmonized in the final programs: the results suggest that this co-designing work was advantageous for the success of the experience and the positive reactions of the children.

5. Conclusions

The present study focuses on the impact of the proposed activities at the group level, before extending investigation to the effects produced on typically and atypically developing children at the individual level. The results suggest that the system can be successfully employed for empowerment of social abilities in group activities. Further studies are envisaged, testing the impact of the activities on children with neurodevelopmental disorders, especially those involving deficits of motor-praxic organization, perceptual and social skills. Children with language and communication disorders are a very interesting target, as they could train some of the perceptual prerequisites for language learning, such as auditory discrimination and rhythmic abilities [57,58]. The activities could have a twofold application: improving weak functions in the at-risk population, either in a mainstream or special educational context or in a rehabilitation context, and empowering developing functions in typical populations in schools or other aggregation/education centers, whilst at the same time promoting greater inclusion and preventing isolation, aggression or bullying behaviors.

Supplementary Materials: The following are available online at http://www.mdpi.com/2076-3417/10/8/2948/s1.

Author Contributions: Conceptualization, M.L.L., M.G., and S.T.; methodology, M.L.L., M.G., and S.T.; software, P.N., S.T. and E.B.; validation, E.B., M.L.L., and G.R..; formal analysis, E.B., M.L.L.; investigation, M.G. and S.T.; resources, P.N., and G.R.; data curation, M.G., S.T., and P.N.; writing—original draft preparation, M.L.L., S.T. and E.B.; writing—review and editing, M.L.L., E.B., M.G., and S.T.; visualization, M.L.L., M.G., S.T., and E.B.; supervision, G.R.; project administration, G.R.; funding acquisition, G.R. All authors have read and agreed to the published version of the manuscript.

Funding: This research was funded by a grant from the Lombardy Region of Italy, in the context of the GIOCOSO project (GIOchi pediatrici per la COmunicazione e la SOcializzazione), ID code 40388780; and by the Italian Ministry of Health (Ricerca Corrente RC-2019 to G.Reni and RC-2020 to E. Biffi).

Acknowledgments: The authors wish to thank the kindergarten "Rosa Spreafico" of the Institute Stoppani in Lecco, Italy, with the director, the teachers and all the families who participated. They also thank all the project partners, with special regard to AERIS, who collaborated in planning project contents and in collecting data t kindergarten.

Conflicts of Interest: The authors declare no conflict of interest.

Data and Software Availability Statement: The data that support the findings of this study and the software specifically developed to perform it are available from the corresponding author, upon reasonable request. Note that the software is compatible with BTS NIRVANA 1 system, but not with further versions of the same device.

References

1. Weiss, P.L.; Katz, N. The potential of virtual reality for rehabilitation. *J. Rehabil. Res. Dev.* **2004**, *41*, 7–10.
2. Rubin, P. *Future Presence: How Virtual Reality Is Changing Human Connection, Intimacy, and the Limits of Ordinary Life*; HarperCollins: New York, NY, USA, 2018.
3. Riva, G.; Bacchetta, M.; Baruffi, M.; Borgomainerio, E.; Defrance, C.; Gatti, F.M.; Galimberti, C.; Fontaneto, S.; Marchi, S.; Molinari, E.; et al. VREPAR projects: The use of virtual environments in psycho-neuro-physiological assessment and rehabilitation. *CyberPsychology Behav.* **1999**, *2*, 69–76. [CrossRef] [PubMed]
4. Laver, K.E.; Lange, B.; George, S.; Deutsch, J.E.; Saposnik, G.; Crotty, M. Virtual reality for stroke rehabilitation. *Cochrane Database Syst. Rev.* **2017**. [CrossRef] [PubMed]
5. Maggio, M.G.; De Luca, R.; Molonia, F.; Porcari, B.; Destro, M.; Casella, C.; Salvati, R.; Bramanti, P.; Calabro, R.S. Cognitive rehabilitation in patients with traumatic brain injury: A narrative review on the emerging use of virtual reality. *J. Clin. Neurosci.* **2019**, *61*, 1–4. [CrossRef] [PubMed]
6. Calabrò, R.S.; Naro, A. Understanding Social Cognition Using Virtual Reality: Are We still Nibbling around the Edges? *Brain Sci.* **2019**, *10*, 17. [CrossRef] [PubMed]
7. Jacobson, R. After the «virtual reality» gold rush: The virtual worlds paradigm. *Comput. Graph.* **1993**, *17*, 695–698. [CrossRef]
8. Bevilacqua, R.; Maranesi, E.; Riccardi, G.R.; Donna, D.V.; Pelliccioni, P.; Luzi, R.; Lattanzio, F. Non-Immersive Virtual Reality for Rehabilitation of the Older People: A Systematic Review into Efficacy and Effectiveness. *J. Clin. Med.* **2019**, *8*, 1882. [CrossRef]
9. Fox, J.; Arena, D.; Bailenson, J.N. Virtual reality: A survival guide for the social scientist. *J. Media Psychol.* **2009**, *21*, 95–113. [CrossRef]
10. Blascovich, J.; Loomis, J.; Beall, A.C.; Swinth, K.R.; Hoyt, C.; Bailenson, J.N. Immersive virtual environment technology as a methodological tool for social psychology. *Psychol. Inq.* **2002**, *13*, 103–124. [CrossRef]
11. Sayma, M.; Tuijt, R.; Cooper, C.; Walters, K. Are We There Yet? Immersive Virtual Reality to Improve Cognitive Function in Dementia and Mild Cognitive Impairment. *Gerontologist* **2019**. [CrossRef]
12. Slater, M.; Wilbur, S. A framework for immersive virtual environments (FIVE): Speculations on the role of presence in virtual environments. *Presence Teleoperators Virtual Environ.* **1997**, *6*, 603–616. [CrossRef]
13. Tieri, G.; Morone, G.; Paolucci, S.; Iosa, M. Virtual reality in cognitive and motor rehabilitation: Facts, fiction and fallacies. *Expert Rev. Med. Devices* **2018**, *15*, 107–117. [CrossRef]
14. Montana, J.I.; Tuena, C.; Serino, S.; Cipresso, P.; Riva, G. Neurorehabilitation of Spatial Memory Using Virtual Environments: A Systematic Review. *J. Clin. Med.* **2019**, *8*, 1516. [CrossRef] [PubMed]
15. De Luca, R.; Torrisi, M.; Piccolo, A.; Bonfiglio, G.; Tomasello, P.; Naro, A.; Calabrò, R.S. Improving post-stroke cognitive and behavioral abnormalities by using virtual reality: A case report on a novel use of nirvana. *Appl. Neuropsychol. Adult* **2017**, *25*, 581–585. [CrossRef] [PubMed]
16. Bohil, C.J.; Alicea, B.; Biocca, F.A. Virtual reality in neuroscience research and therapy. *Nat. Rev. Neurosci.* **2011**, *12*, 752–762. [CrossRef] [PubMed]
17. Grewe, P.; Kohsik, A.; Flentge, D.; Dyck, E.; Botsch, M.; Winter, Y.; Markowitsch, H.J.; Bien, C.G.; Piefke, M. Learning real-life cognitive abilities in a novel 360 degrees -virtual reality supermarket: A neuropsychological study of healthy participants and patients with epilepsy. *J. Neuroeng. Rehabil.* **2013**, *10*, 42. [CrossRef]
18. Loomis, J.M.; Blascovich, J.J.; Beall, A.C. Immersive virtual environment technology as a basic research tool in psychology. *Behav. Res. Methods Instrum. Comput.* **1999**, *31*, 557–564. [CrossRef]
19. Borrego, A.; Latorre, J.; Llorens, R.; Alcañiz, M.; Noé, E. Feasibility of a walking virtual reality system for rehabilitation: Objective and subjective parameters. *J. Neuroeng. Rehabil.* **2016**, *13*, 68. [CrossRef]
20. Meade, M.; Meade, J.G.; Sauzéon, H.; Fernandes, M.A. Active Navigation in Virtual Environments Benefits Spatial Memory in Older Adults. *Brain Sci.* **2019**, *9*, 47. [CrossRef]

21. Cogné, M.; Auriacombe, S.; Vasa, L.; Tison, F.; Klinger, É.; Sauzeon, H.; Joseph, P.-A.; N'Kaoua, B. Are visual cues helpful for virtual spatial navigation and spatial memory in patients with mild cognitive impairment or Alzheimer's disease? *Neuropsychology* **2018**, *32*, 385–400. [CrossRef]
22. Russo, M.; De Luca, R.; Naro, A.; Sciarrone, F.; Aragona, B.; Silvestri, G.; Manuli, A.; Bramanti, A.; Casella, C.; Bramanti, P.; et al. Does body shadow improve the efficacy of virtual reality-based training with BTS NIRVANA? A pilot study. *Medicine* **2017**, *96*, e8096. [CrossRef] [PubMed]
23. Slater, M. Grand challenges in virtual environments. *Front. Robot. AI* **2014**, *1*. [CrossRef]
24. Cho, D.-R.; Lee, S.-H. Effects of virtual reality immersive training with computerized cognitive training on cognitive function and activities of daily living performance in patients with acute stage stroke: A preliminary randomized controlled trial. *Medicine* **2019**, *98*, e14752. [CrossRef] [PubMed]
25. Maggio, M.G.; Maresca, G.; De Luca, R.; Stagnitti, M.C.; Porcari, B.; Ferrera, M.C.; Galletti, F.; Casella, C.; Manuli, A.; Calabrò, R.S. The Growing Use of Virtual Reality in Cognitive Rehabilitation: Fact, Fake or Vision? A Scoping Review. *J. Natl. Med. Assoc.* **2019**, *111*, 457–463. [CrossRef]
26. Lucas, B.R.; Elliott, E.; Coggan, S.; Pinto, R.; Jirikowic, T.; McCoy, S.W.; Latimer, J. Interventions to improve gross motor performance in children with neurodevelopmental disorders: A meta-analysis. *BMC Pediatr.* **2016**, *16*, 193. [CrossRef]
27. Ravi, D.; Kumar, N.; Singhi, P. Effectiveness of virtual reality rehabilitation for children and adolescents with cerebral palsy: An updated evidence-based systematic review. *Physiotherapy* **2017**, *103*, 245–258. [CrossRef]
28. Ghai, S.; Ghai, I. Virtual Reality Enhances Gait in Cerebral Palsy: A Training Dose-Response Meta-Analysis. *Front. Neurol.* **2019**, *10*, 236. [CrossRef]
29. Mesa-Gresa, P.; Gil-Gomez, H.; Lozano, J.A.; Gil-Gómez, J.-A. Effectiveness of Virtual Reality for Children and Adolescents with Autism Spectrum Disorder: An Evidence-Based Systematic Review. *Sensors* **2018**, *18*, 2486. [CrossRef]
30. De Luca, R.; Leonardi, S.; Portaro, S.; Le Cause, M.; De Domenico, C.; Colucci, P.V.; Pranio, F.; Bramanti, P.; Calabrò, R.S. Innovative use of virtual reality in autism spectrum disorder: A case-study. *Appl. Neuropsychol. Child* **2019**. [CrossRef]
31. Kerns, K.A.; Macoun, S.; Macsween, J.; Pei, J.; Hutchison, M. Attention and working memory training: A feasibility study in children with neurodevelopmental disorders. *Appl. Neuropsychol. Child* **2017**, *6*, 120–137. [CrossRef]
32. Akçayır, M.; Akçayır, G. Advantages and challenges associated with augmented reality for education: A systematic review of the literature. *Educ. Res. Rev.* **2017**, *20*, 1–11. [CrossRef]
33. Westermann, G.; Mareschal, D.; Johnson, M.H.; Sirois, S.; Spratling, M.W.; Thomas, M.S.C. Neuroconstructivism. *Dev. Sci.* **2007**, *10*, 75–83. [CrossRef] [PubMed]
34. Kukla, A. *Social Constructivism and the Philosophy of Science*; Routledge: London, UK, 2000.
35. Richard, E.; Tijou, A.; Richard, P.; Ferrier, J.-L. Multi-modal virtual environments for education with haptic and olfactory feedback. *Virtual Real.* **2006**, *10*, 207–225. [CrossRef]
36. Southgate, E.; Smith, S.; Cividino, C.; Saxby, S.; Kilham, J.; Eather, G.; Scevak, J.; Summerville, D.; Buchanan, R.; Bergin, C. Embedding immersive virtual reality in classrooms: Ethical, organisational and educational lessons in bridging research and practice. *Int. J. Child Comput. Interact.* **2019**, *19*, 19–29. [CrossRef]
37. Winn, W. *A Conceptual Basis for Educational Applications of Virtual Reality*; Technical Publication R-93-9; Human Interface Technology Laboratory of the Washington Technology Center, University of Washington: Seattle, WA, USA, 1993.
38. Williamson, B. *Computer Games, Schools, and Young People: A Report for Educators on Using Games for Learning*; Futurelab: Bristol, UK, 2009.
39. Sandford, R.; Williamson, B. *Games and Learning: A Handbook from NESTA FutureLab*; Nesta Futurelab: Bristol, UK, 2005.
40. Gee, J.P. *Situated Language and Learning: A Critique of Traditional Schooling*; Routledge: London, UK, 2004.
41. Ramaley, J.A.; Zia, L. The Real Versus the Possible: Closing the Gaps in Engagement and Learning. In *Educating the net generation*; Oblinger, D., Oblinger, J.L., Lippincott, J.K., Eds.; EDUCAUSE: Boulder, CO, USA, 2005.
42. Papert, S. Teaching children thinking. *Contemp. Issues Technol. Teach. Educ.* **2005**, *5*, 353–365. [CrossRef]

43. Fassbender, E.; Richards, D.; Bilgin, A.; Thompson, W.F.; Heiden, W. VirSchool: The effect of background music and immersive display systems on memory for facts learned in an educational virtual environment. *Comput. Educ.* **2012**, *58*, 490–500. [CrossRef]
44. Reisoğlu, I.; Topu, B.; Yılmaz, R.; Yılmaz, T.K.; Göktaş, Y. 3D virtual learning environments in education: A meta-review. *Asia Pac. Educ. Rev.* **2017**, *18*, 81–100. [CrossRef]
45. Merchant, Z.; Goetz, E.T.; Cifuentes, L.; Keeney-Kennicutt, W.; Davis, T.J. Effectiveness of virtual reality-based instruction on students' learning outcomes in K-12 and higher education: A meta-analysis. *Comput. Educ.* **2014**, *70*, 29–40. [CrossRef]
46. Campos, P.F.; Pessanha, S.; Jorge, J. Fostering collaboration in kindergarten through an augmented reality game. *Int. J. Virtual Real.* **2011**, *10*, 33–39. [CrossRef]
47. Huang, Y.; Li, H.; Fong, W.T.R. Using Augmented Reality in early art education: A case study in Hong Kong kindergarten. *Early Child Dev. Care* **2015**, *186*, 879–894. [CrossRef]
48. Lee, L.-K.; Chau, C.-H.; Chau, C.-H.; Ng, C.-T. Using augmented reality to teach kindergarten students English vocabulary. In Proceedings of the 2017 International Symposium on Educational Technology (ISET), Hong Kong, China, 27–29 June 2017.
49. Lorusso, M.L.; Giorgetti, M.; Travellini, S.; Greci, L.; Zangiacomi, A.; Mondellini, M.; Sacco, M.; Reni, G. Giok the Alien: An AR-Based Integrated System for the Empowerment of Problem-Solving, Pragmatic, and Social Skills in Pre-School Children. *Sensors* **2018**, *18*, 2368. [CrossRef] [PubMed]
50. Lorusso, M.L.; Biffi, E.; Molteni, M.; Reni, G. Exploring the learnability and usability of a near field communication-based application for semantic enrichment in children with language disorders. *Assist. Technol.* **2018**, *30*, 39–50. [CrossRef] [PubMed]
51. Gandolfi, E. Virtual Reality. In *Handbook of Research on K-12 Online and Blended Learning*; Kennedy, K., Richard, E.F., Eds.; Carnegie Mellon University, ETC Press: Pittsburgh, PA, USA, 2018; pp. 545–561.
52. Trawick-Smith, J.; Russell, H.; Swaminathan, S. Measuring the effects of toys on the problem-solving, creative and social behaviours of preschool children. *Early Child Dev. Care* **2011**, *181*, 909–927. [CrossRef]
53. Aran, O.T.; Şahin, S.; Köse, B.; Ağce, Z.B.; Kayihan, H. Effectiveness of the virtual reality on cognitive function of children with hemiplegic cerebral palsy: A single-blind randomized controlled trial. *Int. J. Rehabil. Res.* **2020**, *43*, 12–19. [CrossRef]
54. Shema-Shiratzky, S.; Brozgol, M.; Cornejo-Thumm, P.; Geva-Dayan, K.; Rotstein, M.; Leitner, Y.; Hausdorff, J.M.; Mirelman, A. Virtual reality training to enhance behavior and cognitive function among children with attention-deficit/hyperactivity disorder: Brief report. *Dev. Neurorehabilit.* **2019**, *22*, 431–436. [CrossRef] [PubMed]
55. Gee, J.P. *The Anti-Education Era: Creating Smarter Students through Digital Learning*; St. Martin's Press: New York, NY, USA, 2013.
56. Gee, J.P.; Hayes, E.R. *Language and Learning in the Digital Age*; Routledge: Abingdon-on-Thames, UK, 2011.
57. Knight, A.; Rabon, P. Music for speech and language development in early childhood populations. *Music. Ther. Perspect.* **2017**, *35*, 124–130. [CrossRef]
58. Patscheke, H.; Degé, F.; Schwarzer, G. The effects of training in rhythm and pitch on phonological awareness in four-to six-year-old children. *Psychol. Music.* **2019**, *47*, 376–391. [CrossRef]

applied sciences

MDPI

Article

FCN-Based 3D Reconstruction with Multi-Source Photometric Stereo

Ruixin Wang [1], Xin Wang [1], Di He [1], Lei Wang [2,*] and Ke Xu [1,*]

[1] Collaborative Innovation Center of Steel Technology, University of Science and Technology Beijing, Beijing 100083, China; g20189167@xs.ustb.edu.cn (R.W.); s20191301@xs.ustb.edu.cn (X.W.); hedi8888888888@gmail.com (D.H.)

[2] National Engineering Research Center for Advanced Rolling Technology, University of Science and Technology Beijing, Beijing 100083, China

* Correspondence: 2011wanglei2011@gmail.com (L.W.); xuke@ustb.edu.cn (K.X.); Tel.: +86-10-62332598 (L.W.); +86-10-62332159 (K.X.)

Received: 27 March 2020; Accepted: 21 April 2020; Published: 23 April 2020

Abstract: As a classical method widely used in 3D reconstruction tasks, the multi-source Photometric Stereo can obtain more accurate 3D reconstruction results compared with the basic Photometric Stereo, but its complex calibration and solution process reduces the efficiency of this algorithm. In this paper, we propose a multi-source Photometric Stereo 3D reconstruction method based on the fully convolutional network (FCN). We first represent the 3D shape of the object as a depth value corresponding to each pixel as the optimized object. After training in an end-to-end manner, our network can efficiently obtain 3D information on the object surface. In addition, we added two regularization constraints to the general loss function, which can effectively help the network to optimize. Under the same light source configuration, our method can obtain a higher accuracy than the classic multi-source Photometric Stereo. At the same time, our new loss function can help the deep learning method to get a more realistic 3D reconstruction result. We have also used our own real dataset to experimentally verify our method. The experimental results show that our method has a good effect on solving the main problems faced by the classical method.

Keywords: Photometric Stereo (PS); 3D reconstruction; fully convolutional network (FCN)

1. Introduction

Vision-based 3D reconstruction technology can obtain 3D information on the target object from a 2D image in a non-contact manner, which has the advantages of being less affected by the shape of the actual object and giving a more real and robust reconstruction effect. Vision-based reconstruction methods can be roughly divided into active vision methods and passive vision methods. The reconstruction accuracy of active 3D reconstruction methods is relatively high, such as laser scanning and structured light methods, but their cost and complexity are also higher and their reconstruction speed is slow. The passive vision method can make up for the above shortcomings of the active vision method, but still faces challenges in terms of reconstruction accuracy.

As a 3D reconstruction method based on passive vision, the shape from shading (SFS) [1] can analyze the lightness and darkness information in the image and use the reflected illumination model to recover the normal information of the object from a single image. However, a single image contains less information, so the actual reconstruction effect of this method is average. Therefore, in order to improve the shortcomings of the SFS, RJ Woodhan [2] first proposed the Photometric Stereo, using data redundancy to solve the problem of single image reconstruction in SFS due to factors such as shadows and specular reflections, improving the effect and robustness of the reconstruction. On this basis, some researchers have found that increasing the number of light sources can provide more equations to the solution of unknown parameter parameters [3], thereby compensating for the surface microscopic information missed by the three-dimensional measurement method with three light sources and improving the accuracy of the dimensional measurement, i.e., multi-source Photometric Stereo.

Currently, the research and improvement of Photometric Stereo 3D reconstruction mainly focuses on light source calibration, non-Lambertian reconstruction [4], gradient reconstruction depth [5] and so on. The classic Photometric Stereo method usually assumes that the light intensity on the observation images taken under different illuminations is the same, and the sensor exposure is constant, but these assumptions are difficult to achieve in practical applications. In response to this, Cho et al. [6] developed a method for accurately determining the surface normal direction that is not affected by these factors for situations where the light direction is known but the light intensity is unknown, which improves the accuracy of the Photometric Stereo method in practical applications. Hertzmann et al. [7] proposed a method for calculating the geometry of objects with general reflection characteristics from the image to solve the complex calibration problem of photometric three-dimensional reconstruction, which can be applied to any remote and unknown lighting with almost no calibration operation surroundings.

With the extensive study of deep learning in various fields, neural network frameworks have also been gradually applied to the field of 3D graphics [8,9]. As we all know, the convolutional neural network (CNN) performs well in tasks such as classification and regression. At present, some studies have used CNN to complete three-dimensional tasks. Tang J et al. [10] use the CNN to mix three different three-dimensional shape expressions together, which can bring a better performance to many three-dimensional tasks compared with a single expression. The 3D ShapeNet established by Wu et al. [11] is an earlier proposed 3D reconstruction model of a single image based on voxel representation, using a convolutional depth confidence network to represent geometric 3D graphics as a probability distribution of binary variables on the 3D voxel grid. Its 3D reconstruction was realized by continuously predicting shape types and filling unknown voxels. In a related work, Badrinarayanan et al. [12] established a deep full convolution neural network (FCN) to solve the task of semantics segmentation, which was used to realize the road scene understanding. On the basis of the FCN structure, another network architecture called U-net [13] was established to achieve biomedical image segmentation.

In recent years, the rise of deep learning brings new development direction to the field of machine vision. As a main problem in machine vision, 3D reconstruction has also been widely studied. Eigen et al. [14] adopted a multi-scale deep network with two components, consisting of a coarse-scale network and a fine-grained network, to capture depth information directly. On this basis, a similar neural network architecture was used to process three tasks including depth prediction simultaneously [15], but each task was independently trained by changing its output layer and training objectives. Liu et al. [16] combined the Markov Random Field (MRF) of multi-scale local features and global image features to model the depth of different points and the relationship between them. Other related studies are different from the multi-scale deep network architecture. These include transforming the problem into a classification problem which predicted the likelihood that a pixel would be at any fixed standard depth [17]. Laina et al. [18] used a fully convolutional architecture, encompassing residual learning, to model the ambiguous mapping between monocular images and their corresponding scene depth maps. Xu et al. [19] added a fusion module to the CNN architecture, and the continuous conditional random field (CRF) was used to integrate complementary information on the front-end CNN's multiple side outputs. Li et al. [20] proposed a fast-to-train two-streamed CNN, and the depth and depth gradients were combined either via further convolution layers or directly with an optimization enforcing consistency between the depth and depth gradients. Dechaintre et al. [21] made the result of 3D construction more realistic with a rendering-aware deep network improved by U-net, based on the bidirectional reflectance distribution function (BRDF) [22]. Other related studies include methods based on Bayesian updates and dense [23], the generative adversarial network (GAN) [24], dictionary learning [25], self-augmented convolutional neural networks [26], etc.

For the multi-source Photometric Stereo 3D reconstruction method based on the physical model, using the neural network to simulate the mapping relationship between the real reflection of the object surface and its 3D information is very meaningful research. On the one hand, neural networks can improve the efficiency and accuracy of the multi-source Photometric Stereo, and on the other hand, a lot of existing research on reflection characteristics can also provide a priori knowledge for the neural network algorithms. Although there have been some related studies on learning Photometric Stereo from different perspectives [27–29], the research results in this area are still very limited.

Earlier, Santo H. et al. [29] proposed the use of an FCN in learning Photometric Stereo, and restoring the surface normal of the object from multiple views. After that, Chen G. et al. [28] took the direction of the light source as an input and improved the performance of the algorithm by adding more constraints to the model. Some of the other related studies learnt Photometric Stereo by obtaining the surface normal of the object indirectly. Chen G. et al. [30] proposed a two-stage deep learning structure to solve the uncalibrated Photometric Stereo problem, that is, using a lighting calibration network (LCNet) to recover the light direction and intensity corresponding to the image from any number of images, and then using a normal estimation network (NENet) to predict the normal mapping of the object surface. Compared with the single-stage model, this intermediate supervision effectively reduced the learning difficulty of the network. Moreover, Ikehata S. et al. [31] combined the two-dimensional input image information into an intermediate representation called an observation map to learn Photometric Stereo and used the rotation pseudo-invariance to constrain the network. This method also took the surface normal as the optimization goal. Our method solves the Photometric Stereo 3D reconstruction task from a different perspective. After solving the reflection illumination model, an integration step will be used to restore the three-dimensional topography of the surface, which is also a complicated process. The computational and time cost of this step is also very large, and it may cause cumulative errors and finally cause different degrees of distortion in the reconstructed results. We hope to use depth as the direct optimization goal and obtain the three-dimensional shape of the object surface from end-to-end.

In this paper, we built a U-shaped network structure based on FCN that can obtain the 3D topography of the object surface. By training a parameterized model, we can directly simulate the relationship between physical information such as shadows and reflections on the surface of the object and its depth information. The end-to-end learning can make our method more directly obtain the three-dimensional shape of the object. In addition, we added a regularization constraint on the basis of the general L2 loss function, and the experiments prove that, compared with optimizing the depth value of each pixel directly with the simple L2 loss function [27], this constraint can effectively improve the accuracy of prediction. We also adopted a photometric acquisition setup with a specific configuration to collect a real Photometric Stereo dataset, obtained a high-precision ground truth (GT) using structured light scanning and accurately registered it to the 2D image we collected. The experimental results show that the effectiveness of our method has been verified in a real multi-source Photometric Stereo setup.

The remainder of this study is organized as follows. We first introduce the principle of multi-source Photometric Stereo and the details of our method, including the network structure, our new loss function including two regularization constraints, and the real Photometric Stereo dataset in Section 2. Then the details of our experiments and the experimental results are shown in Section 3. We end with a discussion of our experimental results in Section 4.

2. Materials and Methods

2.1. The Multi-Source Photometric Stereo

The goal of multi-source Photometric Stereo is to recover the original 3D information of the object surface from a set of images with different light source directions. Assume a fixed orthographic camera and directional lighting with multiple equal angle intervals from a fixed latitude line in the upper hemisphere. We assume that a light source from the direction of $\vec{l} \in \mathbb{R}^3$ illuminates a point on the object surface, and that the surface normal of the point is represented by $\vec{n} \in \mathbb{R}^3$. Then, its pixel intensity can be determined as $I = \vec{\rho} E \cdot \vec{n} \cdot \vec{l}$, where $\vec{\rho}$ is the sensitivity coefficient and E is the light source pre-calibrated brightness, which needs to be obtained through a specific light source calibration method. For the k different light directions $L = [\vec{l}_1, \vec{l}_2, \ldots, \vec{l}_k]^{\mathrm{T}} \in \mathbb{R}^{k\times3}$, the light intensity can be expressed as $I = [\vec{I}_1, \vec{I}_2, \ldots, \vec{I}_k]^T \in \mathbb{R}^k$, and so the image formation model can be expressed as

$$\begin{bmatrix} I_1 \\ I_2 \\ \vdots \\ I_k \end{bmatrix} = \vec{\rho} E \cdot \begin{bmatrix} \vec{l}_1 \\ \vec{l}_2 \\ \vdots \\ \vec{l}_k \end{bmatrix} \cdot \vec{n} . \tag{1}$$

By solving the above equations, the surface normal direction $\vec{n} = (n_x, n_y, n_z)$ corresponding to each pixel position will be obtained.

After that, we suppose $(p,\ q, -1) = (n_x/n_z,\ n_y/n_z, -1)$. Here, p and q are the gradients of a point on the three-dimensional surface in the X and Y directions respectively, and will be formed into the matrices $\vec{P}$ and $\vec{Q}$ in the two-dimensional space. Then, the depth of the 3D surface is defined as $Z(x,\ y)$, and the two gradient values in the directions X and Y are $\triangle Z_x$ and $\triangle Z_y$. Lastly, we use the method of Wu Lun et al. [32] for reference to approximate the actual values $\triangle Z_x$ and $\triangle Z_y$ of the gradient with the $\vec{P}$ and $\vec{Q}$ obtained above. Through the classic two-dimensional integration path algorithm (path integration algorithm, PI), we can obtain a three-dimensional surface with the depth $Z(x,\ y)$.

2.2. Network Architecture

We converted the solution of the mapping relationship from image to depth in the multi-source Photometric Stereo method into an end-to-end optimization process with a large number of parameters. The FCN with encoder-decoder architecture has an outstanding performance in the problem of the pixel-level classification of images; its skip structure combined with the results of different depth layers ensures the long-distance dependence between pixels and the robustness and accuracy of the network and improves the accuracy of the feature extraction. Meanwhile, the network structure of the FCN determines that it can perfectly adapt to any size of input, which is exactly what we needed. Therefore, on the basis of the FCN network structure, we adopted U-net as the basis of our network design.

The architecture of the proposed network is shown in Figure 1. The U-shaped network structure could fully combine the simple features of shallow layer in the decoder stage, so it could also adapt to our small dataset. Our network contained twenty-nine layers, including twenty-one convolution layers, four pooling layers and four up-sampling layers. The activation function of all the convolution operations in the network was ReLU, and we took multiple RGB images from different light source directions containing different degrees of shadow and brightness information as the input of the network. In addition, the network outputted the original RGB images synthesized by the proposed network while outputting the predicted depth—that is, the output of the network was a multi-channel output.

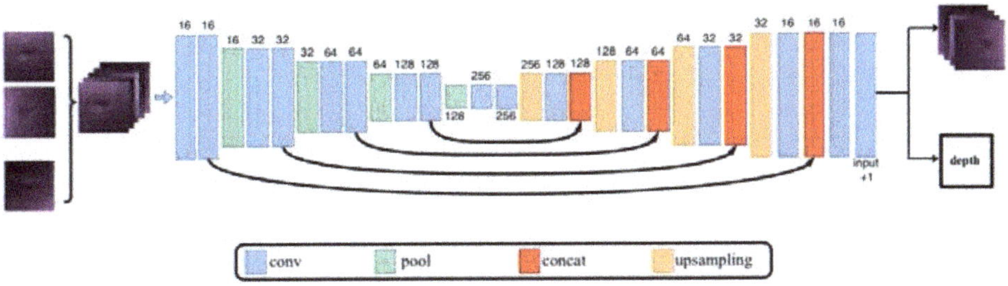

Figure 1. Overview of the proposed network architecture. The network outputted a depth map when given a set of images from the light source direction with different angles as inputs. The kernel sizes for all the convolutional layers were 3 × 3 and for all up-sampling layers 2 × 2. Values above the layers indicate the number of feature channels.

2.3. Loss Function

With the U-shaped network based on the code-decode structure, it was easy to lose some details in the training process, and the result of the final output 3D reconstruction was not accurate enough. We propose a loss function which is suitable for the task optimization based on our network structure—that is, we add two regularization constraints on the basis of the L2 loss function, and the training loss for each sample is set to

$$L_{depth} = ||Z - \widetilde{Z}||^2 + \lambda||I - \widetilde{I}||^2, \tag{2}$$

where Z and $\widetilde{Z}$ denote the predicted depth and the ground truth, and respectively, I and $\widetilde{I}$ are the predicted RGB images and the original RGB images. λ is a custom parameter. Here, we have set it to 1×10^4. As described in Section 2.2, our network structure reconstructed the original image of the corresponding light source while predicting the depth value. In the previous experiments, we found that training the network with L2 loss alone can make the network converge, but its reconstruction effect was not good enough. The defect area of the samples had different reflective characteristics under different angles of light, which was an unavoidable phenomenon in the use of the Photometric Stereo method to solve the three-dimensional reconstruction problem. Therefore, the reconstruction results obtained by simply optimizing the depth of each pixel were largely affected by the highlights in the RGB images, and it was not easy to obtain reasonable reconstruction results. Using two regularization constraints, that is, based on the original depth value as the goal of optimization, the original image is also the optimization goal of the network, which could play the role of additional constraints in the network training so as to weaken the influence of the highlight in the input images and make the reconstruction results closer to the real situation. By minimizing the sum of the deviations between the two prediction targets and the GT, our new loss function could improve the effectiveness of feature extraction. Compared with simply predicting the depth value of each pixel position and calculating their loss, this operation, similar to the image restoration, could help correct the prediction results of the network. In Section 4.2, we further evaluate the effectiveness of our new loss function.

2.4. Dataset

2.4.1. The Real-World Dataset

In order to verify the effectiveness of our method, we hoped to use a real-world sample database to train and test our model. At the beginning, we hoped to match our needs to the currently available datasets. However, due to the practical difficulties in 3D data collection, many datasets are based on synthesis or rendering [27,28] and some of them even have no corresponding GT, and so could not be used to train the neural network [33,34]. We think that there are still great differences between real scene data and rendered simulation data. Therefore, we made a batch of samples by hand and established a real Photometric Stereo experiment platform to collect the dataset we needed.

Our sample database consisted of 100 equal-sized corrugated boards with different degrees of surface damage on them, as shown in Figure 2. The damage on each cardboard was caused by human random. Because the middle part of the corrugated board was partly hollow, the image of the damaged part was very complex under different angles of illumination. The surface features of our samples did not conform to the standard Lambert model, and there were fractures on the surface of the defects which were not a uniform transition. This was not friendly to the classic multi-source Photometric Stereo, as shown in the experimental results.

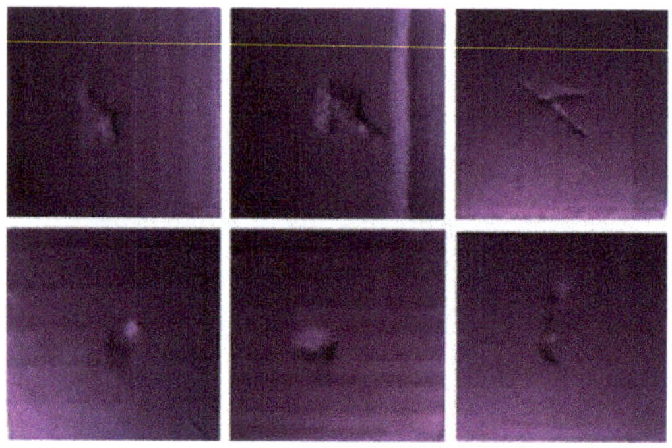

Figure 2. Examples of our real-world dataset.

2.4.2. The Photometric Acquisition Setup

We set up a real photometric stereo experiment platform to collect the images needed for training, as shown in Figure 3a. The camera and the circular light frame were fixed by a frame including clamping devices to ensure that the light conditions of each acquisition were determined and consistent, and the light frame was fixed with the camera (Automation Technology GmbH, Bad Oldesloe, Germany) at its center. The arrangement of the circular light frame is shown in Figure 3b. We designed our light sources as 20 white LED bulbs of the same size (60 degrees) as the scattering angle and fixed them on a circular ring. The angle interval between each adjacent white LED bulb was 18 degrees.

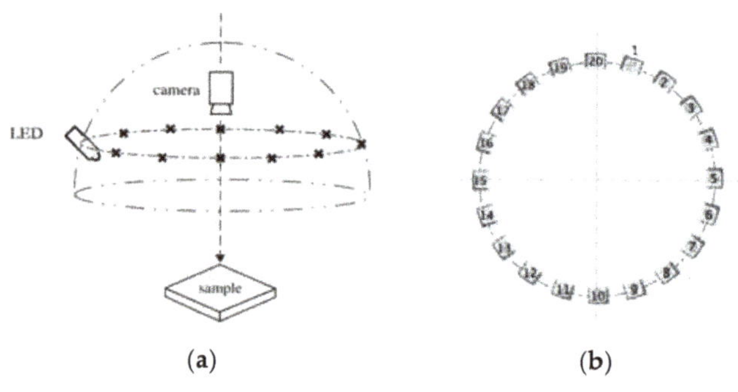

(a) (b)

Figure 3. (**a**) The photometric acquisition setup of the multi-source Photometric Stereo. (**b**) The light configurations of our proposed setup.

2.4.3. Data Capture

Through the program control, we lit up the LED bulbs in each direction in order and collected 2000 images corresponding to 100 samples in turn, all of which were captured in the dark room. We used 95 samples as the training set and the remaining 5 as the test set. The collected samples were cropped and then resized to the pixel size of 256*256, which was convenient for network training and better fitting function. We also obtained the GT of each object by line structured light scanning and accurately registered them on the two-dimensional images we collected, as shown in Figure 4.

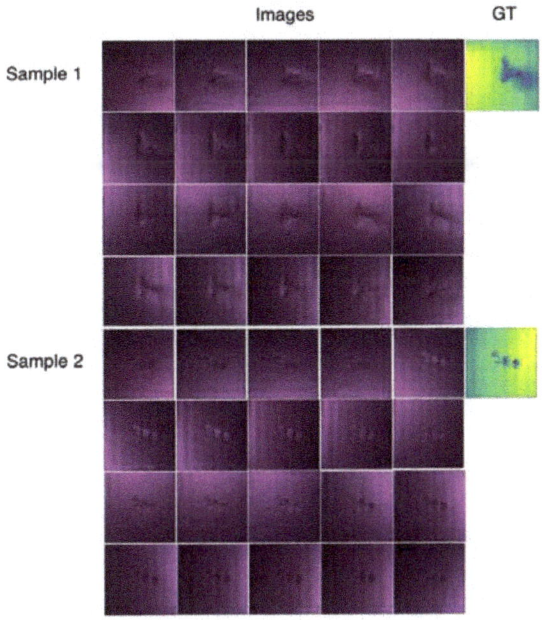

Figure 4. Examples of the collected images and their corresponding ground truth (GT).

3. Results

3.1. Implementation Details

We used a Tensorflow (tensorflow_gpu-1.8.0-cp35-cp35m-win_amd64.whl) with a Nvidia GTX2080 graphics card to implement and train the proposed network. The training process used a batch size of 16 for 100 epochs. The loss function was optimized using the Adagrad Optimizer and the learning rate was 1×10^4. We initialized the weights with a zero-mean Gaussian distribution and a standard deviation of $\sqrt{2/fin}$, where the fin was the number of input units in the weight tensor.

For each sample object, we selected two-dimensional images from the light source direction at 5 equal angle intervals to train our network. That is to say there were 4 kinds of light source combinations for the 20 images collected from each actual sample that could be used as an input for our network. In this way, the size of the training set was 380 (95*4). We used it as a type of data augmentation to train our network. The results predicted by the general loss function optimization network were also evaluated by the same setup. In addition, all 20 images collected for each sample were also used to test the classic multi light source photometric stereo method as a comparative experiment.

3.2. Error Metrics

As shown in Table 1, we used five indices to quantitatively evaluate several methods involved in this experiment which are widely used in the error analysis and accuracy analysis of deep estimation based on deep learning [14,18,27]:

1. root mean squared error(rms)

$$\sqrt{\frac{1}{N}\sum_{i=1}^{N}\left|d_i - d_i^*\right|^2}, \tag{3}$$

2. average relative error(rel)

$$\frac{1}{N}\sum_{i=1}^{N}\frac{\left|d_i - d_i^*\right|}{d_i^*}, \tag{4}$$

3. threshold accuracy(δ)

$$\begin{gathered}\delta = \frac{1}{N}\sum_i \eta_i,\\ \eta_i = \begin{cases} 1 & if\ T < t \\ 0 & if\ T \geq t \end{cases},\\ T = max\left(\frac{d_i}{d_i^*}, \frac{d_i^*}{d_i}\right), t \in [1.25, 1.56, 1.95],\end{gathered} \tag{5}$$

where d_i^* and d_i are the GT and predicted depths respectively of each pixel, and according to the different values of t, the results of δ(t) are divided into three grades.

Table 1. Quantitative evaluation of our method in comparison to the reference method using the L2 norm. Lower is better for rms and rel and higher is better for δ(t).

Methods	rms	rel	δ(1.25)	δ(1.56)	δ(1.95)
L2 Norm	0.3770	0.3552	0.6492	0.9859	0.9906
Ours	0.2797	0.2473	0.2359	0.9727	0.9937

4. Discussion

4.1. Compared with the Classic Multi-Source PS

In some classic multi-source Photometric Stereo 3D reconstruction methods, the effect of highlights on the results is removed by a selecting method—that is, some images that contain severe highlight reflections will not participate in the calculation. However, this loses a lot of meaningful information contained in the highlight position, even reducing the rationality of the prediction. The characteristics of the neural network determined that it could be biased towards learning information from the input that was more relavent to the correct results. Therefore, using the neural network to learn will not lose the useful information of the highlight position itself, but can also help to reduce inaccurate predictions caused by specular reflections and noise. Furthermore, we represented the optimized target as the depth value of each pixel. Compared with other representations such as point clouds or voxel grids, such 2D representations make the computational cost of our network less.

In order to verify the practical significance of the neural network used to learn Photometric Stereo for 3D reconstruction, we compared the results of the classic multi-source Photometric Stereo method (BASELINE) and ours with GT to conduct a qualitative analysis. We reconstructed the target surface with the BASELINE method using 5, 10 and 20 two-dimensional images taken under the illumination of light sources with equal angle intervals, as shown in Figure 5c–e. The BASELINE method had an obvious effect on the reconstruction of the corrugates which excessive smoothly, but the cast shadow and attached shadow caused by the fracture led to an anomaly in the 3D information extraction at the deeper fractures (Sample 3). However, there was a smooth transition in ours at the fracture site, which made our prediction more reasonable. For the defect surface with more small cracks, ours could not reproduce all the details perfectly. In comparison, the BASELINE results lost more surface information, and the smooth inclined position with little feature information could not present a reasonable three-dimensional shape. In addition, because the surface features of the target samples did not conform to the standard Lambert model, the reflection around the defect resulted in different degrees of bulge in the transition from the plane to the defect in the reconstruction surface of the BASELINE method (Sample 2). Our method took GT as the direct optimization goal, which could minimize the influence of the highlights in the input on the correct prediction results.

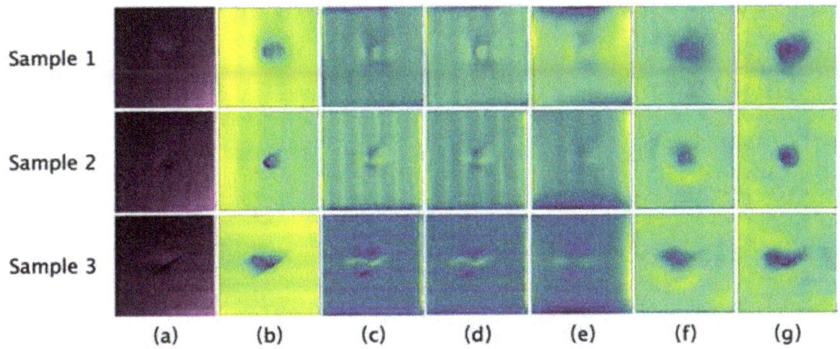

Figure 5. Examples of the results of our method and others. The samples in the example are all from the test set: (**a**) the first column shows the basic 2D information of this sample; (**b**) the corresponding ground truth data (calculated using linear structured light scan) are shown in the second column; (**c**–**e**) the third to fifth columns demonstrate the classic multi-source Photometric Stereo approach using 5,10, and 20 input images; (**f**) the sixth column shows the results using a general L2 loss function; (**g**) the result estimated by our network is shown in final column.

4.2. *Effectiveness of the New Regularization Constraints*

Most of the recent studies use the normal vector solved by the reflected illumination model as the optimization target, but the solution from the normal vector to the depth is also a complex problem. To verify the effectiveness of our new regularization constraints, we used the proposed network and the same configuration, but used a common loss function with a general L2 norm to train the dataset, which was used in the recent related work [27]. By comparing group (f) and group (g) of these three samples in Figure 5, we can find that ours(g) contained more details than the method with the general L2 norm did (f). Since our optimized target also included the original image of the object, generating the input images could help our network to correct the prediction of the depth, so that the reconstruction result was closer to the real. Thus, ours was clearer for the reconstruction of the simple sample surface (corrugates), and the transition of the cracks on the defects was also smoother (Sample 3). As shown in Figure 5 (Sample 2, Sample 3), there was an abnormal bulge around the defects as we can see in group (f), but from the original image and GT corresponding to the sample, this did not conform to the real situation. However, ours had a good effect on the optimization of this special position—that is, the transition from plane to defect was more reasonable. In addition, Table 1 shows the quantitative analysis results of our method and the general L2 norm. It can be seen from the table that our method significantly improved on the parameters rms and rel. However, the threshold accuracy of ours was slightly lower than that of the L2 norm. The main reason for this, we think, was that the restoration of the images made our reconstruction results closer to reality rather than only taking the depth GT as the optimization standard. Therefore, the accuracy of the depth prediction was lower than that of GT, but it could also get the same level of L2 loss within a certain accuracy range.

5. Conclusions

In this paper, we proposed an effective improvement method aimed at problems such as the complex calibration process and low reconstructing speed faced by the traditional multi-source Photometric Stereo method in 3D reconstruction tasks to improve its accuracy and efficiency. Hereto, we trained the neural network model with a large number of parameters in an end-to-end way to simulate the relationship between physical information, such as shadow and reflection on the surface of the object, and depth information in the multi-source Photometric Stereo. In contrast, our method was superior to the classic algorithm in terms of efficiency and accuracy. In addition, we proposed a new regularization constraint, which improved the effectiveness of feature extraction by minimizing the sum of the loss of the two prediction targets, making the prediction closer to reality.

Author Contributions: Conceptualization, D.H. and K.X.; methodology, R.X. and D.H.; software, X.W. and D.H.; validation, L.W. and K.X.; formal analysis, R.W.; investigation, R.W.; resources, R.W., X.W.; data curation, R.W., X.W.; writing—original draft preparation, R.W.; writing—review and editing, K.X.; visualization, L.W.; supervision, K.X.; project administration, L.W.; funding acquisition, K.X. All authors have read and agreed to the published version of the manuscript.

Funding: This research was funded by the National Key R&D Program of China (no.2018YFB0704304), and the National Natural Science Foundation of China (grant number 51674031 and 51874022).

Conflicts of Interest: The authors declare no conflict of interest.

References

1. Horn, B.K.P. Shape from Shading: A Method for Obtaining the Shape of a Smooth Opaque Object from One View. Ph.D. Thesis, Department of Electrical Engineering, MIT, Cambridge, UK, 1970.
2. Woodham, R.J. Photometric method for determining surface orientation from multiple images. *Opt. Eng.* **1980**, *19*, 139–144. [CrossRef]
3. Xu, K.; Wang, L.; Xiang, J.; Zhou, P. Three-dimensional defect detection method of metal surface based on multi-point light source. *China Sci.* **2017**, *12*, 420–424. (In Chinese)
4. Sun, J.; Smith, M.; Smith, L.; Midha, S.; Bamber, J. Object surface recovery using a multi-light photometric stereo technique for non-Lambertian surfaces subject to shadows and specularities. *Image Vis. Comput.* **2007**, *25*, 1050–1057. [CrossRef]
5. Wang, L.; Xu, K.; Zhou, P.; Yang, C. Photometric stereo fast 3D surface reconstruction algorithm using multi-scale wavelet transform. *J. Comput. -Aided Des. Comput. Graph.* **2017**, *29*, 124–129. (In Chinese)
6. Cho, D.; Matsushita, Y.; Tai, Y.W.; Kweon, I. Photometric Stereo Under Non-uniform Light Intensities and Exposures. In Proceedings of the European Conference on Computer Vision, Amsterdam, The Netherlands, 8–16 October 2016.
7. Hertzmann, A.; Seitz, S.M. Example-based photometric stereo: Shape reconstruction with general, varying BRDFs. *IEEE Trans. Pattern Anal. Mach. Intell.* **2005**, *27*, 1254–1264. [CrossRef] [PubMed]
8. Li, X.; Dong, Y.; Peers, P.; Tong, X. Synthesizing 3D Shapes from Silhouette Image Collections using Multi-projection Generative Adversarial Networks. In Proceedings of the IEEE Conference on Computer Vision and Pattern Recognition, Long Beach, CA, USA, 16–20 June 2019; pp. 5535–5544.
9. Sun, C.Y.; Zou, Q.F.; Tong, X.; Liu, Y. Learning Adaptive Hierarchical Cuboid Abstractions of 3D Shape Collections. *ACM Trans. Graph.* **2019**, *38*, 1–13. [CrossRef]
10. Tang, J.; Han, X.; Pan, J.; Jia, K.; Tong, X. A Skeleton-bridged Deep Learning Approach for Generating Meshes of Complex Topologies from Single RGB Images. In Proceedings of the IEEE Conference on Computer Vision and Pattern Recognition, Long Beach, CA, USA, 16–20 June 2019; pp. 4541–4550.
11. Wu, Z.; Song, S.; Khosla, A.; Yu, F.; Zhang, L.; Tang, X.; Xiao, J. 3D ShapeNets: A Deep Representation for Volumetric Shape Modeling. In Proceedings of the 2015 IEEE Conference on Computer Vision and Pattern Recognition (CVPR), Boston, MA, USA, 7–12 June 2015.
12. Badrinarayanan, V.; Kendall, A.; Cipolla, R. SegNet: A Deep Convolutional Encoder-Decoder Architecture for Scene Segmentation. *IEEE Trans. Pattern Anal. Mach. Intell.* **2017**, *39*, 2481–2495. [CrossRef] [PubMed]

13. Ronneberger, O.; Fischer, P.; Brox, T. U-Net: Convolutional Networks for Biomedical Image Segmentation. In Proceedings of the International Conference on Medical Image Computing and Computer-Assisted Intervention, Munich, Germany, 5–9 October 2015.
14. Eigen, D.; Puhrsch, C.; Fergus, R. Depth Map Prediction from a Single Image using a Multi-Scale Deep Network. In *Advances in Neural Information Processing Systems, Proceedings of the 27th International Conference on Neural Information Processing Systems, Montreal, QC, Canada 8–13 December 2014*; Curran Associates, Inc.: New York, NY, USA, 2014.
15. Eigen, D.; Fergus, R. Predicting Depth, Surface Normals and Semantic Labels with a Common Multi-scale Convolutional Architecture. In Proceedings of the 2015 IEEE International Conference on Computer Vision (ICCV), Santiago, Chile, 7–13 December 2015.
16. Liu, F.; Chung, S.; Ng, A.Y. Learning depth from single monocular images. *IEEE Trans. Pattern Anal. Mach. Intell.* **2005**, *18*, 1–8.
17. Ladicky, L.; Shi, J.; Pollefeys, M. Pulling Things out of Perspective. In Proceedings of the 2014 IEEE Conference on Computer Vision and Pattern Recognition (CVPR), Columbus, OH, USA, 23–28 June 2014.
18. Laina, I.; Rupprecht, C.; Belagiannis, V.; Tombari, F.; Navab, N. Deeper Depth Prediction with Fully Convolutional Residual Networks. In Proceedings of the 2016 Fourth International Conference on 3D Vision (3DV), Stanford, CA, USA, 25–28 October 2016.
19. Xu, D.; Ricci, E.; Ouyang, W.; Wang, X.; Sebe, N. Multi-Scale Continuous CRFs as Sequential Deep Networks for Monocular Depth Estimation. In Proceedings of the IEEE Conference on Computer Vision and Pattern Recognition, Honolulu, HI, USA, 21–26 July 2017.
20. Li, J.; Klein, R.; Yao, A. A Two-Streamed Network for Estimating Fine-Scaled Depth Maps from Single RGB Images. In Proceedings of the IEEE International Conference on Computer Vision 2017, Venice, Italy, 22–29 October 2017.
21. Deschaintre, V.; Aittala, M.; Durand, F.; Drettakis, G. Single-Image SVBRDF Capture with a Rendering-Aware Deep Network. *ACM Trans. Graph.* **2018**, *37*, 1–5. [CrossRef]
22. Nicodemus, F.E. Geometrical Considerations and Nomenclature for Reflectance. *NBS Monogr.* **1977**, *160*, 4.
23. Hermans, A.; Floros, G.; Leibe, B. Dense 3D semantic mapping of indoor scenes from RGB-D images. In Proceedings of the IEEE International Conference on Robotics & Automation 2014, Hong Kong, China, 31 May–7 June 2014.
24. Yoon, Y.; Choe, G.; Kim, N.; Lee, J.Y.; Kweon, I. Fine-scale Surface Normal Estimation using a Single NIR Image. In Proceedings of the European Conference on Computer Vision 2016, Amsterdam, The Netherlands, 11–14 October 2016.
25. Xiong, S.; Zhang, J.; Zheng, J.; Cai, J.; Liu, L. Robust surface reconstruction via dictionary learning. *ACM Trans. Graph.* **2014**, *33*, 1–12. [CrossRef]
26. Li, X.; Dong, Y.; Peers, P.; Tong, X. Modeling surface appearance from a single photograph using self-augmented convolutional neural networks. *ACM Trans. Graph.* **2017**, *36*, 45. [CrossRef]
27. Liang, L.; Lin, Q.; Yisong, L.; Hengchao, J.; Junyu, D. Three-Dimensional Reconstruction from Single Image Base on Combination of CNN and Multi-Spectral Photometric Stereo. *Sensors* **2018**, *18*, 764.
28. Chen, G. PS-FCN: A Flexible Learning Framework for Photometric Stereo. In Proceedings of the European Conference on Computer Vision (ECCV) 2018, Munich, Germany, 8–14 September 2018.
29. Santo, H.; Samejima, M.; Sugano, Y.; Shi, B.; Matsushita, Y. Deep Photometric Stereo Network. In Proceedings of the 2017 IEEE International Conference on Computer Vision Workshop (ICCVW), Venice, Italy, 22–29 October 2017.
30. Chen, G.; Han, K.; Shi, B.; Matsushita, Y.; Wong, K.K. Self-calibrating Deep Photometric Stereo Networks. In Proceedings of the IEEE Conference on Computer Vision and Pattern Recognition 2019, Long Beach, CA, USA, 16–20 June 2019.
31. Ikehata, S. CNN-PS: CNN-based Photometric Stereo for General Non-Convex Surfaces. In Proceedings of the European Conference on Computer Vision (ECCV) 2018, Munich, Germany, 8–14 September 2018.
32. Wu, L.; Wang, Y.; Liu, Y. A robust approach based on photometric stereo for surface reconstruction. *Acta Autom. Sin.* **2013**, *39*, 1339–1348. (In Chinese) [CrossRef]

33. Alldrin, N.; Zickler, T.; Kriegman, D. Photometric stereo with non-parametric and spatially-varying reflectance. In Proceedings of the 2008 IEEE Conference on Computer Vision and Pattern Recognition, Anchorage, AK, USA, 23–28 June 2008.
34. Einarsson, P.; Chabert, C.F.; Jones, A.; Ma, W.C.; Lamond, B.; Hawkins, T.; Bolas, M.; Sylwan, S.; Debevec, P. Relighting Human Locomotion with Flowed Reflectance Fields. In *Eurographics Workshop on Rendering, Proceedings of the 17th Eurographics Conference on Rendering Techniques Nicosia, Cyprus, 26–28 June 2006*; Eurographics Association: Goslar, Germany, 2006.

applied sciences

MDPI

Article

Is an ADHD Observation-Scale Based on DSM Criteria Able to Predict Performance in a Virtual Reality Continuous Performance Test?

Débora Areces, Celestino Rodríguez *, Trinidad García and Marisol Cueli

Department of Psychology, University of Oviedo, 33003 Asturias, Spain; arecesdebora@uniovi.es (D.A.); garciatrinidad@uniovi.es (T.G.); cuelimarisol@uniovi.es (M.C.)

* Correspondence: rodriguezcelestino@uniovi.es; Tel.: +34-985-10-95-63

Received: 7 March 2020; Accepted: 25 March 2020; Published: 1 April 2020

Featured Application: This study showed that only the DSM 5 criteria referring to inattention symptoms were able to significantly predict performance in the variables measured by a Continuous Performance Test based on Virtual Reality.

Abstract: The Diagnosis of Attention Deficit/Hyperactivity Disorder (ADHD) requires an exhaustive and objective assessment in order to design an intervention that is adapted to the peculiarities of the patients. The present study aimed to determine if the most commonly used ADHD observation scale—the Evaluation of Attention Deficit and Hyperactivity (EDAH) scale—is able to predict performance in a Continuous Performance Test based on Virtual Reality (VR-CPT). One-hundred-and-fifty students (76% boys and 24% girls) aged 6–16 (M = 10.35; DT = 2.39) participated in the study. Regression analyses showed that only the EDAH subscale referring to inattention symptoms, was a statistically significant predictor of performance in a VR-CPT. More specifically, this subscale showed 86.5% prediction-accuracy regarding performance in the Omissions variable, 80.4% in the Commissions variable, and 74.5% in the Response-time variable. The EDAH subscales referring to impulsivity and hyperactivity were not statistically significant predictors of any variables in the VR-CPT. Our findings may partially explain why impulsive-hyperactive and the combined presentations of ADHD might be considered as unique and qualitatively different sub-categories of ADHD. These results also highlighted the importance of measuring not only the observable behaviors of ADHD individuals, but also the scores in performance tests that are attained by the patients themselves.

Keywords: ADHD; EDAH; assessment; continuous performance test; virtual reality

1. Introduction

ADHD is a common, chronic, and impairing neuropsychiatric disorder, with worldwide prevalence rates ranging from 5% to 7% among the school-age population [1]. ADHD is characterized by a persistent behavioral pattern associated with inattention, overactivity (or hyperactivity), and difficulty in controlling impulses, leading to three presentations: the combined presentation, the predominantly inattentive presentation, and the predominantly impulsive-hyperactive presentation (hereafter I/H) [2]. This disorder relates to significant impairments at home (in family adaptation) and at school (low academic performance) [3]. Additionally, among the long-term consequences of having ADHD symptoms, we could indicate a higher probability of being unemployed, drug abuse, or being imprisoned [4].

In this sense, latent deficits in ADHD are manifested through observable symptoms described in the DSM-5 manual [2], which have been included in different observational scales (completed by teachers or parents). For instance, some of the best known and most widely used observational scales are the following: (1) the Evaluation of Attention Deficit and Hyperactivity (EDAH scale) [5]; (2) the Behavior Assessment System for Children (BASC) [6]; (3) the Child Behavior Checklist (CBCL) [7]; and (4) the Conners' scales [8]. However, the use of these instruments as the sole assessment measure has been harshly criticized because assessment depends on the subjectivity of the observer [9]. This implies an important limitation in the evaluation of ADHD, because terms like "restlessness", or "being a clueless person" could be interpreted differently depending on the person who evaluates a particular case. For example, when parents complete an observational scale, those having more than one child often evaluate the different items by comparing with their other children.

For this reason, some other widely used tests in the assessment of ADHD are the Continuous Performance Tests (CPT) (based on participants' performance), aimed at detecting problems such as deficits in monitoring and updating information in working memory, in inhibiting undesired responses or avoiding to pay attention to irrelevant stimuli and shifting attention away between activities [10]. Among these, Conners' CPT [8], the Children Sustained Attention Task [11], the Integrated Visual and Auditory Test [12], and the Test of Variables of Attention [13] are the most widely used tests in the assessment of ADHD symptoms. These tests are based on the current models about the etiology of ADHD, which state that the dysfunction in executive processes is one important pathway to understanding this disorder [14,15].

CPTs provide different variables, which are associated to the phenotypic behavior of ADHD students [16]. More specifically, a high number of omission errors and the presence of lengthy response times are thought to relate to inattention deficit. On the contrary, a high number of commission errors and higher levels of variability in their patient's responses might indicate the presence of impulsive/hyperactive symptoms. In this sense, the profile obtained for each participant is useful in the differential diagnosis of ADHD and its clinical presentations [17]. However, CPT are also criticized as having low ecological validity, since ADHD symptoms do not always occur in a controlled environment, which differs considerably from real-life conditions [18,19]. Thus, various authors [20–22] consider the inclusion of Virtual Reality in the CPT (VR-CPT) as a solution that would allow a significant increase in ecological validity. VR-CPT offers the possibility of carrying out assessments in more realistic conditions, including distractors present in typical classrooms (i.e., a classmate who speaks to the subject during the execution of the task, a teacher knocking on the door or the sound of an ambulance passing near the window) [23]. This allows clinicians to know in depth how distractors influence attention capacity, as well as what type of distractors interfere significantly in the performance of children and adolescents. Namely, it is possible to measure the influence of the distractors according to the sensory modality in which they are presented. Moreover, data provided by VR-CPTs are more useful in designing an interventional plan than those obtained with a traditional CPTs, which do not provide any information as to the patients´ behavior in daily-life contexts [17].

These findings have been taken into consideration in the current protocols about the assessment of ADHD, which recommend the correct administration of the following diagnosis tools: (1) a structured or semi-structured interview; (2) an observational scale based on DSM criteria; and (3) a CPT, in order to contrast the results and verify the presence of ADHD symptoms [24,25].

Taking all this into consideration, the present study aims at analyzing whether the data collected by the EDAH scale might partially explain the results obtained by a VR-CPT called AULA Nesplora. This objective allows to measure the degree of congruence between what third parties observe and the patient´s own performance will thus be measured, resulting in an important innovation: Although there are some studies that analyze the relationship between performance in a CPT and current and retrospective symptoms in adults and children [16,18], no studies so far analyze the capacity of an observation scale (based on DSM criteria) to predict performance in the variables of VR-CPTs.

2. Materials and Methods

2.1. Participants

The present study made use of a non-probabilistic clinical sample composed of 150 children with ADHD (76% boys and 24% girls) aged 6 to 16 ($M = 10.35$; $SD = 2.93$) and with an average IQ of 109.82 ($SD = 22.53$). Participants have been diagnosed with the combined presentation of ADHD by neuropsychiatrists, according to the Diagnostic and Statistical Manual of Mental Disorders [2].

2.2. Procedure

The study obtained previous approval by the Ethical Committee of the Principality of Asturias (reference: CPMP/ICH/135/95, code: TDAH-Oviedo), and all instructions from the protocol were performed according to institutional guidelines and laws.

Firstly, a member of the research group contacted with local hospitals and clinical services serving children and adolescents diagnosed with ADHD (more particularly, the combined presentation of ADHD). Contact with these services was initially made by phone, and, later, a face-to-face meeting was held with those professionals who agreed to participate in the project [26].

Secondly, different meetings with families/parents were organized in order to explain the main objectives of the present project. Having given previous written consent for the study, the parents completed the observational scale about the Evaluation of Attention Deficit and Hyperactivity Disorder (EDAH scale) [5], which is based on DSM criteria of ADHD symptoms [2]. Then, the children and adolescents performed the Virtual Reality Continuous Performance Test (VR-CPT), called Aula Nesplora CPT. The evaluations were conducted in a laboratory and lasted for 1 h. A member of the research group was always present during the evaluation process, in order to supervise the administration of the tests. Finally, the parents were informed by clinicians about the results obtained in both tools.

2.3. Instruments

Considering the objectives of the present study, the tests used are described below:

The EDAH Scale [5], which was completed by families (the children´s parents). This scale consists of 20 items about symptoms related to Attention Deficit and Hyperactivity/Impulsivity Disorder. It differentiates between ADHD and control groups, as well as between ADHD presentations. The following variables were included in the present study: EDAH-AD (score in the items that measure Attention Deficit), EDAH-I/H (score in Impulsivity/Hyperactivity items), and EDAH-ADHD (the sum of attention deficit plus Impulsivity/Hyperactivity symptoms). The reliability of the instrument, using Cronbach´s Alpha, was 0.74 in the current sample.

AULA Nesplora [23] is a VR-CPT, which evaluates attention, impulsivity, processing speed, and motor activity in children and adolescents aged between 6 and 16. The task is performed in a virtual reality environment, which is shown through Three-Dimensional (3D) glasses equipped with motion sensors and headphones. The virtual environment presented through the glasses is like a standard school classroom. The participant takes the perspective of a student sitting at one of the desks and facing the chalkboard. Head movements (which are related to motor activity) are detected by sensors located in the glasses, since the software updates the field of vision, giving the participant the feeling of actually being in a classroom.

The test consists of three parts, which are gradually explained by a virtual teacher. The first part aims to immerse the participant in a virtual reality environment. More specifically, this task consists of visually locating balloons and popping them. The first part only aims at immersing participants in the virtual reality environment, by visually locating balloons and popping them and, therefore, performance in it is quite irrelevant and the results from this part are not provided by the test. The second task is based on the "x-no" paradigm (traditionally known as "no-go") in which the participant must press a button when he or she does not see or hear the stimulus "apple". This task mainly measures attentional levels, so children or adolescents with inattention problems are expected to make a lot of omission errors in this part. Finally, the last task is based on an "x" paradigm (or "go"): Participants are asked to press a button whenever they see or hear the number "seven". This task aims to measure the inhibitory control capacity, so it is expected that patients with impulsive-hyperactive problems commit a high number of commission errors. Moreover, it is also convenient to highlight that in each part (Go task Vs. No go task) appears different types of distractors (Visual Vs. Auditory distractors) and this offers the possibility of comparing the results from each part in the presence or absence of distractors. This benefit supposes an important innovation in the evaluation of ADHD symptoms because it allows getting a diagnosis with more ecological validity. Moreover, the increase of ecological validity has been shown to be more effective in the diagnosis of ADHD in comparison to other Traditional CPTs, which offer similar variables but without considering the presence of the distractor and with less levels of ecological validity [26].The completion of the test takes approximately 20 min.

To sum up, the variables provided by this test do not differ from those of other CPTs regarding attention deficit and hyperactivity/impulsivity measurements (Omissions, Commissions, Response Time). However, they enhance this information, relating these measurements to sensory modality (visual vs. auditory), presence/absence of distractors, task type (go vs. no-go) and adding a new index called motor activity. Cronbach´s Alpha in this sample was 0.78.

2.4. Data analysis

This study examined the discriminant value of the subscales of EDAH in predicting performance in VR-CPT. The descriptive statistics for the variables under study were analyzed, paying special attention to skewness and kurtosis. Following the criterion of Kline [27], the maximum scores accepted for skewness and kurtosis were limited to a range of 3–10. The results thus allowed us to perform parametric analyses.

In this sense, three regression models were carried out in order to verify the discriminant values of EDAH subscales in predicting the scores in Omissions, Commissions and Response Time provided by a VR-CPT. Percentile scores were used in order to control the effect of age and gender.

SPSS 24 [28] was used in the analysis of data, having $p < 0.05$ as the criterion for reaching statistical significance.

3. Results

As shown in Table 1 and according to the Kline (2011) criteria, it was found that the variables had a normal distribution.

Table 1. Descriptive Statistics for VR-CPT variables and EDAH Subscales.

	M	*SD*	Asymmetry	Kurtosis
Omissions	62.11	25.29	−0.377	−0.924
Commissions	58.03	29.05	−0.229	−1.149
Response Time	49.47	29.10	0.126	−1.169
EDAH. I/H	82.55	16.87	−1.664	3.851
EDAH.AD	82.51	15.38	−1.821	5.468
EDAH.CD	80.40	15.94	−1.584	3.853

Note. *M* = Mean; *SD* = Standard Deviation; EDAH.H = Items of EDAH scale referred to Hyperactive symptoms; EDAH.AD = Items of EDAH scale related to Attention Deficit; EDAH. CD = Items about Conduct Disorder symptoms.

Once the descriptive statistics were analyzed, the three regression models were conducted. The first regression model (Table 2) was statistically significant predictor of the omissions obtained in the VR-CPT, $F(3, 148) = 318.220$, $p < 0.001$. The second regression model was also statically significant for predicting the commissions obtained in the VR-CPT, $F(3, 148) = 198.177$, $p < 0.001$. Similarly, the third regression model was significant in the prediction of response time variable from VR-CPT, $F(3, 148) = 144.804$, $p < 0.001$.

Table 2. Regression models to predict performance in the VR-CPT variable.

Independent variables: EDAH Scale	Dependent Variables: VR-CPT Variables		
	Omissions	Commissions	Response Time
EDAH.AD β (t)	0.411 (2.765 **)	0.615 (3.162 **)	0.782 (3.558 **)
EDAH. H β (t)	0.240 (1.254)	−0.049 (−0.208)	−0.451 (0.652)
EDAH.CD β (t)	0.256 (1.256)	0.334 (1.342)	0.334 (1.342)
R^2	0.865 ***	0.804 ***	0.745 ***

Note. β = Standardized beta coefficient; t = Student t coefficient; R^2 = variance explained; EDAH.AD = Items of EDAH scale related to Attention Deficit; EDAH.H = Items of EDAH scale referred to Hyperactive symptoms; EDAH. CD = Items about Conduct Disorder symptoms.; ** $p < 0.01$;*** $p < 0.001$.

These results indicated that in the regression model for the prediction of scores in the omission variable obtained in VR-CPT, only the score obtained in the Attention Deficit subscale from EDAH was statistically significant. Likewise, regarding the model for predicting the performance of a commission variable from VR-CPT, only the Attention Deficit subscale from EDAH was, again, a statistically significant variable. The same pattern was repeated in the last regression model (for predicting the Response time variable), because only the Attention Deficit subscale was a significant independent variable.

4. Discussion

The present study supports the utility of the inattention subscale (belonging to EDAH scale) in predicting a patient´s performance in a VR-CPT (more specifically, AULA Nesplora). These results coincide with previous investigations, which suggested a strong relationship between the presence of inattention symptoms and a significantly high number of omission errors and slow response time [20–23].

Additionally, the results also showed that the remaining subscales from the EDAH scale (EDAH subscale, related to hyperactive symptoms, and EDAH subscale, related to conduct disorder symptoms) did not significantly predict any particular variable from the VR-CPT. These findings are in the line of previous studies [29], which discussed the difficulty of determining what type of symptoms of ADHD are most dominant. The fact that the EDAH subscale referred to hyperactive symptoms did not predict that any variable from VR-CPT could be due to the fact that CPT are solely based on measuring different Executive Functions (EF). In this sense, there are some clinical studies that support the view that EF deficits, although found in many individuals in groups of children and adolescents who suffer ADHD, are not a necessary feature of ADHD and, therefore, the EDAH subscales based on DSM criteria (and, more particularly, the Hyperactive subscale) are not taking them into account [30,31]. Similarly, this study also resulted in another unexpected finding, since inattention symptoms were capable to predict an 80.4% of performance in the commission variable from the VR-CPT. This result could be partially explained by the fact that EF deficits are mainly related to inattentive rather than impulsive-hyperactive symptoms [32]. Considering the present findings, the following question is posed: is ADHD a single diagnostic category or is it better to talk about two different disorders? Children with the inattentive presentation of ADHD frequently show non-specific attention problems, which are associated with deficient sensory processes, poorly focused attention and less accurate information processing. Understandably, these problems mainly lead to learning disabilities [33]. However, children with predominantly impulsive-hyperactive or combined presentations of ADHD do not have general attention problems like those mentioned in the previous case. These subtypes are more associated with memory retrieval problems, disruptive behavior, and peer rejection [15,33].

Similarly, the results also suggested the importance of carrying out an objective assessment of ADHD, not only considering the symptoms of ADHD contained in DSM-5 manual, but also taking into account the patient's own performance in a CPT, in order to contrast the two different types of measures (symptoms collected by observational scale and variables collected by a CPT). As many protocols recommend [24,25], it is highly relevant to use several assessment tools with the same patient, to ensure the objectivity of the diagnosis process. Moreover, including a CPT based on Virtual Reality increases the ecological validity of the patient´s evaluation and, at the same time, brings out the possibility of analyzing how distractors affect their daily life [20,22].

Therefore, the results obtained in the present study may be useful in guiding clinicians get an objective and reliable assessment of the ADHD symptomatology. However, it is important to highlight some limitations of the study that should be considered in future research lines. In this sense, it would be convenient to include a control group, in order to analyze whether the evidence obtained is maintained. Another important limitation of this study relates to the ADHD sample, as it consists of children and adolescents who have been clinically diagnosed as presenting a combined presentation of ADHD. In this sense, it might also be positive to include the remaining two presentations of ADHD in the ADHD group (that is: the predominantly inattentive presentation and the predominantly impulsive-hyperactive presentation), so as to observe possible differences. Hence, we would have the possibility to compare performance in interesting variables, like motor activity, depending on the ADHD presentation. This would allow us to check whether the inattention presentation presents the lowest level of motor activity and, by contrast, whether the impulsive-hyperactive presentation obtains the expected highest level for this same variable.

Author Contributions: Conceptualization, D.A. and C.R.; methodology, D.A. and C.R.; formal analysis, T.G. and D.A.; investigation, C.R. and D.A; data curation, M.C, T.G. and C.R.; writing—original draft preparation, D.A. and M.C.; writing—review and editing, C.R.; visualization, T.G and D.A.; supervision, C.R.; All authors have read and agreed to the published version of the manuscript.

Funding: This research was funded by the Principality of Asturias regional project with reference FC-GRUPIN-IDI/2018/000199 and a crowdfunding project from the Spanish National Government with reference MINECO-18-FCT-PRECIPITA.

Conflicts of Interest: The authors declare no conflict of interest.

References

1. Polanczyk, G.V.; Willcutt, E.G.; Salum, G.A.; Kieling, C.; Rohde, L.A. ADHD prevalence estimates across three decades: An updated systematic review and meta-regression analysis. *Int. J. Epidemiol.* **2014**, *43*, 434–442. [CrossRef] [PubMed]
2. American Psychiatric Association. *Diagnostic and Statistical Manual of Mental Disorders*, 5th ed.; American Psychiatric Publishing: Arlington, VA, USA, 2013; ISBN 978-0-89-042555-8.
3. Barkley, R.A. *Attention-Deficit Hyperactivity Disorder: A Handbook for Diagnosis and Treatment*, 3rd ed.; Guilford: New York, NY, USA, 2006; ISBN 978-1-59-385210-8.
4. Rodríguez, C.; Núñez, J.C.; Rodríguez, F.J.; Parrales, A.; Bringas, C.; García, T. Attention Deficit Hyperactivity Disorder (ADHD): Prevalence and Sociodemographic Features in Imprisoned Population. *Psicologia, Reflexão Crítica* **2015**, *28*, 698–707. [CrossRef]
5. Farré, A.; Narbona, J. *EDAH: Scale for the Assessment of Attention Deficit Hyperactivity Disorder*; TEA Ediciones: Madrid, Spain, 2001.
6. Reynolds, C.; Kamphaus, R.W. *Behavior Assessment System for Children (BASC): Manual*; TEA Ediciones: Madrid, Spain, 2004; ISBN 978-0-387-79948-3.
7. Achenbach, T.M. *Child Behavior Checklist for Age 4–18*; University of Vermont: Burlington, VT, USA, 1991; ISBN 938-56-50-87.
8. Conners, C.K. *Conners' Continuous Performance Test User's Manual*; Multi-Health Systems: Toronto, ON, Canada, 1995; ISBN 843-5-08-511818-0.

9. García, T.; González-Castro, P.; Areces, D.; Cueli, M.; Rodríguez, C. Executive functions in children and adolescents: The types of assessment measures used and implications for their validity in clinical and educational contexts. *Papeles del Psicólogo* **2014**, *35*, 215–233.
10. Robaey, P.; McKenzie, S.; Schachar, R.; Boivin, M.; Bohbot, V.D. Stop and look! Evidence for a bias towards virtual navigation response strategies in children with ADHD symptoms. *Behav. Brain Res.* **2016**, *298*, 48–54. [CrossRef] [PubMed]
11. Servera, J.; Llabrés, J. *Children Sustained Attention Task (CSAT)*; TEA Ediciones: Madrid, Spain, 2004.
12. Tinius, T.P. The Integrated Visual and Auditory Continuous Performance Test as a neuropsychological measure. *Arch. Clin. Neuropsychol.* **2003**, *18*, 439–454. [CrossRef] [PubMed]
13. Greenberg, L.M. Developmental normative data on the Test of Variables of Attention (TOVA). *J. Child Psychol. Psychiatry* **1993**, *34*, 1019–1030. [CrossRef] [PubMed]
14. Castellanos, F.X.; Sonuga-Barke, E.J.; Milham, M.P.; Tannock, R. Characterizing cognition in ADHD: Beyond executive dysfunction. *Trends Cogn. Sci.* **2006**, *10*, 117–123. [CrossRef]
15. Sagvolden, T.; Johansen, E.B.; Aase, H.; Russell, V.A. A dynamic developmental theory of attention-deficit/hyperactivity disorder (ADHD) predominantly hyperactive/impulsive and combined subtypes. *Behav. Brain Sci.* **2005**, *28*, 397–418. [CrossRef]
16. Epstein, J.N.; Erkanli, A.; Conners, C.K.; Klaric, J.; Costello, J.E.; Angold, A. Relations between continuous performance test performance measures and ADHD behaviors. *J. Abnorm. Child Psychol.* **2003**, *31*, 543–554. [CrossRef]
17. Bart, O.; Raz, S.; Dan, O. Reliability and validity of the Online Continuous Performance Test among children. *Assessment* **2014**, *21*, 637–643. [CrossRef]
18. Areces, D.; García, T.; Cueli, M.; Rodríguez, C. Is a Virtual Reality Test Able to Predict Current and Retrospective ADHD Symptoms in Adulthood and Adolescence? *Brain Sci.* **2019**, *9*, 274–281. [CrossRef] [PubMed]
19. Gioia, G.A.; Kenworthy, L.; Isquith, P.K. Executive function in the real world: BRIEF lessons from Mark Ylvisaker. *J. Head Trauma Rehabil.* **2010**, *25*, 433–439. [CrossRef] [PubMed]
20. Adams, R.; Finn, P.; Moes, E.; Flannery, K.; Rizzo, A.S. Distractibility in attention deficit/hyperactivity disorder (ADHD): The virtual reality classroom. *Child Neuropsychol.* **2009**, *15*, 120–135. [CrossRef] [PubMed]
21. Rizzo, A.A.; Buckwalter, J.G.; Bowerly, T.; Humhrey, L.A.; Neuman, U.; van Rooyen, A.; Kim, L. The virtual classroom: A virtual reality environment for the assessment and rehabilitation of attention deficits. *Rev. Esp. Neuropsicol.* **2001**, *3*, 11–37. [CrossRef]
22. Areces, D.; Rodríguez, C.; García, T.; Cueli, M.; González-Castro, P. Efficacy of a continuous performance test based on virtual reality in the diagnosis of ADHD and its clinical presentations. *J. Atten. Disord.* **2018**, *22*, 1081–1091. [CrossRef]
23. Díaz-Orueta, U.; Garcia-López, C.; Crespo-Eguílaz, N.; Sánchez- Carpintero, R.; Climent, G.; Narbona, J. AULA virtual reality test as an attention measure: Convergent validity with Conners' Continuous Performance Test. *Child Neuropsychol.* **2014**, *20*, 328–342. [CrossRef]
24. Gualtieri, C.T.; Johnson, L.G. ADHD: Is objective diagnosis possible? *Psychiatry (Edgmont)* **2005**, *2*, 44–53. [CrossRef]
25. Areces, D.; Cueli, M.; García, T.; González-Castro, P.; Rodríguez, C. Using brain activation (nir-HEG/Q-EEG) and execution measures (CPTs) in a ADHD assessment protocol. *J. Vis. Exp.* **2018**, *134*, e56796. [CrossRef]
26. Rodríguez, C.; Areces, D.; García, T.; Cueli, M.; González-Castro, P. Comparison between two continuous performance tests for identifying ADHD: Traditional vs. virtual reality. *Int. J. Clin. Health Psychol.* **2018**, *18*, 254–263. [CrossRef]
27. Kline, R.B. *Principles and Practice of Structural Equation Modeling*, 3rd ed.; The Gilford Press: New York, NY, USA, 2011; ISBN 978-1-60-623876-9.
28. Arbuckle, J.L. *SPSS*, version 24.0; SPSS: Chicago, IL, USA, 2016.
29. Toplak, M.E.; Pitch, A.; Flora, D.B.; Iwenofu, L.; Ghelani, K.; Jain, U.; Tannock, R. The unity and diversity of inattention and hyperactivity/impulsivity in ADHD: Evidence for a general factor with separable dimensions. *J. Abnorm. Child Psychol.* **2009**, *37*, 1137–1150. [CrossRef]
30. Burke, J.D. Relationship between conduct disorder and oppositional defiant disorder and their continuity with antisocial behaviors: Evidence from longitudinal clinical studies. In *Externalizing Disorders of Childhood:*

Refining the Research Agenda for DSM-V; Shaffer, D., Leibenluft, E., Rohde, L.A., Eds.; American Psychiatric Publishing: Arlington, VA, USA, 2009.
31. Sonuga-Barke, E.J.; Sergeant, J.A.; Nigg, J.; Willcutt, E. Executive dysfunction and delay aversion in attention deficit hyperactivity disorder: Nosologic and diagnostic implications. *Child Adolesc. Psychiatr. Clin. N. Am.* **2008**, *17*, 367–384. [CrossRef] [PubMed]
32. Sergeant, J.A.; Willcutt, E.; Nigg, J. *How Clinically Functional Are Executive Function Measures of ADHD. Externalizing Disorders of Childhood: Refining the Research Agenda for DSM–V*; American Psychiatric Association: Arlington, VA, USA, 2008.
33. Willcutt, E.G.; Pennington, B.F.; DeFries, J.C. Etiology of inattention and hyperactivity/impulsivity in a community sample of twins with learning difficulties. *J. Abnorm. Child Psychol.* **2000**, *28*, 149–159. [CrossRef] [PubMed]

applied sciences

MDPI

Article

Development and Assessment of a Sensor-Based Orientation and Positioning Approach for Decreasing Variation in Camera Viewpoints and Image Transformations at Construction Sites

Mohsen Foroughi Sabzevar [1,*], Masoud Gheisari [2] and James Lo [1]

1 Department of Civil, Architectural & Environmental Engineering, Drexel University, Philadelphia, PA 19104, USA; james.lo@drexel.edu
2 Rinker School of Construction Management, University of Florida, Gainesville, FL 32611, USA; masoud@ufl.edu
* Correspondence: mf848@drexel.edu; Tel.: +1-601-307-9312

Received: 1 February 2020; Accepted: 18 March 2020; Published: 27 March 2020

Featured Application: In this paper, we propose a position and orientation approach that reduces the image transformation phenomena in advance (i.e., process modification). Thus, this approach which integrates with image matching techniques that have limitations dealing with image transformation (i.e., result modification) could be valuable. The advantage of this approach is that it is not dependent on scene features, and therefore it can be used in situations where the features in a scene change or when extremely image transformations occur. This approach can be used as a supplementary approach to assist the feature-based methods.

Abstract: Image matching techniques offer valuable opportunities for the construction industry. Image matching, a fundamental process in computer vision, is required for different purposes such as object and scene recognition, video data mining, reconstruction of three-dimensional (3D) objects, etc. During the image matching process, two images that are randomly (i.e., from different position and orientation) captured from a scene are compared using image matching algorithms in order to identify their similarity. However, this process is very complex and error prone, because pictures that are randomly captured from a scene vary in viewpoints. Therefore, some main features in images such as position, orientation, and scale of objects are transformed. Sometimes, these image matching algorithms cannot correctly identify the similarity between these images. Logically, if these features remain unchanged during the picture capturing process, then image transformations are reduced, similarity increases, and consequently, the chances of algorithms successfully conducting the image matching process increase. One way to improve these chances is to hold the camera at a fixed viewpoint. However, in messy, dusty, and temporary locations such as construction sites, holding the camera at a fixed viewpoint is not always feasible. Is there any way to repeat and retrieve the camera's viewpoints during different captures at locations such as construction sites? This study developed and evaluated an orientation and positioning approach that decreased the variation in camera viewpoints and image transformation on construction sites. The results showed that images captured while using this approach had less image transformation in contrast to images not captured using this approach.

Keywords: orientation; positioning; viewpoint; image matching; algorithm; transformation

1. Introduction

Formally, the era of computer vision started in the early 1970s [1]. Computer vision is defined as a trick "to extract descriptions of the world from pictures or sequences of pictures" [2]. This technique assists humans in "making useful decisions about real physical objects and scenes based on images" [3]. According to Horn et al. [4], computer vision "analyzes images and produces descriptions that can be used to interact with the environment". In summary, the goal of computer vision is "to describe the world that we see in one or more images and to reconstruct its properties, such as shape, illumination, and color distributions" [1]. One of the fundamental processes in computer vision is called image matching [5]. Image matching is "the process of bringing two images geometrically into agreement so that corresponding pixels in the two images correspond to the same physical region of the scene being imaged" [6]. In other words, during the image matching process, two images that are randomly captured from a scene are compared in order to identify their similarity. "Fast and robust image matching is a very important task with various applications in computer vision." [7]. The process of image matching is required for tracking targets [8], image alignment and stitching [9,10], reconstruction of three-dimensional (3D) models from images [11], object recognition [12], face detection [13,14], data mining [15], robot navigation [8], motion tracking [16,17], and more. These applications are promising in real world problems, and it is possible to leverage them at construction sites to monitor various activities.

1.1. Image Matching Applications in the Construction Industry

In the construction industry, especially in recent years, image matching techniques have shown capabilities for addressing different issues regarding information management. There is abundant research regarding applications of image matching techniques through AEC/FM (architecture, engineering and construction and facilities management). For instance, to solve issues related to difficulties in updating as-built and as-is information on jobsites, some researchers have utilized image matching techniques to create a building information model of the scenes. They have taken images from different angles, stitched them, and attached the data to these models [12]. Others such as Kang et al. [18] reported that in a large-scale indoor environment full of self-repetitive visual patterns, recognizing the location of images captured from different scenes can be confusing. To address this issue, they applied image matching techniques, which analyzed unique features in captured images, to retrieve the location. Kim et al. [19] used image matching techniques to compare virtual images of a construction site with the real construction photographs for the purpose of detecting differences between the actual and planned conditions of the jobsite. Another application of using image matching techniques is to detect changes in a scene by comparing features of pictures captured at different times [20] to estimate the rough progress of a project.

Providing easier access to construction information on a jobsite is another reason to use image matching techniques. For this purpose, some researchers suggested using augmented reality technology to superimpose a layer of data (e.g., text, voice, 3D model, image, etc.) over the locations where access to information is required [21,22]. Marker-based augmented reality (AR) and markerless AR, which both use image matching techniques, can be used for this purpose. For both methods, the image matching algorithms need to detect distinct features between live video frames that are captured from the environment, and a reference image that is already available. In the marker-based approaches, since the algorithms need to detect the features of a label (e.g., Quick Response Code/QR code), the results are very robust [1] in contrast with markerless AR, which needs to use the natural features of the environment that can vary [23,24] (more information regarding AR is presented in Appendix A).

1.2. Problem Statement

In general, there are three main types of algorithms for image matching. The first type is shape matching algorithms, which look for similarities in the shapes of objects in the images [5]. The second type is pixel matching algorithms, which look for similarities in the pixel's intensity [5]. The third type is called feature-based matching algorithms [5]. In this type, the algorithm detects the distinct local features of images, such as corresponding points, lines, and areas [5].

The challenge these algorithms need to deal with is the variation in the context of pictures that were captured from a scene from different viewpoints. When two pictures are not taken from the same viewpoint, the position, orientation, and scale of the features (e.g., objects and background) in the scene are transformed. Thus, these algorithms should detect the similarities between the features that have been displaced and deformed in the images, and then match them.

In previous decades, to deal with these issues of extracting features, researchers have proposed many different techniques. Some of these techniques detect image features regardless of transformations (e.g., translation and rotation) and illumination but not scaling. This group of techniques is known as single-scale detectors. Techniques such as Moravec detector [25], Harris detector [26], SUSAN detector [27], FAST detector [28,29], and Hessian detector [30,31] are examples of single-scale detectors. Other techniques known as multiscale detectors, including Laplacian of Gaussian [32], difference of Gaussian [33], Harris–Laplace [34], Hessian–Laplace [35], and Gabor–Wavelet detector [36] were later created. In addition to rotation, translation, and illumination, these techniques consider the impact of uniform scaling in detecting features, with the assumption that scale is not affected by an affine transformation of the image structures. Thus, to be able to detect the image features as accurately as possible, it was necessary to create techniques that could handle non-uniform scaling (change in scaling in different directions). Scale invariant feature transform (SIFT) is one of the most advanced versions of these algorithms [1]. SIFT can detect and describe image features [1]. In the first step, the SIFT algorithm detects the local distinct points on images. In the second step, these distinct points (keypoints) are converted into histogram vectors based on the image gradient of each point called keypoint descriptors. SIFT gives value to each of these vectors. In the third step, SIFT compares these values to match the keypoints. However, this is not the end of the process, as not all matches conducted by SIFT are correct. There could be some keypoints in two images with equal values but related to different parts of the scene. For instance, a keypoint on the top of a scene could have equal value with a point on the bottom of a scene. In this case, SIFT cannot distinguish between them. Therefore, incorrect matching occurs. These incorrect matches need to be filtered. For the purpose of filtering the incorrect matches, the fourth step is required. In this step, a technique called RANSAC or random sample consensus [37] is widely used. This approach divides the corresponding points into inlier and outlier sets and finds the best portion of points in inlier sets. To ensure this occurs, first, this algorithm randomly samples two keypoints. The width of the inlier boundary is already determined for this algorithm. RANSAC counts what fractions of points are located inside of this inlier boundary. This process is repeated several times for different keypoints. The largest number of points found as inlier is defined as the best matching pattern, and other matches are removed. Figure 1 illustrates the procedures that SIFT detects, describes, and matches the key points, while RANSAC filters incorrect matches.

However, image matching algorithms are not fully successful when image transformation occurs and image viewpoint changes [5,7,38–40]. In fact, increased changes in the image viewpoint can make the matching process unreliable, since the similarity between objects shown on images reduces [5]. For example, an image matching algorithm such as SIFT only works well when the difference between view angles is less than 30 degrees [41]. In addition, if the scaling is too high, the algorithm cannot detect the key points on the frame and the image matching process does not work correctly. For example, three images from a scene are illustrated in Figure 2. The first image (Figure 2a) is the reference image captured. The second image (Figure 2b) is the current frame from the same scene but impacted by the rotation of the camera (more than 30 degrees). The third image (Figure 2c) is also from the same scene but is impacted by high scaling. These scenarios can impact cases such as those using markerless AR that use SIFT and RANSAC during the image matching process. Thus, the algorithm cannot correctly match the features between two images. In addition, when image transformations take place, unrelated and unwanted areas around the scene are also detected. In this situation, change detection algorithms [42] report these areas as a change in the scene. This result is not accurate in construction scenarios where change detection algorithms are used to detect the construction progress based on changes in the image frames.

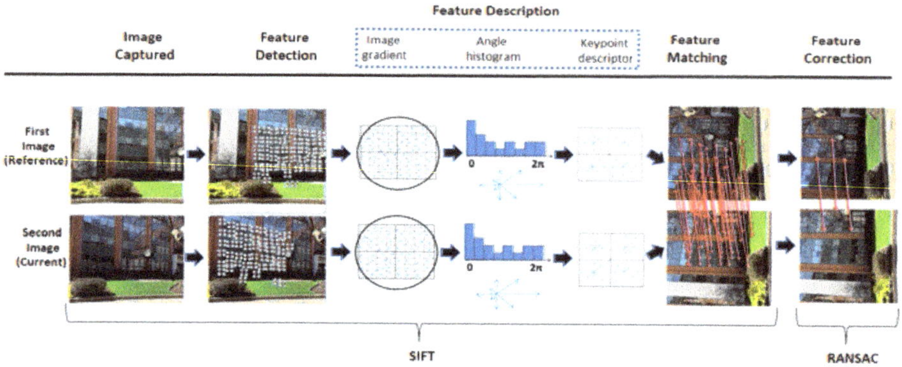

Figure 1. While SIFT detects, describes, and matches the key points, RANSAC filters incorrect matches (Adapted from [1,33]).

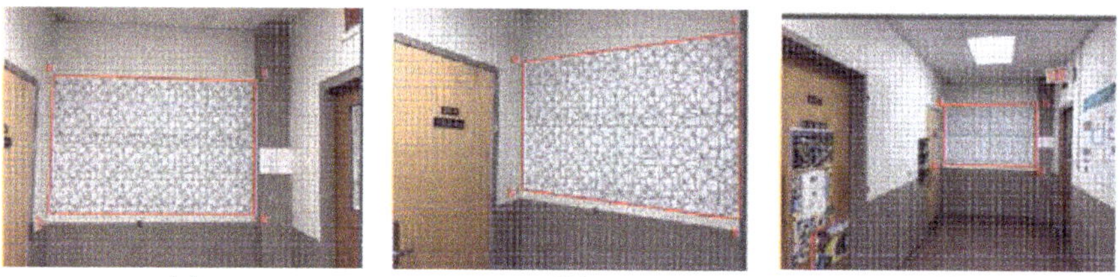

(a) (b) (c)

Figure 2. Differences between the reference and current images when the camera's orientation and position change, which results in transformation of the features of the current images. (**a**) Reference frame; (**b**) Current frame impacted by rotation; (**c**) Current frame impacted by scaling.

In addition to the difficulties posed by image transformation, there is another scenario whereby image matching techniques could fail. For example, when a scene is completely changed during a renovation project, the image matching algorithms cannot match the distinct feature points between two frames (e.g., before and after renovation), therefore, the image matching cannot occur. Figure 3 shows a scene that is completely changed before and after renovation. This scenario can impact markerless AR.

(a) (b)

Figure 3. An example of a scene in which its features have completely changed during renovation. (**a**) Image captured before renovation (reference image); (**b**) Image captured after renovation (current image).

Another scenario in which image matching techniques fail to work accurately is when features in two scenes within a location (e.g., a room) are exchanged during renovation. For instance, before renovation, the reference images were captured from the west wall and north wall. During renovation, the features of the west wall and north wall were switched. Since a feature-based system can only detect environmental features and cannot interpret geographical directions, an image matching technique such as SIFT would fail to generate accurate results during the image matching process. To have a clear understanding of this scenario, two scenes have been sketched, as shown in Figure 4. Figure 4a shows a scene before renovation, a door is attached to the west wall, and a window is attached to the north wall. Figure 4b shows the same room, but this time the window has been moved to the west wall and the door has been moved to the north wall. In fact, feature-based tracking methods detect environmental features but not directions. This scenario can impact use cases like markerless AR and change detection.

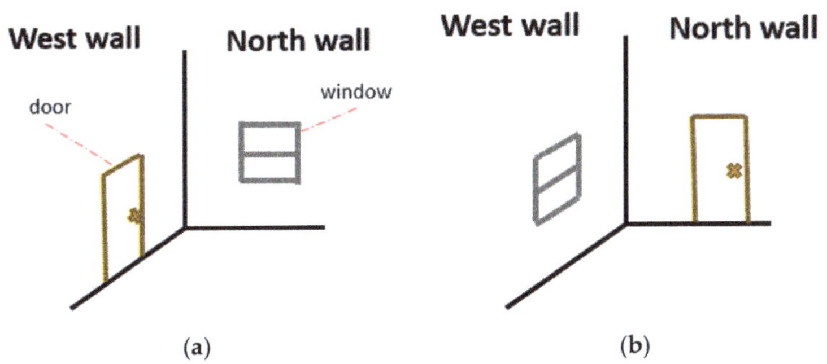

Figure 4. An example of two scenes in one room during renovation. (**a**) West wall captured (before renovation); (**b**) North wall captured (after renovation).

These limitations of the image matching process motivated us to study supplementary ways (e.g., controlling the image capturing process) to support the image matching algorithms in order to prevent sole dependency on natural features in the scene.

1.3. Ways to Control the Image Capture Process (Process Management)

The algorithms explained in the previous section that deal with image transformation conduct a kind of result management, but not process management, on images captured from a scene. Thus, in addition to using image matching algorithms that aim to identify similarities between images captured arbitrarily from different viewpoints, a preprocess should be required to control the viewpoint of the images. With this strategy, changes in viewpoints of pictures are minimized, and the current image matching algorithms can perform more accurately. The key to capturing two pictures from a single viewpoint is to hold a camera in a single position and orientation.

One way to provide this condition is to use a fixed point camera approach [19]. In this approach, for each scene, a camera must be installed with a fixed viewpoint. Therefore, the resulting pictures are from the same point of view. However, this approach is not practical, especially for chaotic locations such as construction sites, which are exposed to the movements of workers, vehicles, and materials that can accidentally block or relocate cameras. Moreover, this method is very costly because a camera is needed for each scene (Figure 5a).

The second way is to embed a benchmark (point of reference) for each scene on the jobsite and use the total station approach (i.e., installing the camera on a tripod) when taking pictures. In this way, crews can retrieve the position and orientation of the camera in different trials. However, the feasibility of implementing such an idea in a location that is under construction and exposed to different disturbances, such as the movement of workers and equipment, dust, floor washing liquids, or demolishing and replacing floor covers, which could remove any marks and nails, makes this option unreliable (Figure 5b).

Another way is to use a system that can navigate crews to locate the camera in a reference location and viewpoint without using a physical reference point or installing a fixed-point camera for each scene. To locate a camera on a single location and viewpoint, the position and orientation parameter values of the camera need to be retrieved remotely. However, the question is, "Is there any way to repeat and retrieve the camera's position and orientation parameter values remotely on messy, dusty, and temporary locations like construction sites for the purpose of decreasing image transformation?"

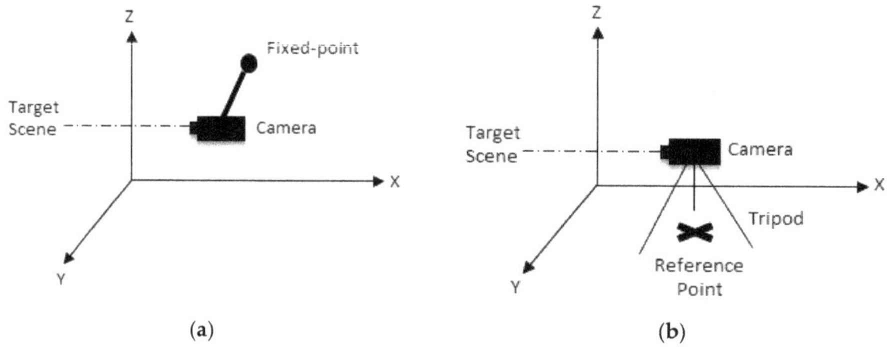

Figure 5. Holding a camera in a single position and orientation on a jobsite. (**a**) Using a fixed-point camera for each scene on a jobsite; (**b**) Embedding a benchmark (point of reference) for each scene on a jobsite.

1.4. Research Objectives

This study aims to answer the research question using the following objectives: (1) Identify different scenarios in which image transformation can taking place due to changes in the viewpoint of the camera, (2) propose an approach based on localization systems to repeat and retrieve the camera's position and orientation in different trials to decrease image transformation, (3) prototype this approach, and (4) evaluate how this new approach versus the traditional method could reduce image transformation in terms of accuracy and precision. Measuring precision is necessary because it shows whether or not the participants can produce and reproduce a constant pattern for taking pictures from a scene under different conditions. Measuring accuracy is essential because it shows whether the participants could produce and reproduce pictures close to a reference picture that was randomly (from different position and orientation) captured. The primary contribution of this paper to the body of knowledge is to identify a method that can reduce transformation errors in images captured from a scene at a construction site. This method should support image matching techniques and improve their chance of success.

1.5. Research Methodology

To achieve the first objective, an illustrative case study has been conducted to identify different scenarios in which image transformation can take place due to changes in the viewpoint of the camera. In addition, a literature review has been conducted to identify advanced types of image transformation and related features. To achieve the second objective, sensor-based tracking systems were reviewed, and the required position and orientation sensors were identified. A system architecture was proposed to show how these systems can be integrated and implemented for the purpose of this study. To achieve the third objective, a prototype based on the system architecture was developed. To achieve the fourth objective, an experiment was designed and conducted. The following two sections explain the required background information and investigative methods.

2. Background Information for Method Development

2.1. Image Transformation

According to Szeliski [1], the first step in matching two images is to detect or extract the distinct features of these images. However, this is not easy. Feature detection is challenging because when two images (e.g., the key reference frame and current frame) have been captured from a scene at different viewpoints, their features such as position, orientation, and scale are not exactly the same. This phenomenon is called image transformation. Thus, image transformation is impacted based on the position and orientation of the camera. The position and orientation of a camera depends on six spatial degrees of freedom, including three degrees of freedom for position (i.e., X, Y, and Z), and three degrees of freedom for orientation (i.e., pitch, roll, and yaw/head) [43]. Figure 6 illustrates the coordinate system that can be defined based on six degrees of freedom.

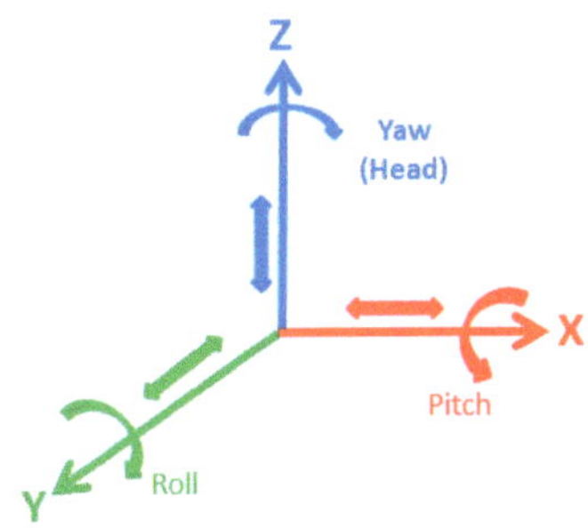

Figure 6. Coordinate system including six degrees of freedom, three linear and three angular (adapted from [44]).

2.2. Image Transformation Scenarios: Illustrative Case Study (i.e., Examples of Image-Based Scene Transformations)

To have a better understanding of camera position and orientation and their impact on image transformation, an illustrative case study has been conducted. In this case study, a camera was installed on a tripod with six degrees of freedom. In this first step, a reference picture was captured from a scene with a fixed camera's orientation and position. In the second step, the secondary pictures were captured from a different camera's position and orientation. For each capture, only one degree of freedom was applied. In other words, three pictures were captured when the position of the camera changed in the X, Y, or Z directions with a fixed orientation, and three pictures were captured while the position was fixed and the orientation changed in the X, Y, or Z directions. The six images captured in these ways were aligned over the reference picture separately to identify the transformation impacts and based on the observations, six conceptual diagrams were created, as shown in Figure 7.

The first type is a linear transformation that occurs on the X-axis. This type occurs when the relative position of a camera changes in the X direction while producing two images. The second type is a linear transformation on the Z-axis. This transformation occurs when the camera is repositioned in the Z direction. The third type of linear transformation occurs on the Y-axis. In this type, which is correlated with scaling, the picture is captured when the position of the camera in the Y direction is changed. In this type, the size of objects in the image changes. The fourth type is an angular transformation that occurs around the X-axis. In this type, the orientation of the camera changes, and the camera is rotated around in the X direction. The fifth type is an angular transformation that occurs around the Y-axis. In this type, the camera rotates in the Y direction. The sixth type is an angular transformation that occurs on the Z-axis. In this type, the camera is rotated around in the Z direction.

In the first and second types of transformations, only the locations of objects in the images change. In the third type, in addition to the locations of objects, the sizes of the objects change. In the fourth type of transformation, the locations of objects change. In the fifth and sixth types of transformations, due to changes in the orientation of the camera, the shapes of objects in the image change. In addition to these changes, in all these transformations, due to changes in the position of the camera or changes in orientation, some objects that are captured on the first image disappear and some new objects are captured.

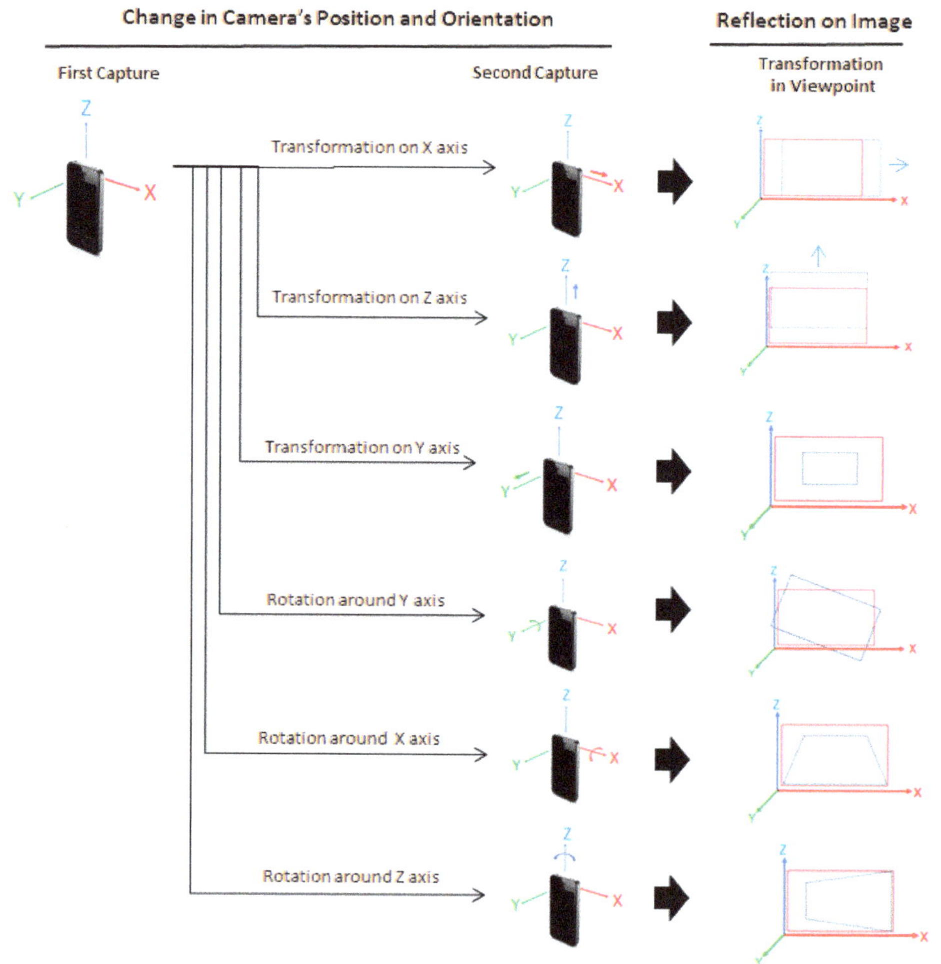

Figure 7. Changes in the camera's position and orientation can cause image transformation.

Advanced Transformations

In real situations, without using a tripod, these fundamental transformations combine and create new types of transformations. For instance, if transformations on the X- and Z-axis coincide, it is called translation. In this type, it is assumed that factors such as image orientation, lengths, angles, parallelism, and straight lines remain unchanged. In other words, this type of transformation only has two degrees of freedom. If relative translation and the rotation of the camera lens regarding the Y-axis occur together, it is called Euclidean (rigid). In this type, factors such as lengths of edges, angles, parallelism, and straight lines remain unchanged. In other words, this type of transformation has three degrees of freedom.

The third type, similarity, occurs when the relative rotation and scale of the second image changes in relation to the first image. This means that the second picture, in addition to the rotation of the camera around the Y-axis, was captured from a different position relative to the scene (on the Y-axis). In this type, angles, parallelism, and straight lines remain unchanged. In other words, this type of transformation has four degrees of freedom. The fourth type, called affine, occurs when a camera that takes the second picture rotates around two coordination axes such that parallelism and straight lines remain unchanged. In other words, this type of transformation has six degrees of freedom.

The fifth type, which is called projective (homography), occurs when a camera rotates around one or more coordination axes such that only the straight lines remain unchanged. In other words, this type of transformation has eight degrees of freedom.

The image matching algorithms need to deal with these image transformations and bring them into agreement with the reference picture. To have a better understanding, Szeliski [1] suggested a diagram to visualize these different types of transformations (Figure 8).

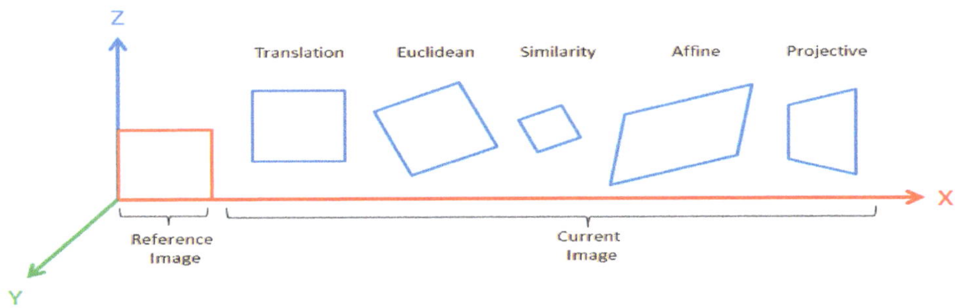

Figure 8. Two-dimensional (2D) geometric image transformations (adapted from Szeliski [1]).

2.3. *Propose an Approach Based on Localization Systems to Remotely Repeat and Retrieve the Camera's Position and Orientation to Decrease Image Transformation (Sensor-Based Tracking Systems)*

As was previously indicated, at temporary and messy places such as construction sites, one way to potentially decrease the impact of image transformation is a system that navigates the crews to hold the camera in a single position and orientation without using a tripod or a fixed-point camera. For this purpose, an accurate positioning and orientation system is required. Sensor-based techniques, independent from vision techniques, could be suitable candidates. In other words, sensor-based approaches use non-vision sensors to track a scene. Mechanical sensors, magnetic sensors, GPS (Global Positioning System), and ultrasonic and inertia sensors are some examples of non-vision tracking sensors. The following paragraph introduces the limitations of these types of sensors.

GPS has low user coverage in an indoor environment (4.5%) [45]. It requires direct lines of sight from a user's receiver to at least three orbital satellites [46,47] and its signal accuracy is degraded by occlusion. Wi-Fi has high user coverage indoors (94.5%) [45], with 15 to 20 m accuracy in indoor environments [45]. Bluetooth has 75% accuracy for partial coverage and 98% accuracy for full coverage in a room, while target devices need to be stationary for long periods of time [48]. Ultrasonic sensors are sensitive to temperature, occlusion, and ambient noise, require significant infrastructure, and have a low update rate [47]. Infrared is short range and limited because of line-of-sight requirements, as seen in Active Badge [49]. Radio frequency (type of signals, IEEE 802.11, WLAN) has a median accuracy of 2 to 3 m [50]. Inertial sensors are prone to drift and require constant recalibration [51]. Radio frequency (type of signals, UWB) emits ultra-wideband signals that can pass through walls and have high accuracy [52,53].

Required Position and Orientation Sensors

From these different tracking sensors, the most accurate positioning system could be the system that works with ultra-wideband (UWB) [54]. The accuracy of this system claims to be (±)10 cm [54]. According to [54], "the accuracy achieved with this technology is several times better than traditional positioning systems based on WIFI, Bluetooth, RFID or GPS signals." [54]. Some companies are developing UWB positioning sensors. One of them is called Pozyx. The sensors produced by this company include a tag and some anchors (at least four anchors are required). The tag sends and receives signals to anchor modules through a wireless radio technology called ultra-wideband (UWB) [54]. These signals can penetrate walls in an indoor environment. The anchor modules play the role of reference points for the tag. In this system, to calculate the position, the distance of one tag module to each anchor module is calculated based on time-of-flight (TOF) of the waves between the tag and anchors, where [54]:

Distance = time of flight × speed of light
Speed of light = 299,792,458 m/s

Then, through a method called multilateration [55], the position of the tag module with regard to anchor modules is calculated. For 3D orientation purposes, some sensors such as acceleration, magnetic field, and angular velocity are embedded in the tag module, which handles orientation responsibility. According to the sensor manual [54], each of these sensors has its own limitations, but through combining the outputs from different types of sensors, 3D orientation is computed accurately.

3. Methods

3.1. System Architecture: Positioning and Orientation

To better understand how these 3D positioning and orientation systems can be integrated and implemented for the purpose of this study, a system architecture was proposed. As shown in Figure 9, for the positioning estimations, the tag communicates with four anchors (i.e., reference points) through ultra-wideband RF signals. For orientation estimations, there are three sensors, acceleration, magnetic, and angular velocity, that can work together to estimate the tag orientation. The tag needs to be connected with a computing device such as a tablet to transfer the received data for analyzing and displaying to users. Using this information, the user can monitor the position and orientation of the tag. The first challenge is how can the tag be used for navigating the camera lens? The second challenge is how can data generated from the tag be displayed through a user interface for the purpose of monitoring the camera's position and orientation? To meet these challenges, a prototype was developed.

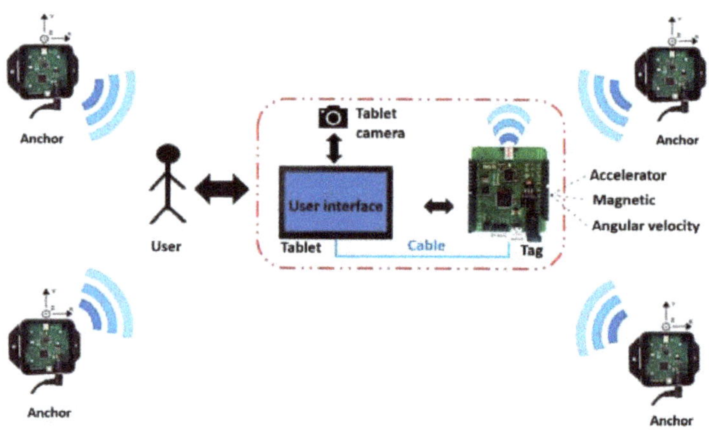

Figure 9. Positioning and orientation system architecture.

3.2. Prototype Development

The prototype development included two phases. In the first phase, the positioning and orientation sensors were integrated with a tablet camera such that the sensors can detect any change in the position and orientation of the camera. In the second phase, a user interface was created to display information regarding the position and orientation of the camera so that users could monitor the position and orientation of the camera. The following paragraphs explain these phases in detail:

Phase 1 To use the tag module for navigating the camera's lens, the simplest way could be attaching the tag module to the backside of the tablet in a fixed condition. To execute this idea, tape holders were used to attach the tag without any degrees of freedom. In this study, since the position and orientation of the tag relative to the tablet camera remained fixed, the position and orientation of the tag and camera lens were assumed to be the same. Figure 10 shows how the tag module was attached to the tablet.

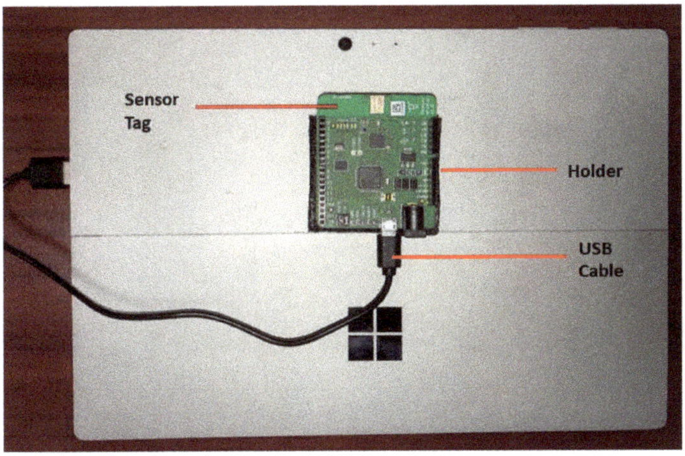

Figure 10. Physical integration of tag and tablet to be used on jobsite.

Phase 2 To be able to display and monitor the positioning and orientation sensors outputs, a user interface was designed and prototyped (Figure 11). This user interface could collect the data regarding the position and orientation of the sensor tag and visualize that data in the form of dynamic diagrams simultaneously. The programming language Python was used to prototype this user interface, with Microsoft Windows selected as the operating system and Surface [56] selected as the handheld device to run this user interface. These systems were selected due to their compatibility with the sensors. Figure 11 illustrates the created user interface. The user interface included indicators that could display the position and orientation of the camera lens in the room. As shown in Figure 11, on the left side, two positioning indicators were designed. The first one could show the position of the tablet in the room on the X-Y axes. The second one could show the position of the tablet on the Z-axis.

On the right side, the orientation indicators are shown (Figure 11). The first one is related to the rotation of the tablet around the Z-axis, which is called the head. The second one is related to the rotation of the tablet around the Y-axis, which is called roll. The third one is related to the rotation of the tablet around the X-axis, which is called the pitch. The zero point on indicators occurred when the red point stopped at the center of the indicator. The fourth one is not an indicator. It was designed to illustrate the Cartesian coordinating system axes. This diagram was designed and displayed on the user interface next to indicators to make sure the participants were aware of the room's coordination system during the experiment.

The user, by moving left and right, and forward and backward, could change the XY indicator; by moving up and down, the Z indicator; and by rotating the tablet, the pitch, roll, and head (i.e., yaw) indicators. After investigating the reference location and orientation, the user could look at the scene through the user interface screen and click the shutter button to capture a picture.

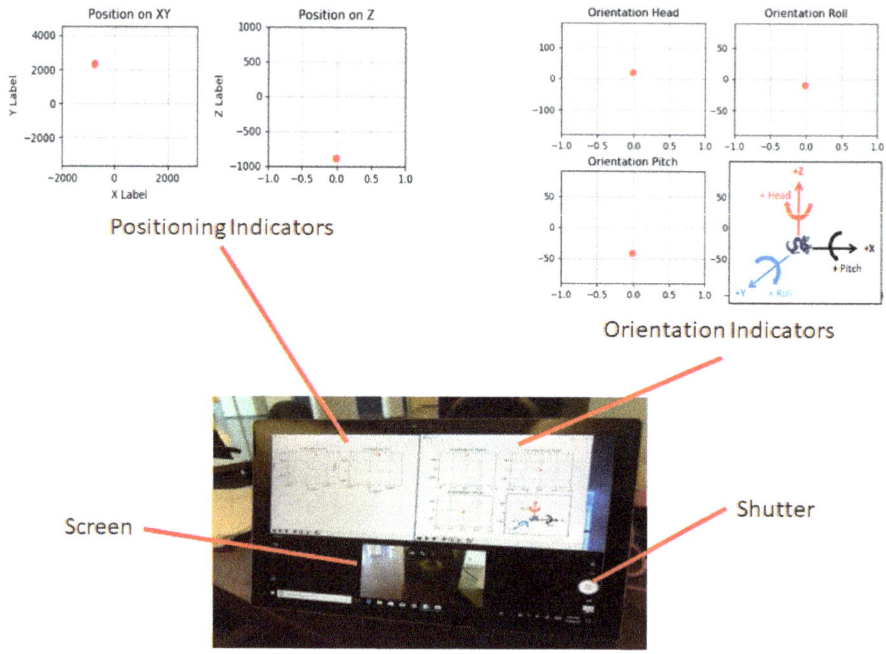

Figure 11. User interface prototype.

3.3. Experimental Testing of the Prototype System

To evaluate how the prototype approach (i.e., sensor-based approach) versus the traditional approach (i.e., non-sensor-based approach) could result in reducing image transformations in terms of accuracy and precision, an experiment was designed and conducted. The following sections explain the experiment design and the associated tasks.

3.3.1. Experimental Design

The experiment included two tasks. The first task was taking a picture from a scene without using positioning and orientation sensors, whereas the second task was taking a picture from the same scene but with the assistance of these sensors. The experiment was a within-subject experiment. In other words, each participant needed to conduct both tasks.

In this experiment, the pictures captured by participants were evaluated based on accuracy and precision parameters. Accuracy was defined as the capability of each approach to reproduce pictures that resemble a reference picture (i.e., error in accuracy = average transformations values − reference value). Our experimenter captured the reference picture from the scene before the experiment. The camera's position and orientation to capture the reference picture were decided based on the common sense of the experimenter. In a real situation, this picture could be the first picture captured from a scene, and therefore other pictures need to be taken from the same viewpoint later. To measure errors regarding the accuracy of the captured images in contrast with the reference image, the average transformation values for each linear direction and angular orientation needed to be calculated separately. The results show how close the images are to the reference image.

In this experiment, precision was defined as the capability of each approach to reproduce pictures that resemble with each other (i.e., error in precision = standard deviation). To measure the error in precision, the standard deviation of transformation for each direction and angle needed to be calculated separately. The results show how close the transformation values are to one another.

For this experiment, a scene with unique features was selected (Figure 12). This scene was an image (172 × 120 cm) installed on a wall in a lab. This scene followed two criteria: (1) The design of the scene should not provide any measurement tool to the participants when they are conducting the experiment tasks. Measurements are only for data collection and analysis by the experimenter. For this purpose, instead of using a checkerboard that has straight lines and potentially could assist the participants in taking pictures and create bias in individual rating behavior, an image was used that looked like a broken window without any recognizable assists (e.g., straight lines) in its context. Although there was not any assist in the context of this image for participants, the design of this image was symmetrical. The experimenter could use this feature for data collection, measurements, and analysis purposes. (2) The image was installed inside a large room with an open zone. Therefore, the participants had enough space to conduct their tests without any physical barriers that could impact their behavior when capturing pictures.

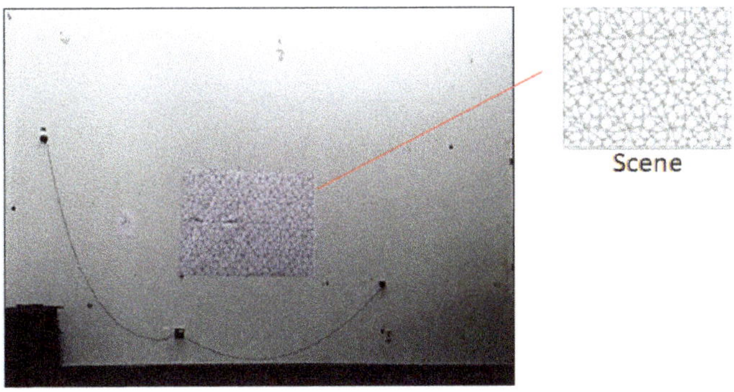

Figure 12. The scene in which participants were asked to capture pictures.

3.3.2. Experiment Tasks

To conduct the experiment tasks, paper-based instructions were created and given to the participants before each task. These instructions included two separate parts. The first part explained the experiment process for the first task. To conduct the first task, each participant needed to read the first part of the instructions. Then, the participant received a tablet to take a picture from the scene. For the first task, the participants needed to use their common sense regarding the position and orientation of the tablet camera.

The second part of the instruction explained the process for the second task. To conduct the second task after completing the first task, the participants needed to read the second part of the instructions. Concurrently, the experimenter needed to equip the tablet with the sensor tag and run the associated python code to activate positioning so that orientation sensors could make the user interface indicators available to the participant. Thus, this time, the participant could monitor the position and orientation of the camera by viewing the indicators. Using these indicators, the participants needed to look for a reference viewpoint with the following features:

Position → XY = [0], Z = [0]
Orientation → Head = [0], Pitch = [0], Roll = [0].

To achieve the defined position, the participants could walk and change their position in different directions to find the reference position where XY = (0) and Z = (0). In addition, they could rotate the tablet around different directions to find the reference orientation where (pitch, roll, head) = (0, 0, 0). As was previously mentioned, this point of view (position and orientation) was defined based on the common sense of experimenter. For this reason, this point of view could not be predictable for participants. It was not located on a position at the center zone of the room or on an orientation angle perpendicular to the scene. It was the best image that the experimenter sensed could capture from the scene. During the second test, the experimenter monitored the participants to ensure they captured the pictures when the red points in all these indicators stopped on zero (0). For each task, the participants were allowed to generate only one picture. There were not any time limitations when participants read the guidelines and conducted the tasks for the experiment. Figure 13 illustrates the coordination system of the scene.

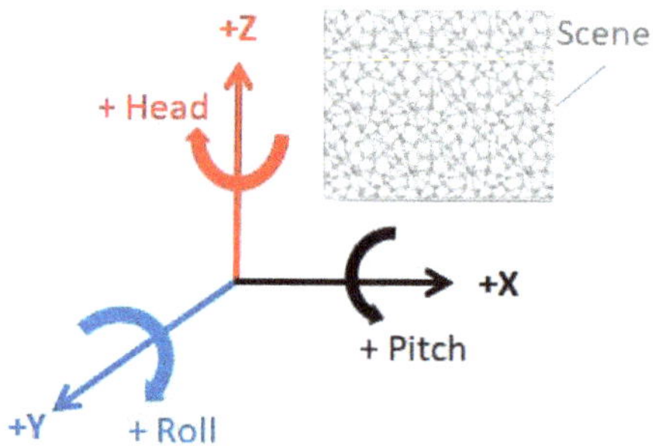

Figure 13. Cartesian coordination system of scene.

4. Results and Discussion

To conduct the experiment, 37 graduate and undergraduate students were randomly selected. For each task, 37 pictures were collected. The experiment was conducted at one location within similar indoor environment conditions and laboratory settings. The reference position and orientation value were similar for both tasks. To avoid learning effect, the first task was non-sensor-based for all the participants, because the second task, which involved the sensors, directed the participants to the defined reference position and orientation. The participants were tested individually to ensure they would not learn from each other. The pictures collected from the participants were divided into two groups (Appendix B). The first group included pictures related to the first task and the second group from the second task. The first group contained 37 pictures taken without using the sensor system, and the second group contained 37 pictures that were taken with assistance from the sensors.

4.1. Limitations

From 37 images in the second group, ten pictures were discarded due to systematic errors that the experimenter reported during the experiment. In addition, two pictures from the first group were discarded due to the extreme camera rotation (90 degrees) around the Z-axis. This skewed rotation changed the coordination system for two pictures, and therefore the results were not calculable. Furthermore, in this experiment, the indicator regarding the Z direction (positioning only) was decided to be off to increase the speed of the system. The initial tests showed that the average transformations in the Z direction were very similar to transformations in the X direction. Considering this point and due to technical limitations, the indicator related to Z was decided to be put in the "off" stetting during the experiment.

A tripod was used for the initial study to understand the relationship between the camera's orientation and the ratios. The tripod was equipped with leveling and protractor tools. It was better if we used a digital one.

It could have been better if we used the original tablet sensors instead of the tag sensors to monitor the orientation. However, the main issue that we observed in both systems (tag and tablet sensors) needed recalibration. Any time that the orientation sensors were used, the yaw had a slight error. Therefore, we decided to use the tag that generated data for both position and orientation.

The UWB system is a promising method that can penetrate walls. However, this experiment was conducted inside an open area. Thus, the potential impact of barriers concrete walls, steel structures, and building infrastructure (e.g., stairs, furniture, machines, etc.) that could have block line-of-sight were not considered.

The intrinsic parameters of the camera for both tasks were the same. For both tasks, the same camera with the same zooming level was used. The participants could not change the zooming level. For extrinsic calibration, manual approaches were used, as are explained in the text. Minor errors could have be included, but since the same methods were implemented for both groups of pictures (i.e., sensor-based and non-sensor-based), the results were unbiased.

The scene included a flat image of 172 × 120 cm instead of a 3D object. The reason for this simplicity was reducing errors in calculations. This 2D scene could be enough to evaluate the precision and accuracy of the two approaches (sensor-based versus non-sensor-based) in retrieving the position and orientation of the camera.

4.2. Measuring Changes in Camera's Position and Orientation

The features in two images can transfer (i.e., displace) if the position and orientation of the camera that captured those pictures change. In this research study, to understand what type of transformation results in a certain type of change in a camera's position and orientation, some methods were determined. These methods could assist the authors in assessing the causes of image transformation in different pictures. These methods are described in the following paragraphs:

Change in the position of the camera in the Y direction To be able to measure any change in the position of the camera in the Y direction, the method illustrated in Figure 14 was used. In this method, the position of the scene is fixed, but the camera's position changes. The distance between the current images in the scene can be estimated where ($y - y' = y \times i/i'$). In this equation, y is the distance between the scene to the camera that captured the reference picture, i is the distance between two points in the reference image, i' is the distance between similar points in the current image, and y' is the distance between the reference camera and the current one.

After estimating the distance of the camera to the scene for all pictures, to find the change in position of the camera, the average distances should be compared with the distance measured regarding the reference picture (i.e., accuracy) and also with each other (i.e., precision).

Change in the position of the camera in the X direction To measure the change in position of the camera in the X direction, a reference point was selected at the center of the board installed on the scene. The distance between this point and the center of the camera lens (i.e., the center of the picture) was measured for all pictures (Figure 15). Since the scales of the pictures were different, these distances were converted into a single scale to become comparable. To find the change in position of the camera, the average distances were compared with the distance measured in the reference picture (i.e., accuracy) and also with each other (i.e., precision).

Change in the orientation of the camera around the Y-axis (roll) To measure the orientation of the camera around the Y-axis, a horizontal line that crossed the center of image was drawn. Then, a protractor was used, and the angle that this line made with the image was measured as rotation around the Y-axis. To find the change in orientation of the camera around the Y-axis, the average rotations for each group were compared with the reference rotation (i.e., accuracy) and also with each other (i.e., precision).

Change in the orientation of the camera around the Z-axis (head) Since the images were two-dimensional, the rotation angle of the camera around the Z-axis was not easy to measure. Therefore, other variables were considered. These variables are the length of the left and right sides of an image that change when the rotation around the X-axis occurs. Turning the camera to the left expands the left side and reduces the right side, and vice versa (Figure 16). Knowing these principles, the left and right sides for all pictures from both groups were measured, and then the ratio for each one was measured (i.e., ratio = smaller vertical side/larger vertical side).

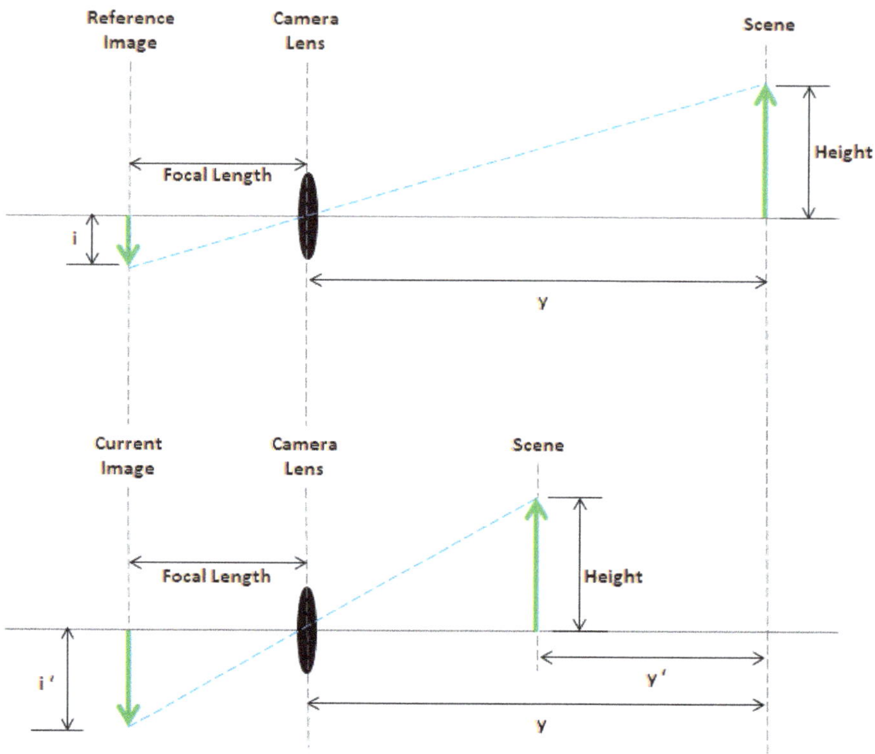

Figure 14. The method used to calculate the camera distance to the scene (adapted from [57]).

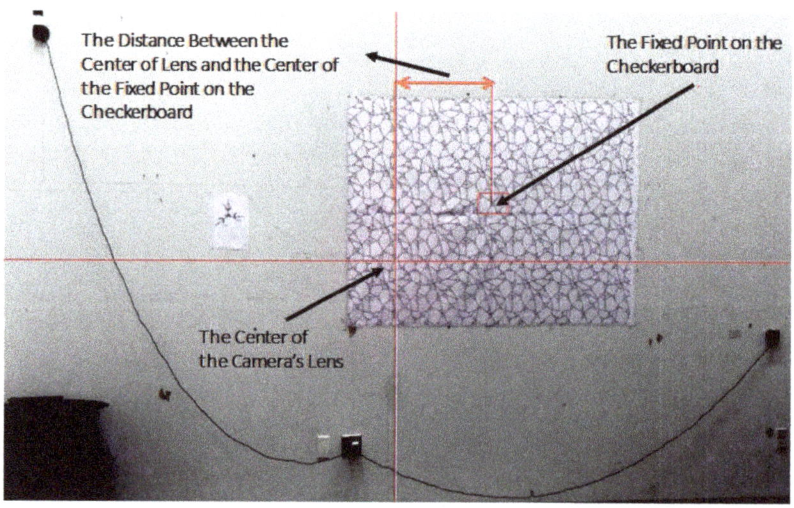

Figure 15. Distance between center of lens and center of the board installed on the wall.

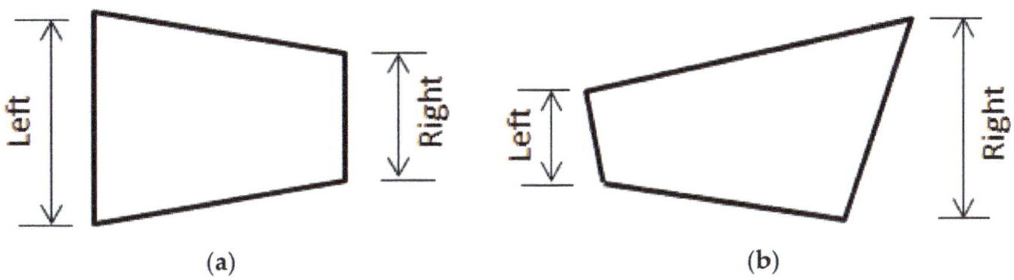

Figure 16. How to measure the length of vertical sides. (**a**) If the vertical sides are parallel; (**b**) If the vertical sides are not parallel.

To understand the relationship between the camera's orientation and these ratios, a separate test was conducted. In this test, a scene was provided, and a camera was installed on a tripod. The camera lens was leveled and located in a parallel position to the scene. The camera had only one degree of freedom around the Z-axis. In this condition, the first picture was captured. Next, the camera was rotated 10 degrees around the Z-axis, and the second picture was captured. This process was repeated, and the results were recorded. Using the results, a graph with a regression line was created (Figure 17). This graph was used to convert the image ratios related to the experimental data to meaningful rotation degrees.

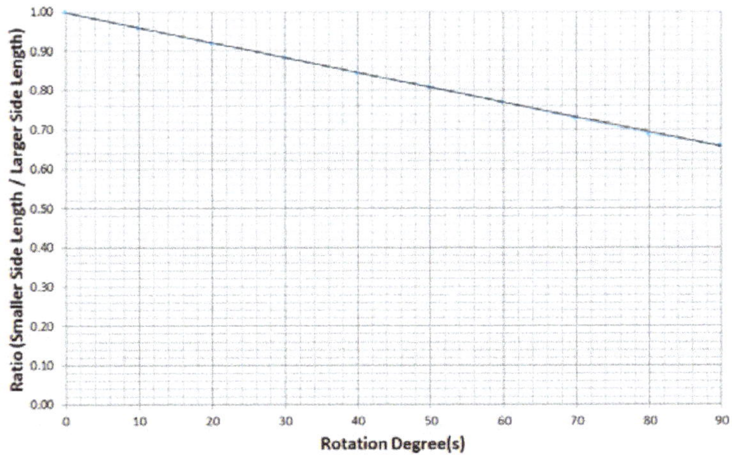

Figure 17. Relationship between the image sides' ratios and the camera's degree of rotation.

To find the change in orientation of the camera around the Z-axis, the average rotation values of each group of pictures were compared with the reference rotation (i.e., accuracy) and also with each other (i.e., precision).

Change in the orientation of camera around the X-axis (pitch) To measure the orientation of the camera around the X-axis, the same principle and graph used for the Y-axis were applied for this axis. However, this time, turning the camera around the X-axis could impact the length of the upper and lower sides of the images. Therefore, the ratio for each image was measured (i.e., ratio = smaller horizontal side/larger horizontal side) (Figure 18).

To measure the degree of rotation around the X-axis, the graph illustrated in Figure 17 was used. To find the change in orientation of the camera around the X-axis, the average rotation values for each group of pictures were compared with the reference rotation (i.e., accuracy), and also with each other (i.e., precision).

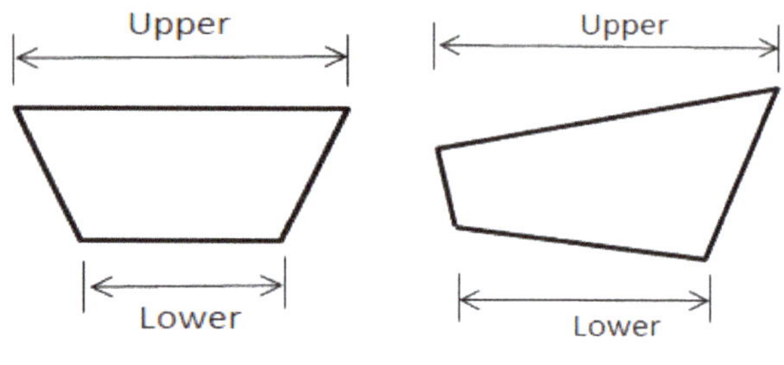

Figure 18. How to measure the length of horizontal sides. (**a**) If horizontal sides are parallel; (**b**) If horizontal sides are not parallel.

4.3. Results of Camera's Positioning Accuracy and Precision in the X and Y Directions

The results of the experiment to discuss the accuracy of the two approaches for producing pictures resembling the reference picture regarding the X and Y directions are presented in Table 1 and are explained as follows:

Accuracy in the X direction The results of the experiment regarding accuracy (i.e., producing pictures resembling the reference picture) showed that when the images were captured with the assistance of positioning sensors, they more accurately resembled the reference picture. In other words, when the participants did not use the sensors during the first task, the average error, in terms of accuracy in the X direction, was (30 cm), but when they used the sensors during the second task, the average error decreased to (0.3 cm).

Accuracy in the Y direction The same situation occurred in the Y direction. In the Y direction, which reflected scaling, the average accuracy error decreased from (33 cm) to (6.8 cm) when the participants used a positioning sensor during the second task.

Table 1. Comparison between accuracy and precision of the two approaches regarding linear transformation.

Type of Image	The Distance between the Center of Lens and Center of the Fixed Point on Scene	In the X Direction (cm)	In the Y Direction (cm)	In the Z Direction (cm)
Reference image		44.71	330.5	N/A
Group 1 (without sensor)	Avg.	14.7	297	N/A
	Min.	0	154	N/A
	Max.	77	582	N/A
	Precision Error (SD)	15.5	112	N/A
	Accuracy Error	30	33	N/A
	Range	77	428	N/A
Group 2 (with sensor)	Avg.	45	337	N/A
	Min.	11	312	N/A
	Max.	101	366	N/A
	Precision Error (SD)	21	13.3	N/A
	Accuracy Error	0.3	6.8	N/A
	Range	90	54	N/A

So far, this sensor-based approach could produce more accurate results than the non-sensor-based approach by reducing image transformation in both X and Y directions. The results of the experiment to discuss the precision of two approaches in producing pictures resembling each other regarding the X and Y directions are explained as follows:

Precision in the X direction The results regarding precision (i.e., producing pictures resembling each other) in the X direction showed an exciting result. In this direction, precision decreased when the participants used the sensor-based approach. In other words, the pictures captured during the first task without sensors had less standard deviation (15.5 cm) as compared with those that were captured during the second task (21 cm). This result indicates that when the participants wanted to take pictures from the scene without sensors, based on their common sense, they selected locations in the X direction that better resembled the reference image (the X direction was parallel to the scene). Therefore, the degree of repeatability increased. In contrast, since the sensors inherently generated error, the participants were navigated to the locations in the X direction that were less close to each other. The results are shown in Table 1.

Precision in the Y direction The results of the experiment regarding precision determined when the images were captured with the assistance of positioning sensors showed that they were more precise in the Y direction. In other words, when participants used the sensors during the second task, the standard deviation in the Y direction decreased (from 112 cm to 13.3 cm). This result showed that in the Y direction, the positioning sensor during the second task could navigate the participants to distances that resembled the reference image more than the first task when they used their common sense.

Thus, in the Y direction, the standard deviation when participants used their common sense in selecting a location in the perpendicular direction to the scene (i.e., Y) was almost six times more than this value in the parallel direction with the scene (i.e., X). Furthermore, according to the sensor's standard manual, the positioning sensor is expected to have errors between +10 and −10 cm. In this experiment, the average standard deviation of the positioning sensor was measured (21 cm) in the X direction and (13.3 cm) in the Y direction.

Another useful result that could be extracted from Table 1 is range (range = max − min). Subtracting the maximum from the minimum revealed that the maximum range occurred in the Y direction (582 − 154 = 505 cm) when the non-sensor-based approach was used. In contrast, the range value for the sensor-based approach was 366 − 312 = 54 cm. This result shows that in the worst-case scenario, the separation of data for the sensor-based approach is ten times better than that of the non-sensor-based approach. Regarding the X direction, when the sensor-based approach was used, the maximum range occurring in the X direction (101 − 11 = 90 cm) which was slightly more than when the non-sensor-based approach was used (77 − 0 = 77 cm).

4.4. Results of Camera's Orientation, Accuracy, and Precision around the X, Y, and Z Directions

The results of the experiment to discuss the accuracy of two approaches in producing pictures resembling the reference image regarding the camera's orientations around the X, Y, and Z axes are presented in Table 2 and are explained as follows:

Accuracy around the X-axis (pitch) The results of the experiment regarding accuracy (i.e., producing pictures resembling the reference picture) showed that the results in both approaches are very similar. While the orientation of the camera around the X-axis for the reference image was measured as 7 degrees, the average reference was 5 degrees for the first group of pictures and 2 degrees for the second group of pictures. Thus, the average accuracy error for pictures captured without using a sensor is slightly less (2 degrees vs. 5 degrees) than when the images were captured with the assistance of orientation sensors. This result shows that using the sensor did not improve the accuracy for rotation around the X-axis (pitch).

Accuracy around the Y-axis (roll) The results showed that the average accuracy around the Y-axis for both approaches is the same. While the orientation of the camera for the reference image around the Y-axis measured 0, the average orientation for Groups 1 and 2 (with and without sensors) measured the same (1 degree). This result showed that when participants wanted to take pictures from a scene using their common sense, they could hold the tablet camera almost in the same orientation as when they used the orientation sensors (Table 2). However, as was indicated in the limitation section, two of the pictures captured by participants had 90 degrees rotation around the Y-axis of the tablet. Although these two exceptional pictures were discarded because of the high statistical skews that could affect the calculations, this could occur in real situations if crews are not warned in advance.

Accuracy around the Z-axis (yaw) The results showed the average accuracy around the Z-axis for the sensor-based approach is slightly better than the non-sensor-based approach. While the orientation of the camera around the Z-axis for the reference image measured 10 degrees, the average for the non-sensor-based approach was 17 degrees and the sensor-based approach was 15 degrees. Therefore, the orientation accuracy error for the sensor-based approach (5 degrees) was slightly less than the non-sensor-based approach (7 degrees). This result indicates that the participants, using their common sense, can generate results closer to the reference than when they use sensors.

Table 2. Comparison between accuracy and precision of sensor-based and non-sensor-based approaches regarding the camera's orientation factors (i.e., pitch, roll, and yaw/head).

Type of Image	Rotation	Pitch (Ratio), Degree	Roll Degree	Yaw or Head (Ratio), Degree
Reference image		(0.97), 7	0	(0.96), 10
Group 1 (without sensor)	Avg.	(0.98), 5	1	(0.93), 17
	Min.	(1), 0	0	(1), 0
	Max.	(0.85), 38	6.5 [90 *]	(0.69), 80
	Precision Error (SD)	(0.03), 7	1.5	(0.08), 20
	Accuracy Error	(0.01), 2	1	(0.03), 7
	Range	38	6.5 [90 *]	80
Group 2 (with sensor)	Avg.	(0.99), 2	1	(0.94), 15
	Min.	(1), 0	0	(0.99), 2
	Max.	(0.96), 10	4.5	(0.85), 38
	Precision Error (SD)	(0.01), 2	1.2	(0.03), 7
	Accuracy Error	(0.02), 5	1	(0.02), 5
	Range	10	4.5	36

* Two of the pictures captured by participants had a 90 degrees rotation around the Y-axis of the tablet. Although these two exceptional pictures were discarded because of the high statistical skews that could impose on the affect calculations, this can occur again in real situations if the crews are not warned in advance.

In general, the degree of resemblance of the pictures produced by both approaches is very close to the reference picture. The precision of the two approaches in producing pictures resembling each other regarding orientations around the X, Y, and Z axes is presented in Table 2 and explained as follows:

Precision around the X-axis (pitch) The result regarding the standard deviation for the sensor-based approach was less than the non-sensor-based approach (7 degrees vs. 2 degrees). Therefore, precision (i.e., producing pictures that resemble each other) around the X-axis improved when the participants used the sensor-based approach. While the precision error for the non-sensor-based approach was 7 degrees, this value decreased to 2 degrees when they used the sensor-based approach. Therefore, the degree of repeatability of the camera's orientation and picture resemblance for pitch increased.

Precision around the Y-axis (roll) The standard deviation for both sensor-based and non-sensor-based approaches was almost the same (1.5 degrees vs. 1.2 degrees). Therefore, the results regarding the average precision error around the Y-axis (roll) were almost the same.

Precision around the Z-axis (yaw) The standard deviation around the Z-axis reduced from 20 degrees to 7 degrees when participants used the sensor-based approach. This means the precision error for the sensor-based approach is less, as the participants could repeat the orientation of the camera regarding (yaw) with less error when using the sensor-based approach.

The other interesting results presented in Table 2 could be the range values (range = max − min) of the changes in the camera's position and orientation when the sample pictures were captured.

The maximum range in orientation occurred around Y (90 − 0 = 90 degrees) and Z (80 − 1 = 79 degrees) when the non-sensor-based approach was used. When the sensor-based approach was used, the maximum range in orientation occurred around Z (38 − 2 = 36 degrees). As was previously indicated, the SIFT algorithms cannot correctly identify the distinct points if the orientation is more than (30 degrees). Therefore, based on the results, the sensor-based approach can prevent this issue during the image capturing phase. Logically, the analysis of the captured images during the image matching phase by image matching algorithms is errorless.

5. Summary and Conclusions

Due to the wide use of image matching techniques in the construction industry, and the vulnerability of these techniques to correctly detect and match scene features when extreme transformations in images occur, this study aimed to investigate how to reduce image transformations. For this purpose, different scenarios in which image transformation can take place were visualized. It was shown how these transformations could occur when the position and orientation of a camera change in three linear directions and three angular orientations. As was illustrated, to reduce image transformations, changes in the viewpoint (i.e., position and orientation) of the camera needed to be reduced. For this purpose, different techniques were reviewed, and the most accurate one was selected. This technique included positioning sensors that worked based on UWB waves, and orientation sensors such as acceleration, magnetic, and angular velocity. To apply these sensors for the purpose of reducing image transformation, a system architecture was defined, and a prototype was developed. The development of the prototype included two phases. In the first phase, the positioning and orientation modules (i.e., tag and anchors) were integrated with a tablet camera such that these sensors could detect any change in the position and orientation of the camera. In the second phase, a user interface was created to display information regarding the position and orientation of the camera such that users could monitor the location and viewpoint of the camera.

To compare how using the sensor-based approach could be different than a non-sensor-based approach, in terms of decreasing changes in position and orientation of the camera, an experiment was designed and conducted. The experiment included two tasks. For the first task, the participants were asked to use their common sense to capture the best picture possible from a scene. For the second task, they were asked to capture a picture from the same scene but with the assistance of positioning and orientation sensors. The images participants generated for these two tasks were evaluated in terms of accuracy (i.e., producing pictures that resemble the reference picture), and precision (i.e., producing pictures that resemble each other). The results of the experiment demonstrated that when participants used the sensor-based approach, a significant reduction in accuracy errors in the X and Y directions, and also the precision error in the Y direction, was achieved. The precision error in the X direction was slightly higher when the participants used the sensor-based approach. Regarding the orientation, the average results for both approaches did not show a significant difference. While accuracy error was slightly better for the non-sensor-based approach for pitch, it was slightly worse for yaw, and the same for roll (however, if the two samples with 90 degree rotations were not discarded from data related to the non-sensor-based approach, the error for this approach increased significantly). For the sensor-based approach, precession errors were slightly lower for pitch and roll and moderately lower for yaw.

In conclusion, these results showed that applying the sensor-based approach can control the camera's overall position and orientation and reduce image transformation. This can be important for feature detection algorithms used in applications such as augmented reality and change detections that use features of the environment in temporary and messy locations such as construction sites, where using a tripod or fixed-point camera is not possible. This research had technical limitations. The accuracy and precision of the sensor-based approach could improve. For instance, in this experiment, only four anchors were used. Using more anchors and even tags could improve the results. By using more powerful tablets, the time for data processing could be reduced, and the positioning system in the Z direction, which for this experiment was off, would be functioning. In future studies, the pictures produced by these two approaches could be tested by the image matching process to evaluate how the accuracy of the related algorithms could improve. If these limitations are eliminated by a sensor-based approach, failure scenarios such as extreme rotation and scaling, eliminated scene, and scene displacement can be improved.

Author Contributions: All authors contributed to the idea and concept of this study; Data curation, M.F.S. and J.L.; Formal analysis, M.F.S.; Funding acquisition, J.L.; Methodology, M.F.S. and J.L.; Project administration, J.L.; Software, M.F.S.; Supervision, M.G. and J.L.; Validation, M.G. and J.L.; Visualization, M.F.S.; Writing—original draft, M.F.S.; Writing—review & editing, M.G. and J.L. All authors have read and agreed to the published version of the manuscript.

Funding: This research was funded by the National Science Foundation (NSF), award number 1562515.

Conflicts of Interest: The authors declare no conflict of interest.

Appendix A

Augmented Reality: One technique that can link/combine paper-based and digital-based environments is augmented reality (AR). AR is "a technology that superimposes a computer-generated image (model) on a user's view of the real world, thus providing a composite view" [58]. AR is a part of the reality–virtuality continuum [59] (Figure A1). According to Azuma [60], AR "allows the user to see the real world, with virtual objects superimposed upon or composited with the real world. Therefore, AR supplements reality, rather than completely replacing it." In other words, AR is a combination of real-world and digital information through a single interface [21]. Thus, AR is an appropriate technology that can be used to access detailed information.

Figure A1. Concept of augmented reality [59].

There are two types of AR techniques, i.e., marker-based and markerless. The following paragraphs explain the differences between these two techniques:

Marker-based AR (feature-based, artificial markers) In this approach, an artificial marker needs to be located in the scene or environment as a reference. Then, information about the marker is interpreted by a handheld computing device (smartphone/tablet) application. Artificial markers are printed and attached to the locations [1]. Some examples of artificial markers are dot-based markers [61], QR code markers [62,63], circular markers [64], square markers [65], and alphabetic combination markers [65]. Due to fiducial marker use in the environment, and the fact that these markers are distinguishable in the environment (physical world), the marker-based tracking approach is very robust with high accuracy [66–68].

Markerless AR (feature-based, natural features) This type of AR system uses the natural features of the environment as references [24]. Depending on the algorithm used for this system, these features could be edges, corners, segments, or points [23]. In this online approach, features extracted from current video frames taken from the scene are compared with features extracted from an initial key frame. Then, correspondence between feature pairs is created. This loop continues until the best match between features has been computed [1]. If enough numbers of matches are identified, the virtual data stored in repository is queried and appears on the screen of the computing device, such as a smartphone or tablet.

Appendix B

B.1. First Group of Samples

The first group of pictures was taken by 37 participants without using the positioning and orientation system during the first task. Figure A2 shows the collected data from the first task.

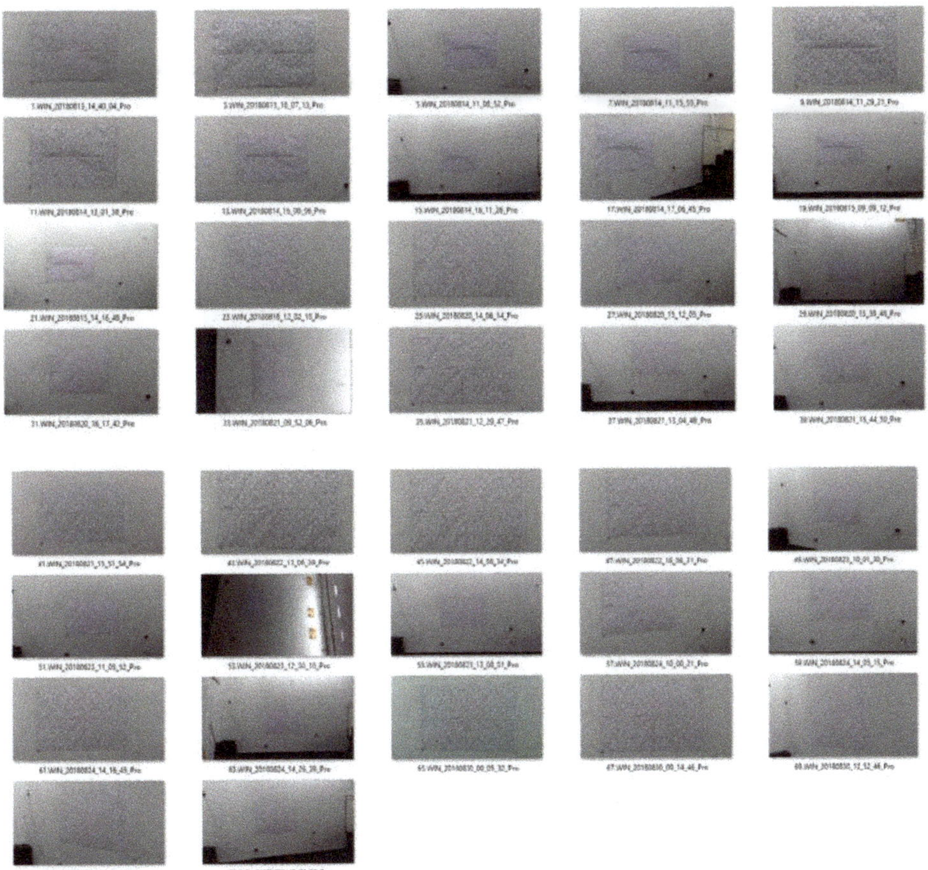

Figure A2. The first sample group of pictures.

B.2. Second Group of Samples

The second group of pictures was taken by 37 participants using the positioning and orientation system during the second task. Figure A3 shows the collected data from the second task.

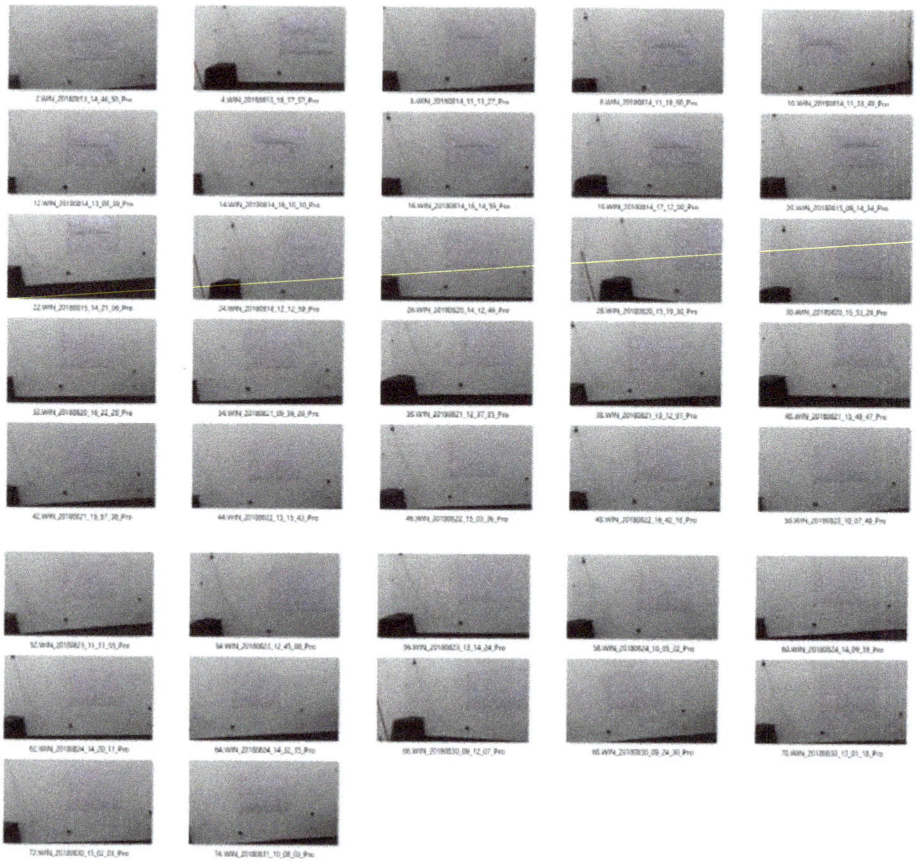

Figure A3. The second sample group of pictures.

References

1. Szeliski, R. Computer Vision: Algorithms and Applications. *Computer (Long. Beach. Calif.)* **2010**, *5*, 832.
2. Forsyth, D.A.; Ponce, J. *Computer Vision, A Modern Approach*; Printice Hall: Upper Saddle River, NJ, USA, 2003.
3. Shapiro, L.G.; Stockman, G.C. *Computer Vision: Theory and Applications*; Prentice Hall: Upper Saddle River, NJ, USA, 2001.
4. Horn, B.; Klaus, B.; Horn, P. *Robot Vision*; MIT Press: Cambridge, NY, USA, 1986; ISBN 0262081598.
5. Chen, M.; Shao, Z.; Li, D.; Liu, J. Invariant matching method for different viewpoint angle images. *Appl. Opt.* **2013**, *52*, 96–104. [CrossRef]
6. Dai, X.L.; Lu, J. An object-based approach to automated image matching. In Proceedings of the IEEE 1999 International Geoscience and Remote Sensing Symposium. IGARSS'99 (Cat. No. 99CH36293), Hamburg, Germany, 28 June–2 July 1999; Volume 2, pp. 1189–1191.
7. Karami, E.; Prasad, S.; Shehata, M. Image Matching Using SIFT, SURF, BRIEF and ORB: Performance Comparison for Distorted Images. In Proceedings of the 2015 Newfoundland Electrical and Computer Engineering Conference, St. John's, NL, Canada, 14–15 April 2015; p. 4.
8. Sinha, S.N.; Frahm, J.M.; Pollefeys, M.; Genc, Y. Feature tracking and matching in video using programmable graphics hardware. *Mach. Vis. Appl.* **2011**, *22*, 207–217. [CrossRef]
9. Brown, M.; Lowe, D.G. Automatic panoramic image stitching using invariant features. *Int. J. Comput. Vis.* **2007**, *74*, 59–73. [CrossRef]

10. Szeliski, R.; Shum, H.-Y. Creating full view panoramic image mosaics and environment maps. In Proceedings of the 24th Annual Conference on Computer Graphics and Interactive Techniques, Los Angeles, CA, USA, 3–8 August 1997; pp. 251–258.
11. Kratochvil, B.E.; Dong, L.X.; Zhang, L.; Nelson, B.J. Image-based 3D reconstruction using helical nanobelts for localized rotations. *J. Microsc.* **2010**, *237*, 122–135. [CrossRef]
12. Lu, Q.; Lee, S. Image-based technologies for constructing as-is building information models for existing buildings. *J. Comput. Civ. Eng.* **2017**, *31*, 4017005. [CrossRef]
13. Moghaddam, B.; Pentland, A. Probabilistic visual learning for object representation. *IEEE Trans. Pattern Anal. Mach. Intell.* **1997**, *19*, 696–710. [CrossRef]
14. Rowley, H.A.; Baluja, S.; Kanade, T. Neural network-based face detection. *IEEE Trans. Pattern Anal. Mach. Intell.* **1998**, *20*, 23–38. [CrossRef]
15. Pérez-Lorenzo, J.; Vázquez-Martín, R.; Marfil, R.; Bandera, A.; Sandoval, F. *Image Matching Based on Curvilinear Regions*; IntechOpen: London, UK, 2007.
16. Takacs, G.; Chandrasekhar, V.; Tsai, S.; Chen, D.; Grzeszczuk, R.; Girod, B. Rotation-invariant fast features for large-scale recognition and real-time tracking. *Signal Process. Image Commun.* **2013**, *28*, 334–344. [CrossRef]
17. Tang, S.; Andriluka, M.; Schiele, B. Detection and tracking of occluded people. *Int. J. Comput. Vis.* **2014**, *110*, 58–69. [CrossRef]
18. Kang, H.; Efros, A.A.; Hebert, M.; Kanade, T. Image matching in large scale indoor environment. In Proceedings of the 2009 IEEE Computer Society Conference on Computer Vision and Pattern Recognition Workshops, CVPR 2009, Miami, FL, USA, 20–25 June 2009; pp. 33–40.
19. Kim, H.; Kano, N. Comparison of construction photograph and VR image in construction progress. *Autom. Constr.* **2008**, *17*, 137–143. [CrossRef]
20. Jabari, S.; Zhang, Y. *Building Change Detection Using Multi-Sensor and Multi-View- Angle Imagery*; IOP Conference Series: Earth and Environmental Science; IOP Publishing: Halifax, NS, Canada, 2016; Volume 34.
21. Gheisari, M.; Foroughi Sabzevar, M.; Chen, P.; Irizzary, J. Integrating BIM and Panorama to Create a Semi-Augmented-Reality Experience of a Construction Site. *Int. J. Constr. Educ. Res.* **2016**, *12*, 303–316. [CrossRef]
22. Foroughi Sabzevar, M.; Gheisari, M.; Lo, L.J. Improving Access to Design Information of Paper-Based Floor Plans Using Augmented Reality. *Int. J. Constr. Educ. Res.* **2020**, 1–21. [CrossRef]
23. Belghit, H.; Zenati-Henda, N.; Bellabi, A.; Benbelkacem, S.; Belhocine, M. Tracking color marker using projective transformation for augmented reality application. In Proceedings of the 2012 International Conference on Multimedia Computing and Systems, Tangier, Morocco, 10–12 May 2012; pp. 372–377.
24. Yuan, M.L.; Ong, S.-K.; Nee, A.Y.C. Registration using natural features for augmented reality systems. *IEEE Trans. Vis. Comput. Graph.* **2006**, *12*, 569–580. [CrossRef]
25. Moravec, H.P. Techniques towards Automatic Visual Obstacle Avoidance. In Proceedings of the International Joint Conference on Artificial Intelligence, Cambridge, MA, USA, 22–25 August 1977; p. 584.
26. Harris, C.G.; Stephens, M. A combined corner and edge detector. In Proceedings of the Alvey Vision Conference, Manchester, UK, 31 August–2 September 1988; pp. 147–151.
27. Smith, S.M.; Brady, J.M. SUSAN—A new approach to low level image processing. *Int. J. Comput. Vis.* **1997**, *23*, 45–78. [CrossRef]
28. Rosten, E.; Drummond, T. Fusing points and lines for high performance tracking. In Proceedings of the Tenth IEEE International Conference on Computer Vision (ICCV'05) Volume 1, Washington, DC, USA, 17–21 October 2005; Volume 2, pp. 1508–1515.
29. Rosten, E.; Drummond, T. Machine learning for high-speed corner detection. In Proceedings of the European Conference on Computer Vision, Graz, Austria, 7–13 May 2006; pp. 430–443.
30. Beaudet, P.R. Rotationally invariant image operators. In Proceedings of the 4th International Joint Conference on Pattern Recognition, Tokyo, Japan, 7–10 November 1978.
31. Lakemond, R.; Sridharan, S.; Fookes, C. Hessian-based affine adaptation of salient local image features. *J. Math. Imaging Vis.* **2012**, *44*, 150–167. [CrossRef]
32. Lindeberg, T. Scale selection properties of generalized scale-space interest point detectors. *J. Math. Imaging Vis.* **2013**, *46*, 177–210. [CrossRef]
33. Lowe, D.G. Distinctive image features from scale-invariant keypoints. *Int. J. Comput. Vis.* **2004**, *60*, 91–110. [CrossRef]

34. Mikolajczyk, K.; Schmid, C. Scale & affine invariant interest point detectors. *Int. J. Comput. Vis.* **2004**, *60*, 63–86.
35. Bay, H.; Ess, A.; Tuytelaars, T.; Van Gool, L. Speeded-up robust features (SURF). *Comput. Vis. Image Underst.* **2008**, *110*, 346–359. [CrossRef]
36. Yussof, W.N.J.H.W.; Hitam, M.S. Invariant Gabor-based interest points detector under geometric transformation. *Digit. Signal Process.* **2014**, *25*, 190–197. [CrossRef]
37. Fischler, M.A.; Bolles, R.C. Random sample consensus: A paradigm for model fitting with applications to image analysis and automated cartography. *Commun. ACM* **1981**, *24*, 381–395. [CrossRef]
38. Morel, J.-M.; Yu, G. ASIFT: A new framework for fully affine invariant image comparison. *SIAM J. Imaging Sci.* **2009**, *2*, 438–469. [CrossRef]
39. Yu, G.; Morel, J.-M. A fully affine invariant image comparison method. In Proceedings of the 2009 IEEE International Conference on Acoustics, Speech and Signal Processing, Taipei, Taiwan, 19–24 April 2009; pp. 1597–1600.
40. Yu, Y.; Huang, K.; Chen, W.; Tan, T. A novel algorithm for view and illumination invariant image matching. *IEEE Trans. Image Process.* **2011**, *21*, 229–240.
41. Wu, J.; Cui, Z.; Sheng, V.S.; Zhao, P.; Su, D.; Gong, S. A comparative study of SIFT and its variants. *Meas. Sci. Rev.* **2013**, *13*, 122–131. [CrossRef]
42. Dellinger, F.; Delon, J.; Gousseau, Y.; Michel, J.; Tupin, F. Change detection for high resolution satellite images, based on SIFT descriptors and an a contrario approach. In Proceedings of the 2014 IEEE Geoscience and Remote Sensing Symposium, Quebec City, QC, Canada, 13–18 July 2014; pp. 1281–1284.
43. Höllerer, T.; Feiner, S. Mobile augmented reality. In *Telegeoinformatics: Location-Based Computing and Services*; Karimi, H.A., Hammad, A., Eds.; CRC Press: Boca Raton, FL, USA, 2004; ISBN 0-4153-6976-2.
44. Sebastian Richard Hitting the Spot. Available online: http://spie.org/x26572.xml (accessed on 8 August 2019).
45. LaMarca, A.; Chawathe, Y.; Consolvo, S.; Hightower, J.; Smith, I.; Scott, J.; Sohn, T.; Howard, J.; Hughes, J.; Potter, F. Place lab: Device positioning using radio beacons in the wild. In Proceedings of the International Conference on Pervasive Computing, Munich, Germany, 8–13 May 2005; pp. 116–133.
46. Khoury, H.M.; Kamat, V.R. Evaluation of position tracking technologies for user localization in indoor construction environments. *Autom. Constr.* **2009**, *18*, 444–457. [CrossRef]
47. Rolland, J.P.; Davis, L.D.; Baillot, Y. A survey of tracking technologies for virtual environments. In *Fundamentals of Wearable Computers and Augmented Reality*; CRC Press: Boca Raton, FL, USA, 2001; pp. 67–112.
48. Bargh, M.S.; de Groote, R. Indoor localization based on response rate of bluetooth inquiries. In Proceedings of the First ACM International Workshop on Mobile Entity Localization and Tracking in GPS-Less Environments, San Francisco, CA, USA, 19 September 2008; pp. 49–54.
49. Want, R.; Hopper, A.; Falcao, V.; Gibbons, J. The active badge location system. *ACM Trans. Inf. Syst.* **1997**, *4*, 42–47. [CrossRef]
50. Bahl, P.; Padmanabhan, V.N. RADAR: An in-building RF-based user location and tracking system. In Proceedings of the Proceedings IEEE INFOCOM 2000. Conference on Computer Communications. Nineteenth Annual Joint Conference of the IEEE Computer and Communications Societies (Cat. No. 00CH37064), Tel Aviv, Israel, 26–30 March 2000; Volume 2, pp. 775–784.
51. Karlekar, J.; Zhou, S.Z.Y.; Nakayama, Y.; Lu, W.; Chang Loh, Z.; Hii, D. Model-based localization and drift-free user tracking for outdoor augmented reality. In Proceedings of the 2010 IEEE International Conference on Multimedia and Expo, ICME 2010, Singapore, 19–23 July 2010; pp. 1178–1183.
52. Deak, G.; Curran, K.; Condell, J. A survey of active and passive indoor localisation systems. *Comput. Commun.* **2012**, *35*, 1939–1954. [CrossRef]
53. Gezici, S.; Tian, Z.; Giannakis, G.B.; Kobayashi, H.; Molisch, A.F.; Poor, H.V.; Sahinoglu, Z. Localization via ultra-wideband radios: A look at positioning aspects for future sensor networks. *IEEE Signal Process. Mag.* **2005**, *22*, 70–84. [CrossRef]
54. Pozyx. Available online: https://www.pozyx.io/ (accessed on 8 August 2017).
55. Popa, M.; Ansari, J.; Riihijarvi, J.; Mahonen, P. Combining cricket system and inertial navigation for indoor human tracking. In Proceedings of the 2008 IEEE Wireless Communications and Networking Conference, Las Vegas, NV, USA, 31 March–3 April 2008; pp. 3063–3068.
56. Microsoft. Available online: https://www.microsoft.com/en-us/surface (accessed on 10 December 2017).

57. Distance to Objects Using Single Vision Camera. Available online: https://www.youtube.com/watch?v=Z3KX0N56ZoA (accessed on 1 March 2020).
58. Soanes, C. *Oxford Dictionary of English*; Oxford University Press: New York, NY, USA, 2003; ISBN 0198613474.
59. Milgram, P.; Kishino, F. A taxonomy of mixed reality visual displays. *IEICE Trans. Inf. Syst.* **1994**, *77*, 1321–1329.
60. Azuma, R.T. A survey of augmented reality. *Presence Teleoperators Virtual Environ.* **1997**, *6*, 355–385. [CrossRef]
61. Bergamasco, F.; Albarelli, A.; Rodola, E.; Torsello, A. Rune-tag: A high accuracy fiducial marker with strong occlusion resilience. In Proceedings of the CVPR 2011, Providence, RI, USA, 20–25 June 2011; pp. 113–120.
62. Kan, T.-W.; Teng, C.-H.; Chou, W.-S. Applying QR code in augmented reality applications. In Proceedings of the 8th International Conference on Virtual Reality Continuum and its Applications in Industry, Yokohama, Japan, 14–15 December 2009; pp. 253–257.
63. Ruan, K.; Jeong, H. An augmented reality system using Qr code as marker in android smartphone. In Proceedings of the 2012 Spring Congress on Engineering and Technology, Xi'an, China, 27–30 May 2012; pp. 1–3.
64. Naimark, L.; Foxlin, E. Circular data matrix fiducial system and robust image processing for a wearable vision-inertial self-tracker. In Proceedings of the Proceedings. International Symposium on Mixed and Augmented Reality, Darmstadt, Germany, 1 October 2002; pp. 27–36.
65. Han, S.; Rhee, E.J.; Choi, J.; Park, J.-I. User-created marker based on character recognition for intuitive augmented reality interacion. In Proceedings of the 10th International Conference on Virtual Reality Continuum and Its Applications in Industry, Hong Kong, China, 11–12 December 2011; pp. 439–440.
66. Pucihar, K.Č.; Coulton, P. Exploring the Evolution of Mobile Augmented Reality for Future Entertainment Systems. *Comput. Entertain.* **2015**, *11*, 1–16. [CrossRef]
67. Tateno, K.; Kitahara, I.; Ohta, Y. A nested marker for augmented reality. In Proceedings of the 2007 IEEE Virtual Reality Conference, Charlotte, NC, USA, 10–14 March 2007; pp. 259–262.
68. Yan, Y. *Registration Issues in Augmented Reality*; University of Birmingham: Edgbaston, Birmingham, UK, 2015. Available online: https://pdfs.semanticscholar.org/ded9/2aa404e29e9cc43a08958ca7363053972224.pdf (accessed on 1 February 2020).

applied sciences

MDPI

Article

Generative Adversarial Network for Image Super-Resolution Combining Texture Loss

Yuning Jiang and Jinhua Li *

College of Data Science and Software Engineering, Qingdao University, Qingdao 266071, China; 2017021427@qdu.edu.cn

* Correspondence: lijh@qdu.edu.cn

Received: 15 January 2020; Accepted: 28 February 2020; Published: 3 March 2020

Abstract: Objective: Super-resolution reconstruction is an increasingly important area in computer vision. To alleviate the problems that super-resolution reconstruction models based on generative adversarial networks are difficult to train and contain artifacts in reconstruction results, we propose a novel and improved algorithm. Methods: This paper presented TSRGAN (Super-Resolution Generative Adversarial Networks Combining Texture Loss) model which was also based on generative adversarial networks. We redefined the generator network and discriminator network. Firstly, on the network structure, residual dense blocks without excess batch normalization layers were used to form generator network. Visual Geometry Group (VGG)19 network was adopted as the basic framework of discriminator network. Secondly, in the loss function, the weighting of the four loss functions of texture loss, perceptual loss, adversarial loss and content loss was used as the objective function of generator. Texture loss was proposed to encourage local information matching. Perceptual loss was enhanced by employing the features before activation layer to calculate. Adversarial loss was optimized based on WGAN-GP (Wasserstein GAN with Gradient Penalty) theory. Content loss was used to ensure the accuracy of low-frequency information. During the optimization process, the target image information was reconstructed from different angles of high and low frequencies. Results: The experimental results showed that our method made the average Peak Signal to Noise Ratio of reconstructed images reach 27.99 dB and the average Structural Similarity Index reach 0.778 without losing too much speed, which was superior to other comparison algorithms in objective evaluation index. What is more, TSRGAN significantly improved subjective visual evaluations such as brightness information and texture details. We found that it could generate images with more realistic textures and more accurate brightness, which were more in line with human visual evaluation. Conclusions: Our improvements to the network structure could reduce the model's calculation amount and stabilize the training direction. In addition, the loss function we present for generator could provide stronger supervision for restoring realistic textures and achieving brightness consistency. Experimental results prove the effectiveness and superiority of TSRGAN algorithm.

Keywords: super-resolution reconstruction; generative adversarial networks; dense convolutional networks; texture loss; WGAN-GP

1. Introduction

With the popularization of Internet and the development of information technology, the amount of information accepted by human is growing at an explosive rate. Images, videos and audio are the main carriers of information transmission. Related research [1] has pointed out that the information humans receive through vision accounts for 60%~80% of all media information, so visible images are an important way to obtain information. However, the quality of an image is often restricted by hardware equipment such as imaging system and the bandwidth during image transmission process. A low-resolution (LR) image with missing details is eventually presented. The reduction of image resolution will cause a serious decrease in image quality. It will greatly affect people's visual experience and cannot meet the requirements for image quality performance indicators in industrial production. Therefore, how to obtain high-resolution (HR) images has become an urgent issue.

At present, there are mainly two ways to improve image resolution. The first is to upgrade hardware devices such as image sensors and optics, but this method is too costly and difficult to promote in practical applications. The other is Images Super-Resolution Reconstruction (ISRR) technology which inputs LR images and generates HR images by using machine learning algorithms and digital image processing technology. It has been widely used in fields such as the medical field, communication field, public safety field and remote sensing imaging field for their low cost and practical application values.

The core of original ISRR algorithms are to use the information of neighboring pixels to estimate the pixels of HR images. Typical algorithms include nearest-neighbor interpolation [2], bilinear interpolation [3] and bicubic interpolation [4]. Their disadvantage is that they do not take into account the semantics of entire image, resulting in the lack of high-frequency details in the reconstructed images.

Subsequently, the reconstruction-based ISRR algorithm has been researched and developed. It introduces image priors or constraints between HR and LR images and uses sample information to infer the distribution of real data. Common ISRR algorithms based on reconstruction include convex set projection method [5], iterative back projection method [6] and maximum posterior probability estimation method [7]. Such methods are subject to computational resources and prior conditions when reconstructing images and are unable to produce satisfactory high-quality images.

In order to obtain higher quality reconstructed images, the learning-based ISRR algorithm has been proposed and developed rapidly. It makes full use of information in image sample library to learn the mapping relationship between HR and LR image. According to different design strategies, it is mainly divided into ISRR algorithms based on sparse representation and deep learning. Yang et al. [8] have applied sparse representation theory to ISRR. Tang et al. [9] have proposed a refined local learning scheme to reduce the image artifacts and further improve the image visual quality. Similar algorithms for reconstructing images by learning mapping relationships include Bayesian process estimation [10], statistical learning [11] and linear regression representation algorithm [12].

The matrix or tensor decomposition algorithms that yield low-rank approximations have been developed for various image completion and resolution up-scaling problems. Hatvani et al. [13] have introduced tensor-factorization-based approach which offers a fast solution without the use of known image pairs or strict prior assumptions to solve ISRR task. To tackle the obstacles of low-rank completion methods, Zdunek et al. [14] have proposed to model the incomplete images with overlapping blocks of Tucker decomposition representations.

In recent years, methods based on deep learning have developed rapidly. Since Dong et al. [15] proposed Super-Resolution Convolutional Neural Network (SRCNN) model which first applied Convolutional Neural Networks (CNN) to ISRR, various network architecture designs and training strategies based on CNN [16–19] have been developed. However, these methods tend to output over-smoothed results without sufficient high-frequency details. In response to the above problem, Johnson et al. [20] have presented to calculate the super-resolution model's perceptual loss in feature space instead of pixel space. [17,21] have introduced Generative Adversarial Network (GAN) [22] to encourage network to generate more realistic and natural images. Lim et al. [23] have enhanced the deep residual network by removing the Batch Normalization (BN) layers in SRGAN (Generative Adversarial Network for Image Super-Resolution) model. Xintao Wang et al. [24] have used Residual Dense Block (RDB) to constitute the main body of generator network. Although the effect of reconstructed images has been improved, unfortunately, these methods still existed unpleasant artifacts in generated images.

In order to further improve the quality of reconstructed images, this paper presents TSRGAN (Super-Resolution Generative Adversarial Networks Combining Texture Loss) model which is based on GAN. Firstly, we use RDB as the basic unit of generator network and adopt Visual Geometry Group (VGG)19 network as the basic framework of discriminator network. This measure can strengthen the reuse of forward features, reduce the amount of training parameters and control the training direction of reconstruction images. Secondly, four losses are introduced to constitute the total objective function of generator. We propose texture loss to encourage local information matching, enhance perceptual loss by employing the features before activation layer to calculate, optimize adversarial loss based on WGAN-GP (Wasserstein GAN with Gradient Penalty) theory and use content loss to ensure the accuracy of low-frequency information. Experimental results show that the model in this paper has achieved good results, which can generate images with more realistic textures.

2. Related Work

2.1. Generative Adversarial Networks

GAN is a new network framework proposed by Ian Goodfellow et al. [22], it estimates generative model through adversarial process. The zero-sum game is the basic idea of GAN model, the generator (G) and discriminator (D) constitute the main framework of the model. GAN trains network through adversarial learning to achieve Nash equilibrium [25], achieving the goal of estimating data's potential distribution and generating new data samples.

G and D can be represented by any differentiable function, taking random variable z and real data x as input, respectively. G(z) represents the result generated by G that obeys the distribution of real samples (p_{data}) as much as possible. If D's input is the real sample, D outputs 1, otherwise D outputs 0. D actually acts as a two-classifier. The goal of G is to fool D, so that D could finally give an evaluation result which is closer to 1. G and D oppose each other and iteratively optimize until D can't distinguish whether the input sample is from G or real data, then it can be considered that the target G has been obtained. The basic framework described in this process is shown in Figure 1. The objective function of GAN is as follows:

$$\min_{G}\max_{D} V(D,G) = E_{x\sim p_{data}}(x)[\log D(x)] + E_{z\sim p_z(z)}(z)[\log(1-D(G(z)))] \quad (1)$$

where G minimizes the objective function to generate samples that can better confuse D, D maximizes the objective function so that D can better distinguish the authenticity of input samples.

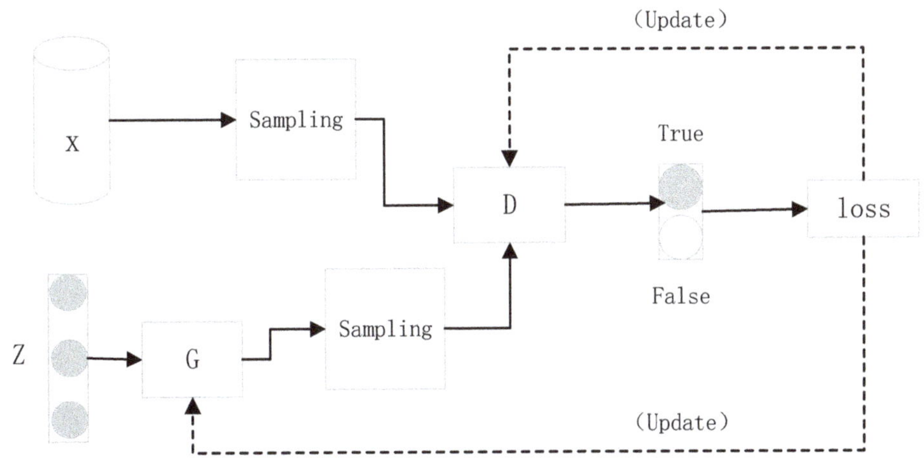

Figure 1. The basic framework of Generative Adversarial Network (GAN).

2.2. Dense Convolutional Network

In deep learning networks, the problem of gradient disappearance and gradient dispersion will become more serious as the increase of network layers. The ResNets proposed in [26], the Highway Networks proposed in [27] and the Stochastic depth structure proposed in [28] are all improved networks for the above problems. Although the proposed algorithms are different in network structure and training process, their key point is to create a short path from the forward feature layer to the backward one. Considering the need to ensure the maximum degree of information transmission between different layers, Huang et al. [29] have proposed dense convolutional network (DenseNet), each layer in DenseNet must obtain additional feature inputs from its all feedforward layers and transfer its own feature map to all subsequent layers for effective training. DenseNet has created a deeper and more efficient convolutional network, its dense connection mechanism is shown in Figure 2. The network has obvious advantages in mitigating the disappearance of gradient. Moreover, the structural design that enhances feature propagation and feature reuse can greatly reduce the number of parameters. DenseNet has been widely used in semantic cutting [30], speech recognition [31] and image classification [29].

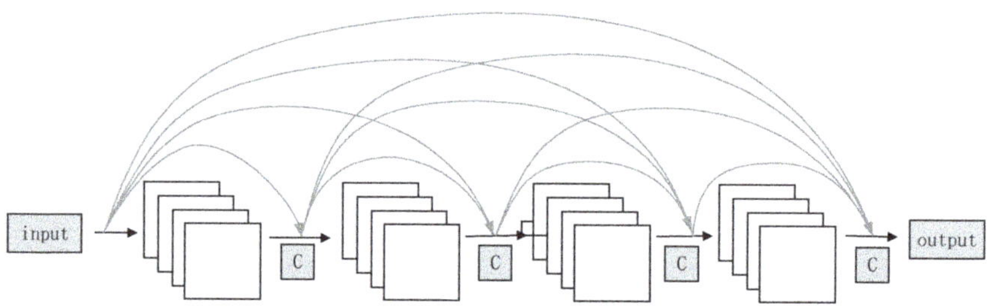

Figure 2. DenseNet's dense connection mechanism.

3. Proposed Methods

This paper uses generative adversarial networks as the main frame, including generator network and discriminator network. The overall structure of TSRGAN is shown in Figure 3. LR image is the generator network's input, then the convolutional layers are responsible for extracting features. Subsequently, the feature map inputs residual model for non-linear mapping. Then the image is reconstructed through the upsampling layer and convolutional layer. Next, the network outputs the reconstruction result. Finally, we input the fake and real HR images into discriminator network separately, which is responsible for discriminating the authenticity of image.

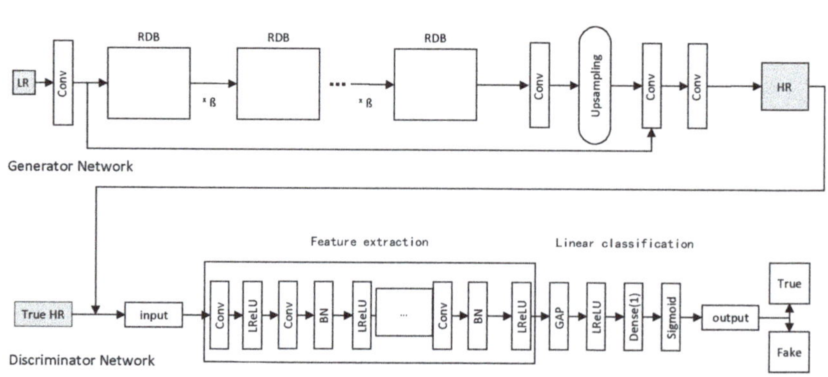

Figure 3. Architecture of generator and discriminator network.

3.1. Network Architecture

3.1.1. Generator Network

In order to further improve the quality of image reconstruction, this paper improves the network based on SRGAN model. Firstly, all BN layers are removed in SRGAN. BN is easy to introduce artifacts and limit the generalization ability of network. Studies have shown that removing the BN layers can improve reconstruction performance and reduce the computational complexity, such as SR task [23] and deblurring task [32]. Secondly, Leaky Rectified Linear Unit (LeakyReLU) is used instead of Rectified Linear Unit (ReLU) as the network's non-linear activation function to avoid gradient vanishing problem:

$$y = \max(0, x) + a * \min(0, x) \tag{2}$$

where x is the input, y is the output and a is a constant between 0 and 1. Finally, based on the researches in [31,33,34], it is shown that deep networks and multi-level connections can improve the performance of algorithm. Therefore, we use RDB instead of Residual Block (RB) which is used in SRGAN as the basic network element. RDB has a deeper and more complex structure than RB, it has the advantages of both residual networks and dense connections. It increases the depth of network while improving the reuse of image feature information. Ultimately, it improves the qualities of reconstructed images. The specific structure is shown in Figure 4. Our generator network is a deep model with 36 RDB, it has larger capacity and stronger ability to capture semantic information. Therefor it can reduce the noises of reconstructed images and generate images with more realistic textures.

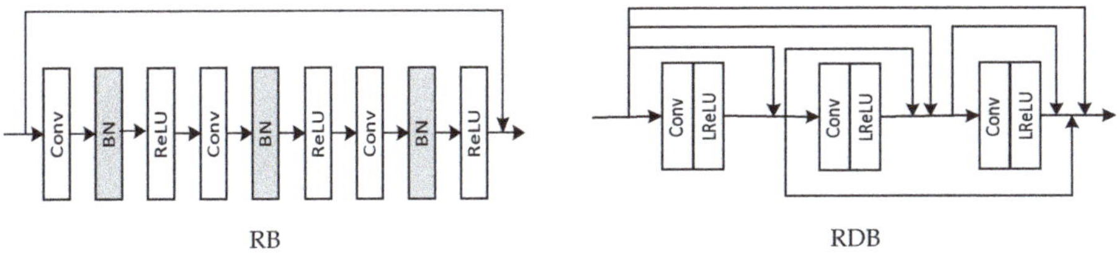

Figure 4. The structure of Residual Block (RB) and Residual Dense Block (RDB).

3.1.2. Discriminator Network

As for the discriminator network, this paper uses the classic VGG19 network as basic architecture, which can be simplified into two modules: feature extraction and linear classification. Feature extraction module includes 16 convolutional layers, after each convolutional layer we use LeakyReLU as the activation function. In addition, the BN layer is used after each convolutional layer except the first one to avoid gradient vanishing problem and enhance the model's stability. Then the discriminator network needs to judge the input sample image. We use Global Average Pooling (GAP) [35] instead of fully connected layer which is used in most image classification models for fear of reducing the training speed of model and increasing the risk of overfitting. GAP is responsible for calculating the pixel average value of each feature map, and then all the values are sent into sigmoid activation function after linear fusion. Ultimately, network outputs D's judgement result for the input sample. Training discriminator network helps generator network to restore results that are closer to the ground-truth images.

3.2. Loss Functions

Loss function is an important factor that affects the quality of image reconstruction. In order to restore the high-frequency information and improve the intuitive visual experience of image, this paper uses content loss L_{con}, adversarial loss L_{adv}, perceptual loss L_{per} and texture loss L_{tex} as the objective function of the generator network:

$$L_G = L_{\text{per}} + L_{tex} + \lambda L_{adv} + \eta L_{con} \tag{3}$$

where λ and η are the coefficients which are used to balance different loss functions.

3.2.1. Content Loss

Mean Square Error (MSE) loss is used as the model's content loss for the sake of ensuring the consistency of low-frequency information between reconstructed image and LR image. It is in charge of optimizing the squared error between pixels corresponding to the generated and real HR images. Reducing the distance between pixels can more quickly and effectively ensure the accuracy of the reconstructed image information, so that the results could get a higher value of peak signal to noise ratio.

$$L_{\text{con}} = L_{MSE}(\theta) = \frac{1}{N}\sum_{i=1}^{N}\left\|I_i^H - G\left(I_i^L, \theta\right)\right\|^2 \tag{4}$$

where I_i^H represents the real HR image, I_i^L represents the LR image, N represents the number of training samples and $G(x,\theta)$ represents the mapping function between LR and HR images learned by the generator network.

3.2.2. Adversarial Loss

Based on the adversarial game mechanism between generator and discriminator network, the discriminator network needs to product the probability of image which is output by generator network being true or false. To maximize the probability that the reconstructed image deceives D, we adopt the adversarial loss proposed in WGAN-GP [36] model to replace the one proposed in GAN model. Improved L_{adv} penalizes D for the gradient of input, it can help stable training of GAN architecture and generate higher quality samples with faster convergence speed with little need for tuning of hyperparameters.

$$L_{adv} = E_{x\sim p_G}[D(x)] - E_{x\sim p_{data}}[D(x)] + \lambda E_{x\sim penalty}\left[\left(\left\|\nabla_x D(x)\right\| - 1\right)^2\right] \tag{5}$$

3.2.3. Perceptual Loss

In order to generate images with more accurate brightness and realistic textures, L_{per} based on VGG network is set to be calculated using feature layer information before activation layer instead of after it. It is defined on the activation layer of the pre-trained deep network to minimize the Euclidean distance between two activation features:

$$L_{per} = \frac{1}{W_{ij}H_{ij}}\sum_{x=1}^{W_{ij}}\sum_{y=1}^{H_{ij}}\left(\phi^{ij}\left(I^{HR}\right)_{x,y} - \phi^{ij}\left(G\left(I^{LR}\right)\right)_{x,y}\right)^2 \tag{6}$$

where, W_{ij} and H_{ij} describe the dimensions of the respective feature maps within the VGG network, $\varnothing^{ij}$ indicates the feature map obtained by the j-th convolution (after activation) before the i-th maxpooling layer within the network. The improved L_{per} overcomes two drawbacks of the original design: First, the activated features are very sparse, especially after a very deep network, the sparse activation provides weak supervision and thus leads to inferior performance. Second, using features after activation also causes inconsistent reconstructed brightness compared with the ground-truth image.

3.2.4. Texture Loss

Although perceptual loss can improve the quality of reconstructed image as a whole, it still has the problem of introducing unnecessary high-frequency structures. We propose to incorporate texture loss presented in [21] to constitute the total loss function of G. L_{tex} encourages local matching of texture information, it extracts feature maps generated by the intermediate layer of convolutional network of generator and discriminator network. Then it calculates the corresponding gram matrix. Finally, L2 loss function is used to calculate texture loss for the obtained Gram matrix values:

$$L_{tex} = \left\| G(\phi(I^{gen}) - G\left(\phi\left(I^{HR}\right)\right) \right\|_2^2, \tag{7}$$

where I^{gen} indicates images that are reconstructed by generator, G indicates the Gram matrix, $G(F) = FF^T$. Texture loss provides strong supervision to further reduce visually incredible artifacts and produce more realistic textures.

4. Experiments and Results

4.1. Experimental Details

The experimental platform we use is NVIDIA GeForceMX150, Intel (R) Core (TM) i7-8550U CPU@2.00GHz, 8 GB RAM, the compilation software we use are pycharm2017 and MATLAB 2018a, and the pytorch deep learning toolbox is used to build and train the network. This paper uses DIV2K dataset, which consists of 800 training images, 100 validation images and 100 testing images. We augment the training data with random horizontal flips and 90 rotations. We perform experiments on three widely used benchmark datasets Set5 [37], Set14 [38] and BSD100 [39]. All experiments are performed with a scale factor of 4× between low- and high-resolution images. The mini-batch size is set to 16. The spatial size of cropped HR patch is 128 × 128.

The training process is divided into two stages. First, we train a generative model with L_1 loss as the objective function. Then, we use the initially trained model as the initialization of G. The generator is trained using the loss function in Equation (3). The initial learning rate is set to 1×10^{-4}. For optimization, we use Adam with $\beta 1 = 0.9$, $\beta 2 = 0.999$. We alternately update the generator and discriminator network until the model converges. In addition, we introduce a residual scaling [40] strategy which scales down the residuals by multiplying a constant β between 0 and 1 before adding them to the main path to prevent instability. β is set to 0.2 in this paper.

For accurately evaluating image quality and proving the effectiveness of algorithm, Peak Signal to Noise Ratio (PSNR) and Structural Similarity Index (SSIM) are adopted as image quality evaluation indicators. μ_X and μ_Y represent the mean values of images X and Y, σ_X and σ_Y represent the standard deviations of images X and Y and σ_{XY} represents the covariance of images X and Y. PSNR is responsible for measuring the distortion of images from the difference in pixels, and SSIM is responsible for measuring the similarity of the images from the brightness, contrast and structure. The larger the two values, the closer the reconstruction result is to the ground-truth image.

$$PSNR = 10 \times \log_{10} \frac{255^2 \times W \times H \times C}{\sum_{i=1}^{W} \sum_{j=1}^{H} \sum_{z=1}^{C} \left[\overline{x}(i,j) - x(i,j)\right]^2 + 1 \times 10^{-9}} \tag{8}$$

$$SSIM(X,Y) = \frac{(2\mu_X\mu_Y + C_1)(2\sigma_{XY} + C_2)}{(\mu_X{}^2 + \mu_Y{}^2 + C_1)(\sigma_X{}^2 + \sigma_Y{}^2 + C_2)} \tag{9}$$

4.2. Experimental Results

4.2.1. Quantitative Evaluation

We have performed super-resolution experiments on Set5 and Set14 to analyze the effects of introducing RDB structure, L_{tex} and improving initial L_{adv}, L_{per} on super-resolution performance. The PSNR values of different model variants are shown in Table 1. It can be observed that each of the above four enhanced measures can improve the super-resolution performance of the network, and the effect is the best when all of them are used. In addition, we have adopted different values for λ and η in Equation (3) and performed experiments on Set5. The results have shown that the reconstruction effect is the best when $\lambda = 3 \times 10^{-3}$ and $\eta = 2 \times 10^{-2}$. Table 2 presents the average PSNR results on Set5 dataset.

Table 1. Average PSNR (dB) of different super-resolution models on Set5 and Set14 datasets.

RDB	L_{adv}	L_{per}	L_{tex}	Set5	Set14
×	×	×	×	30.37	27.02
√	×	×	×	31.22	27.98
√	√	×	×	31.54	28.27
√	√	√	×	31.83	28.39
√	√	√	√	32.38	28.73

Table 2. The average PSNR (dB) results on Set5 dataset when λ and η take different values.

η \ λ	1×10^{-3}	2×10^{-3}	3×10^{-3}	4×10^{-3}	5×10^{-3}
1×10^{-2}	32.31	32.34	32.36	32.35	32.36
2×10^{-2}	32.32	32.35	32.38	32.36	32.35
3×10^{-2}	32.31	32.33	32.36	32.34	32.32

For fair comparison, the SISR methods in comparison are Bicubic [4], ScSR [8], SRGAN [17], EDSR [23] and ESRGAN [24], all these methods are tested on Set5, Set14 and BSD100, respectively. Average PSNR/SSIM values on different datasets with those methods are recorded in Table 3, and the total running time with those methods on different datasets is recorded in Table 4. It can be seen from Table 3 that the performance of TSRGAN on PSNR is generally better than other algorithms. Except that the SSIM value is slightly lower than ESRGAN 0.009 on Set14, it is also superior than other algorithms. Note that Table 4 shows the results that Bicubic consumes the shortest time for it only has interpolation operations. ScSR spends longer time for learning sparse representation dictionaries between the LR and HR image patch pairs. SRGAN, EDSR, ESRGAN and TSRGAN models all need longer time to train for they have extensive convolutional layers. SRGAN has the slowest reconstruction speed because the BN layer is not removed in the network structure, while TSRGAN is slightly slower than EDSR and ESRGAN due to the introduction of deeper network and texture loss. Synthesizing Tables 3 and 4, TSRGAN obviously improves PSNR and SSIM indicators for measuring the quality of image reconstruction without losing too much speed, which verifies its effectiveness and superiority.

Table 3. Average PSNR (dB)/SSIM comparison of different SR algorithms on Set5, Set14 and BSD100 datasets.

Algorithm	Set5	Set14	BSD100
Bicubic	30.07/0.862	27.18/0.786	26.68/0.729
ScSR	30.29/0.868	27.69/0.790	26.94/0.730
SRGAN	30.36/0.873	27.02/0.772	26.51/0.724
EDSR	31.53/0.882	28.02/0.793	27.23/0.732
ESRGAN	32.05/0.895	28.49/0.819	27.58/0.747
TSRGAN	32.38/0.967	28.73/0.810	27.67/0.764

Table 4. Total running time of different algorithms on Set5, Set14 and BSD100 datasets.

Algorithm	Bicubic/s	ScSR/s	SRGAN/s	EDSR/s	ESRGAN/s	TSRGAN/s
Set5	1.725	2.376	3.763	3.005	3.247	3.750
Set14	1.816	2.693	4.098	3.729	3.862	3.899
BSD100	12.519	20.067	28.686	26.103	27.034	27.935

4.2.2. Qualitative Evaluation

In order to ensure the contrast effect, we select an image from datasets Set5 and Set14, respectively. The actual reconstruction results of each algorithm are shown in Figures 5 and 6. Comparing the reconstruction results, it can be observed that the reconstruction details of Bicubic and ScSR are too few, and the generated images are very blurred. Although SRGAN and EDSR have restored some high-frequency information, the edge sharpening effect is poor. The overall effect of ESRGAN is better, but it has introduced unpleasant artifacts and noises. The reconstruction results of TSRGAN are superior to other algorithms in terms of sharpness and detail. As can be seen from enlarged details in Figure 5, TSRGAN can generate a clearer and more natural hat textures. According to Figure 6, it can be observed that TSRGAN has generated image with more accurate brightness information and more pleasing texture details.

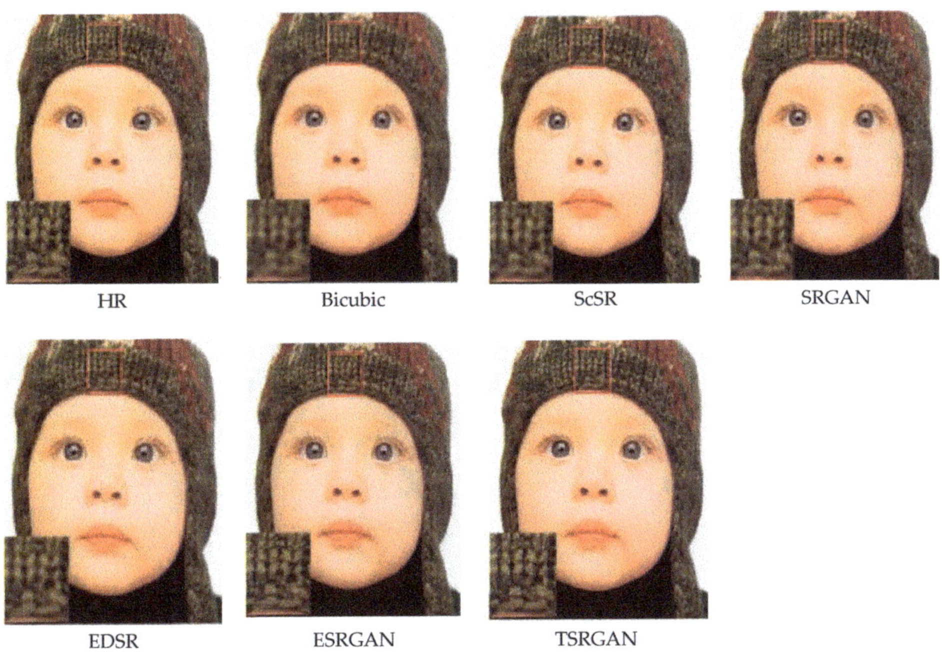

Figure 5. Reconstruction effects from the selected algorithms on Set5 dataset.

Figure 6. Reconstruction effects from the selected algorithms on Set14 dataset.

5. Conclusions

Based on the generative adversarial network framework, we have described a super-resolution model TSRGAN. We have designed the method of removing BN layers and introducing residual dense blocks to deepen the structure of generator network. In addition, we have used WGAN-GP to improve adversarial loss to provide stronger and more effective supervision for model training. Moreover, we have enhanced perceptual loss by using the features before activation layer, which offer stronger supervision and thus restore more accurate brightness and realistic textures. Finally, we have cited texture loss which encourages to match local texture details to achieve better outcomes. The experimental results show that our method makes the average PSNR of reconstructed images reach 27.99 dB and the average SSIM reach 0.778 without losing too much speed, which is superior to other comparison algorithms in objective evaluation index. TSRGAN has significantly improved subjective visual evaluations such as brightness information and texture details, this further proves that our algorithm can reconstruct more realistic images. In future research work, we will consider super-resolution reconstruction of images in specific fields or scenes to improve the quality of image generation.

Author Contributions: Writing—original draft, Y.J.; Writing—review & editing, Y.J. and J.L. All authors have read and agreed to the published version of the manuscript.

Funding: This research was funded by Key Research and Development Plan - Major Scientific and Technological Innovation Projects of ShanDong Province (2019JZZY020101).

Acknowledgments: This study is undertaken within the framework of SRGAN. Furthermore, the authors wish to thank Zhao Junli and Yang Chun for their helpful comments and encouragement for many aspects of the paper, and we wish to thank Dong Yuehang for helping with the ISRR experimental environment support.

Conflicts of Interest: We declare that we have no conflict of interest.

References

1. Gonzalez, R.; Woods, R.E. Digital Image Processing. *Up. Saddle River Nj Pearson Hall.* **2002**, *28*, 290–291.
2. Schultz, R.R.; Stevenson, R.L. A Bayesian approach to image expansion for improved definition. *IEEE Trans. Image Process.* **1994**, *3*, 233–242. [CrossRef] [PubMed]
3. Gribbon, K.T.; Bailey, D.G. A novel approach to real-time bilinear interpolation. In Proceedings of the DELTA, Second IEEE International Workshop on Electronic Design, Test and Applications, Perth, WA, Australia, 28–30 January 2004; pp. 126–131.
4. Zhang, L.; Wu, X. An edge-guided image interpolation algorithm via directional filtering and data fusion. *IEEE Trans. Image Process.* **2006**, *15*, 2226–2238. [CrossRef] [PubMed]
5. Jung, S.W.; Kim, T.H.; Ko, S.J. A novel multiple image deblurring technique using fuzzy projection onto convex sets. *IEEE Signal Process. Lett.* **2009**, *16*, 192–195. [CrossRef]
6. Nayak, R.; Harshavardhan, S.; Patra, D. Morphology based iterative back-projection for super-resolution reconstruction of image. In Proceedings of the 2014 2nd International Conference on Emerging Technology Trends in Electronics, Communication and Networking, Surat, India, 26–27 December 2014; pp. 1–6.
7. Sun, D.; Gao, Q.; Lu, Y.; Huang, Z.; Li, T. A novel image denoising algorithm using linear Bayesian MAP estimation based on sparse representation. *Signal Process.* **2014**, *100*, 132–145. [CrossRef]
8. Yang, J.; Wright, J.; Huang, T.; Ma, Y. Image super-resolution via sparse representation. *IEEE Process. IEEE Trans. Image Process.* **2010**, *19*, 2861–2873. [CrossRef] [PubMed]
9. Tang, S.; Xiao, L.; Liu, P. Single image super-resolution method via refined local learning. *J. Shanghai Jiaotong Univ. (Sci.)* **2015**, *20*, 26–31. [CrossRef]
10. He, L.; Qi, H.; Zaretzki, R. Beta process joint dictionary learning for coupled feature spaces with application to single image super-resolution. In Proceedings of the IEEE Conference on Computer Vision and Pattern recognition, Portland, ON, USA, 23–28 June 2013; pp. 345–352.
11. Peleg, T.; Elad, M. A statistical prediction model based on sparse representations for single image super-resolution. *IEEE Trans. Image Process.* **2014**, *23*, 2569–2582. [CrossRef] [PubMed]
12. Hu, Y.; Wang, N.; Tao, D.; Gao, X.; Li, X. SERF: A simple, effective, robust, and fast image super-resolver from cascaded linear regression. *IEEE Trans. Image Process.* **2016**, *25*, 4091–4102. [CrossRef] [PubMed]
13. Hatvani, J.; Basarab, A.; Tourneret, J.Y.; Gyöngy, M.; Kouamé, D. A Tensor Factorization Method for 3-D Super Resolution with Application to Dental CT. *IEEE Trans. Med. Imaging* **2018**, *38*, 1524–1531. [CrossRef] [PubMed]
14. Zdunek, R.; Sadowski, T. Image Completion with Hybrid Interpolation in Tensor Representation. *Appl. Sci.* **2020**, *10*, 797. [CrossRef]
15. Dong, C.; Loy, C.C.; He, K.; Tang, X. Image Super-Resolution Using Deep Convolutional Networks. *IEEE Trans. Pattern Anal. Mach. Intell.* **2016**, *38*, 295–307. [CrossRef] [PubMed]
16. Kim, J.; Kwon Lee, J.; Mu Lee, K. Accurate image super-resolution using very deep convolutional networks. In Proceedings of the IEEE Conference on Computer Vision and Pattern Recognition, Las Vegas, NV, USA, 27–30 June 2016; pp. 1646–1654.
17. Ledig, C.; Theis, L.; Huszar, F.; Caballero, J.; Cunningham, A.; Acosta, A.; Aitken, A.; Tejani, A.; Totz, J.; Wang, Z.H.; et al. Photo-realistic single image super-resolution using a generative adversarial network. In Proceedings of the IEEE Conference on Computer Vision and Pattern Recognition, Honolulu, HI, USA, 21–26 July 2017; pp. 4681–4690.
18. Lai, W.S.; Huang, J.B.; Ahuja, N.; Yang, M. Deep laplacian pyramid networks for fast and accurate super-resolution. In Proceedings of the IEEE Conference on Computer Vision and Pattern Recognition, Honolulu, HI, USA, 21–26 July 2017; pp. 624–632.
19. Tai, Y.; Yang, J.; Liu, X. Image super-resolution via deep recursive residual network. In Proceedings of the IEEE Conference on Computer Vision and Pattern Recognition, Honolulu, HI, USA, 21–26 July 2017; pp. 3147–3155.
20. Johnson, J.; Alahi, A.; Fei-Fei, L. Perceptual losses for real-time style transfer and super-resolution. In Proceedings of the European Conference on Computer Vision, Amsterdam, The Netherland, 8–16 October 2016; Springer: Berlin/Heidelberg, Germany, 2016; pp. 694–711.

21. Sajjadi, M.S.M.; Scholkopf, B.; Hirsch, M. Enhancenet: Single image super-resolution through automated texture synthesis. In Proceedings of the IEEE International Conference on Computer Vision, Venice, Italy, 22–29 October 2017; pp. 4491–4500.
22. Goodfellow, I.; Pouget-Abadie, J.; Mirza, M.; Xu, B.; Warde-Farley, D.; Ozair, S.; Courville, A.; Bengio, Y. Generative adversarial nets. In Proceedings of the Advances in Neural Information Processing Systems, Montreal, QC, Canada, 8–13 December 2014; pp. 2672–2680.
23. Lim, B.; Son, S.; Kim, H.; Nah, S.; Lee, K.M. Enhanced deep residual networks for single image super-resolution. In Proceedings of the IEEE Conference on Computer Vision and Pattern Recognition Workshops, Honolulu, HI, USA, 21–26 July 2017; pp. 136–144.
24. Wang, X.; Yu, K.; Wu, S.; Gu, J.; Liu, Y.; Dong, C.; Loy, C.C.; Qiao, Y.; Tang, X. Esrgan: Enhanced super-resolution generative adversarial networks. In Proceedings of the European Conference on Computer Vision (ECCV), Munich, Germany, 8–14 September 2018.
25. Ratliff, L.J.; Burden, S.A.; Sastry, S.S. Characterization and computation of local nash equilibria in continuous games. In Proceedings of the 2013 51st Annual Allerton Conference on Communication, Control, and Computing (Allerton), Monticello, IL, USA, 2–4 October 2013; IEEE: Piscataway, NJ, USA, 2013; pp. 917–924.
26. He, K.; Zhang, X.; Ren, S.; Sun, J. Identity mappings in deep residual networks. In Proceedings of the European Conference on Computer Vision, Amsterdam, The Netherland, 8–16 October 2016; Springer: Berlin/Heidelberg, Germany, 2016; pp. 630–645.
27. Srivastava, R.K.; Greff, K.; Schmidhuber, J. Training very deep networks. In Proceedings of the Advances in Neural Information Processing Systems, Montreal, QC, Canada, 7–12 December 2015; pp. 2377–2385.
28. Huang, G.; Sun, Y.; Liu, Z.; Sedra, D.; Weinberger, K. Deep networks with stochastic depth. In Proceedings of the European Conference on Computer Vision, Amsterdam, The Netherland, 8–16 October 2016; Springer: Berlin/Heidelberg, Germany, 2016; pp. 646–661.
29. Huang, G.; Liu, Z.; Van Der Maaten, L.; Weinberger, K.Q. Densely connected convolutional networks. In Proceedings of the IEEE Conference on Computer Vision and Pattern Recognition, Honolulu, HI, USA, 21–26 July 2017; pp. 4700–4708.
30. Jégou, S.; Drozdzal, M.; Vazquez, D.; Romero, A.; Bengio, Y. The one hundred layers tiramisu: Fully convolutional densenets for semantic segmentation. In Proceedings of the IEEE Conference on Computer Vision and Pattern Recognition Workshops, Honolulu, HI, USA, 21–26 July 2017; pp. 11–19.
31. Park, S.; Jeong, Y.; Kim, H.S. Multi-resolution DenseNet based acoustic models for reverberant speech recognition. *Phon. Speech Sci.* **2018**, *10*, 33–38. [CrossRef]
32. Nah, S.; Hyun Kim, T.; Mu Lee, K. Deep multi-scale convolutional neural network for dynamic scene deblurring. In Proceedings of the IEEE Conference on Computer Vision and Pattern Recognition, Honolulu, HI, USA, 21–26 July 2017; pp. 3883–3891.
33. Zhang, Y.; Tian, Y.; Kong, Y.; Zhong, B.; Fu, Y. Residual densenet work for image super-resolution. In Proceedings of the IEEE Conference on Computer Vision and Pattern Recognition, Salt Lake City, UT, USA, 18–23 June 2018; pp. 2472–2481.
34. Zhang, Y.; Li, K.; Li, K.; Wang, L.; Zhong, B.; Fu, Y. Image super-resolution using very deep residual channel attention networks. In Proceedings of the European Conference on Computer Vision (ECCV), Munich, Germany, 8–14 September 2018; pp. 286–301.
35. Szegedy, C.; Liu, W.; Jia, Y.; Sermanet, P.; Reed, S.; Anguelov, D.; Erhan, D.; Vanhoucke, V.; Rabinovich, A. Going deeper with convolutions. In Proceedings of the IEEE Conference on Computer Vision and Pattern Recognition, Boston, MA, USA, 7–12 June 2015; pp. 1–9.
36. Gulrajani, I.; Ahmed, F.; Arjovsky, M.; Dumoulin, V.; Courville, A. Improved training of wasserstein gans. In Proceedings of the Advances in Neural Information Processing Systems, Long Beach, CA, USA, 4–9 December 2017; pp. 5767–5777.
37. Bevilacqua, M.; Roumy, A.; Guillemot, C.; Alberi Morel, M.-L. Low-Complexity Single-Image Super-Resolution based on Nonnegative Neighbor Embedding. In Proceedings of the Acoustics, Speech and Signal Processing (ICASSP), Kyoto, Japan, 25–30 March 2012.
38. Zeyde, R.; Elad, M.; Protter, M. On single image scale-up using sparse-representations. In Proceedings of the International Conference on Curves and Surfaces, Avignon, France, 24–30 June 2010; Springer: Berlin/Heidelberg, Germany, 2010; pp. 711–730.

39. Martin, D.; Fowlkes, C.; Tal, D.; Malik, J. A database of human segmented natural images and its application to evaluating segmentation algorithms and measuring ecological statistics. In Proceedings of the Eighth IEEE International Conference on Computer Vision, ICCV, Vancouver, BC, Canada, 7–14 July 2001; Volume 2, pp. 416–423.
40. Szegedy, C.; Ioffe, S.; Vanhoucke, V.; Alemi, A. Inception-v4, inception-resnet and the impact of residual connections on learning. In Proceedings of the Thirty-First AAAI Conference on Artificial Intelligence, San Francisco, CA, USA, 4–9 February 2017.

applied sciences

MDPI

Article

The Imperial Cathedral in Königslutter (Germany) as an Immersive Experience in Virtual Reality with Integrated 360° Panoramic Photography

Alexander P. Walmsley and Thomas P. Kersten *

HafenCity University Hamburg, Photogrammetry & Laser Scanning lab, Überseeallee 16, 20457 Hamburg, Germany; Alexander.Walmsley@hcu-hamburg.de

* Correspondence: Thomas.Kersten@hcu-hamburg.de

Received: 6 January 2020; Accepted: 7 February 2020; Published: 23 February 2020

Abstract: As virtual reality (VR) and the corresponding 3D documentation and modelling technologies evolve into increasingly powerful and established tools for numerous applications in architecture, monument preservation, conservation/restoration and the presentation of cultural heritage, new methods for creating information-rich interactive 3D environments are increasingly in demand. In this article, we describe the development of an immersive virtual reality application for the Imperial Cathedral in Königslutter, in which 360° panoramic photographs were integrated within the virtual environment as a novel and complementary form of visualization. The Imperial Cathedral (Kaiserdom) of Königslutter is one of the most important examples of Romanesque architecture north of the Alps. The Cathedral had previously been subjected to laser-scanning and recording with 360° panoramic photography by the Photogrammetry & Laser Scanning lab of HafenCity University Hamburg in 2010. With the recent rapid development of consumer VR technology, it was subsequently decided to investigate how these two data sources could be combined within an immersive VR application for tourism and for architectural heritage preservation. A specialised technical workflow was developed to build the virtual environment in Unreal Engine 4 (UE4) and integrate the panorama photographs so as to ensure the seamless integration of these two datasets. A simple mechanic was developed using the native UE4 node-based programming language to switch between these two modes of visualisation.

Keywords: 3D modelling; 3D representation; game engine; laser scanning; panoramic photography; virtual reality

---

1. Introduction

Virtual reality has recently become a much broader field, finding applications in medicine, architecture, military training, and cultural heritage, among other fields. With this growth has come some discrepancy in the definition of the medium: while in some fields VR is used to refer to 360° immersive panoramas and videos, in other fields it refers to fully-realised interactive CGI environments. These two "kinds" of VR have traditionally been approached very differently, owing to highly diverging workflows and the different data sources required. However, there are currently no effective ways of bringing together these two kinds of data (each of which have their own complementary advantages in visualisation) into a single VR application. This is particularly important for applications in cultural heritage, where documentation often takes the form of multiple different kinds of complementary data (e.g., written, photographic, 3D, video and field recordings, among other forms).

Virtual reality is defined by Merriam Webster as "an artificial environment which is experienced through sensory stimuli (such as sights and sounds) provided by a computer and in which one's actions partially determine what happens in the environment" [1]. This very broad definition allows for most modern applications of VR to be taken into account. Additional definitions may be found in literature by Dörner et al. [2], Freina and Ott [3], and Portman et al. [4].

In the following we present the development workflow for a room-scale virtual reality experience of a cultural heritage monument which integrates a high-resolution CGI environment with 360° panoramic photography, allowing the user to "toggle" between the virtual and the real environments from within the VR headset. This implementation has the advantage of exploiting the potential for the interactivity of a real-time game engine environment with the high-fidelity of high dynamic range image (HDRI) panoramic photography.

2. Related Work

While much credit for the generalization of VR technology and its increasing accessibility is due to the video game industry, which has invested heavily in pushing the industry forward [5], VR is now being employed in a wide range of disciplines. To date, VR has been successfully used for, among other applications, virtual surgery, virtual therapy, and flight and vehicle simulations. In the field of cultural heritage, VR has been instrumental in the development of the field of virtual heritage [6–8]. At the HafenCity University Hamburg, several VR projects concerning this subject have already been realized. The town museum in Bad Segeberg, housed in a 17th-century townhouse, was digitally constructed for a VR experience using the HTC Vive Pro [9]. Three historical cities (as well as their surrounding environments) have been developed as VR experiences: Duisburg in 1566 [10], Segeberg in 1600 [11], and Stade in 1620 [12]. In addition, two religious and cultural monuments are also available as VR experiences: the Selimiye Mosque in Edirne, Turkey [13], and a wooden model of Solomon's Temple [14].

The amount of work specifically regarding the real-time VR visualization of cultural heritage monuments is currently limited but growing. Recent museum exhibits using real-time VR to visualize cultural heritage include Batavia 1627 at the Westfries Museum in Hoorn, Netherlands [15], and Viking VR, developed to accompany an exhibit at the British Museum [16]. A number of recent research projects also focus on the use of VR for cultural heritage visualization [17–20], as well as on aspects beyond visualisation, including recreating the physical environmental stimuli [21]. The current paper contributes to this growing discussion by seeking to integrate 360° panorama photographs within an immersive real-time visualization of a cultural heritage monument. At this stage, only very limited work regarding panoramic photography integration in real-time VR is known to the authors [22].

3. The Imperial Cathedral (Kaiserdom) in Königslutter, Germany

The town of Königslutter, some 20 km east of Braunschweig (Lower Saxony, Germany), is dominated by the Imperial Cathedral, known in German as the Kaiserdom (Figure 1). One of the most impressive examples of Romanesque architecture north of the alps, the cathedral's construction was begun under the direction of Kaiser Lothar III, German emperor from 1133 onwards [23,24]. The church was built in the form of a three-aisled cross-shaped column basilica. The cathedral is notable particularly for its repeated architectural references to northern Italian architectural styles of the time, indicating that it might be the work of an Italian architect or indeed someone who was well-travelled in those regions. Among the most important features of the cathedral is an ornamental hunting frieze, which hugs the external wall of the main aisle (see Figure 1 centre). Between 2002 and 2006, restoration was carried out on the exterior of the cathedral, followed by the interior between 2006 and 2010. The cathedral measures 75 m in length, 42 m in width, and 56 m in height.

Figure 1. Panoramic view of the Imperial Cathedral in Königslutter, Germany (**top**), the hunting frieze on the external wall of the main apsis of the cathedral with story-telling figures (**centre**), and a panoramic view of the interior of the cathedral (**bottom**).

4. Methodology

4.1. Project Workflow

The overall workflow for the production of the VR experience of the Kaiserdom is schematically represented in Figure 2. Special focus was given to achieving a realistic 1:1 representation of the cathedral, including the integration of panoramic photos in the VR experience (see Section 4.6). The project was divided into five major phases of development (Figure 2): (1a) data acquisition by terrestrial laser scanning with one Riegl VZ-400 scanner (outside) and two Zoller + Fröhlich IMAGER 5006 scanners (inside), (1b) registration and geo-referencing of scans using RiScan Pro and LaserControl, (1c) segmentation of point clouds into object tiles, (2a) 3D solid modelling with AutoCAD using segmented point clouds, (2b) generation of panoramic images using PTGui, (3) texture mapping of polygon models using Autodesk Maya and Substance Painter, (4a) placement of meshes and building the scene within the UE4 game engine, (4b) integration of motion control and interactions in UE4, (4c) integration of 360° panoramic imagery, and (5) immersive and interactive visualisation of the cathedral in the VR system HTC Vive Pro using Steam VR 2.0 as an interface between the game engine and the Head Mounted Display (HMD).

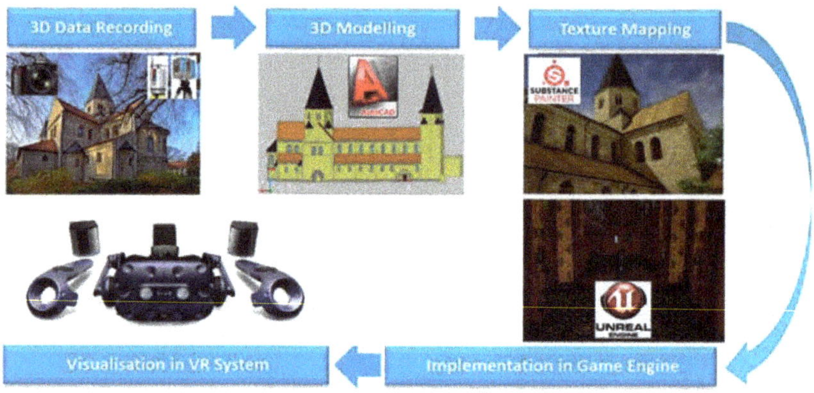

Figure 2. Workflow for the development of the virtual reality (VR) experience.

4.2. Data Acquisition

The data acquisition was already described in 2012 in Kersten and Lindstaedt [25] and is summarised in the following. The laser scan data for the Kaiserdom was acquired at 55 stations inside the cathedral by two Zoller + Froehlich IMAGER 5006 (www.zf-laser.com) terrestrial laser scanners, and at 8 stations outside the cathedral by one Riegl VZ-400 (www.riegl.com) on 5 January and 23 June 2010 (Figure 3). In total, the scanning took 15 h. The scanning resolution was set to high (6 mm @ 10 m) for the IMAGER 5006 and to 5 mm at object space for the Riegl VZ-400. The precision of the geodetic network stations was 2.5 mm, while the precision of the control points for laser scanning was 5 mm. In order to later colourise the point cloud, as well as for the building of the virtual tour, 360° panoramic photos were taken at each IMAGER 5006 scan station and at a few supplementary stations using a Nikon DSLR camera (see Section 4.4).

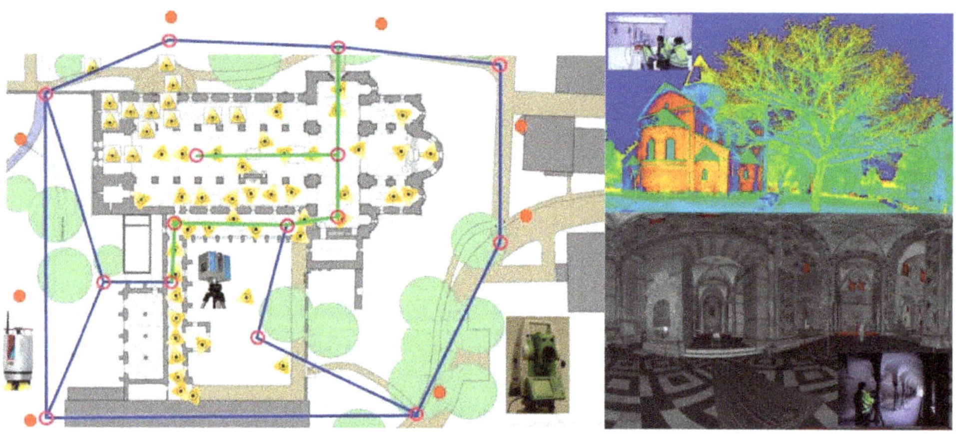

Figure 3. Geodetic 3D network (blue and green lines) and position of the scan stations (IMAGER 5006 = yellow triangles, Riegl VZ-400 = red dots) at the cathedral (**left**), Riegl VZ-400 point cloud of (**top right**) and 2D presentation of an IMAGER 5006 scan (**bottom right**).

4.3. 3D Modelling

The 3D modelling was also described in 2012 in Kersten and Lindstaedt [25] and is briefly summarised in the following. The generated point cloud, being too large to import directly into a CAD program, was first segmented and then transferred to AutoCAD using the plugin PointCloud. Once imported, the cathedral was blocked out manually with a 3D mesh by extruding polylines along the surfaces and edges of the point cloud structure. This method has the advantage of not generating too large a file, while retaining visual control of the built model using a superimposed point cloud. Figure 4 shows the final constructed 3D CAD model of the entire cathedral in four different perspective views.

Figure 4. Constructed 3D model of the imperial cathedral in Königslutter, Germany—View of the four fronts in AutoCAD.

For some smaller details on the cathedral, the automated meshing functions in Geomagic were used to quickly generate a mesh directly from the point cloud (Figure 5). This works by means of a simple triangulation algorithm, which works better for more complex and irregular shapes and surfaces.

Figure 5. Generation of small complex objects of the cathedral using the meshing function in Geomagic for the segmented point clouds.

4.4. Panoramic Photography

In order to subsequently colourise the point cloud, as well as to generate a virtual online tour of the cathedral, a series of 360° panoramic photos were taken at each IMAGER 5006 scan station using a Nikon DSLR camera with a nodal point adapter. Supplementary panoramic photos were also taken at 10 additional locations outside the cathedral, as well as 19 further points within the cathedral. These were taken without any laser-scanning targets or extraneous objects present in the shot. The acquisition and processing of the panoramic photography was also described in 2012 in Kersten and Lindstaedt [25]. For better understanding of the whole workflow, the processing of the panoramic photography is briefly summarised in the following. Each set of photographs consists of 16 images—one pointing towards the sky, three towards the ground and 12 photos for the 360° circle in the horizontal position. The software PTGui automatically generated a spherical panorama with 11,700 × 5850 pixels (ca. 43 MB) for each camera station. These panorama images were converted into a set of six cube images (in total ca. 5 MB). The panorama viewing software KRpano (https://krpano.com) was initially used to generate an interactive virtual tour for web browsers (Figure 6). The tour can be viewed at https://www.koenigslutter-kaiserdom.de/virtuelleTour/tour.html (must have Adobe Flash 9/10 enabled). In this browser-based tour, all spherical panoramas are linked to each other via hotspots or via the overview map (bottom-right corner). This provides a quick and convenient way of navigating through the panoramas, simply by clicking on the relevant map icon.

Figure 6. Interactive virtual tour through the imperial cathedral using full spherical panorama photography on several stations inside and outside of the building, including an overview map of stations (bottom-right corner).

4.5. Game Engine Unreal and VR System HTC Vive

A game engine is a simulation environment where 2D or 3D graphics can be manipulated through code. Developed primarily by the video games industry, they provide ideal platforms for the creation of VR experiences for other purposes (e.g., cultural heritage), as many of the necessary functionalities are already built in, eliminating the need to engineer these features independently. While there are dozens of appropriate game engines that could be used, the most popular for small studios and production teams tend to be the Unity engine (Unity Technologies, San Francisco, California, USA), CryEngine (Crytek, Frankfurt am Main, Germany) and Unreal Engine (Epic Games, Cary, North Carolina, USA). For this project, the Unreal Engine was chosen for its advantage in the built-in blueprints visual coding system, which allows users to build in simple interactions and animations without any prior knowledge of C++, the programming language on which the engine is built [26].

The specific hardware required to run VR is a VR headset, two "lighthouse" base stations, two controllers, and a VR-ready PC. For this project, the HTC Vive Pro was chosen as a headset. The lighthouses are needed to track the user's movement in 3D space (Figure 7), while the controllers are used for mapping interactions in the virtual world. Tracking is achieved with a gyroscope, accelerometer, and laser position sensor within the VR headset itself, and can detect movements with an accuracy of 0.1° [27]. Figure 7 shows the setup of the VR system HTC Vive Pro, including the interaction area (blue) for the user.

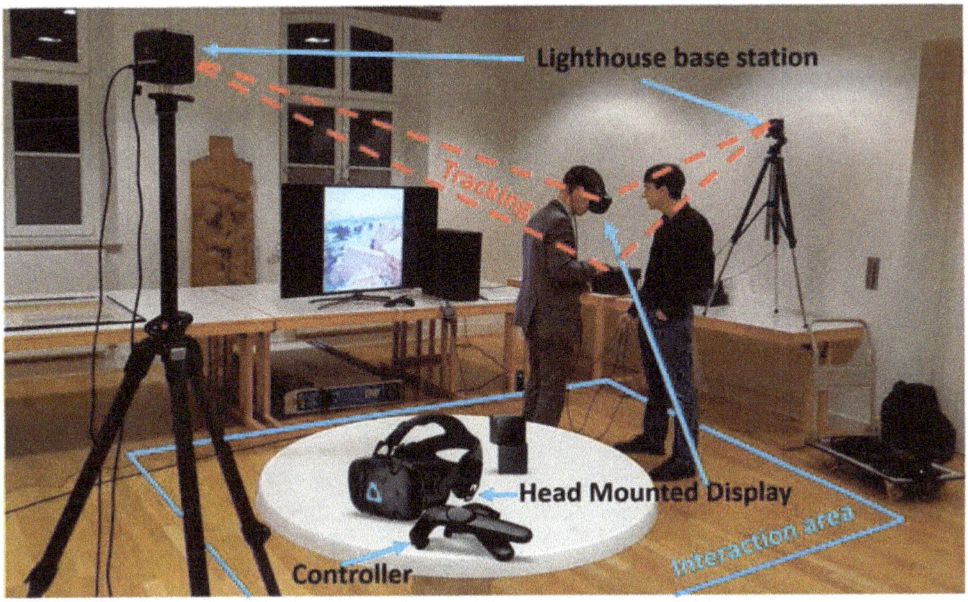

Figure 7. Components and schematic setup of the VR system HTC Vive Pro with interaction area (blue).

4.6. Implementation In Virtual Reality

In order to bring the model into virtual reality, some changes had to be made to the mesh and textures in order to make them run more efficiently within the game engine. The strict performance criteria of VR mean that every effort needs to be made to optimize the models and ensure that a sufficiently high frame rate (ideally 90 frames per second, though for many applications above 40 is sufficient) can be achieved. Much of this part of the workflow was done manually.

First, the mesh was split into different parts in order to make the data volume of the files smaller and therefore speed up the time taken for each iteration of the texturing process. Because UE4's built-in render engine renders only those meshes and textures that are within the screen-space of the viewer at any one time, a logical approach is to separate the interior from the exterior meshes, so as to unload the exterior data when the user is inside the cathedral and vice versa when they are outside. The two principal parts of the Kaiserdom—the central nave and the cloisters—were also processed separately for the same reason. In a few areas of the model, such as the southern side of the cloister, additional modelling was done in order to supplement the scan data. A low-poly city model provided by the company CPA Software GmbH (Siegburg, Germany) was used as a basis to model low-poly buildings in the area around the Cathedral. As these buildings were not central to the experience, they were modelled only in low detail so as not to take up too much rendering space on the GPU. Buildings further away from the Cathedral, which were only visible on the periphery of the virtual environment, were left in their raw form (simple grey rectangular meshes) to avoid any extraneous modelling work.

Much of the work in the VR optimization process was dedicated to the production of high-quality textures suitable for real-time VR. There is a fundamental trade-off here between the quality of the textures needed to appear photorealistic at close range and the data streaming limit of the engine (which varies due to hardware and software specifications). As a rule, creating a photorealistic environment for VR requires high-quality textures in order to boost the experience of immersion. While the Unreal Engine automatically implements level-of-detail algorithms to reduce the load on the render engine, a certain amount of manual optimization must be done in addition to achieve performance goals. As such, texture resolution was varied depending on how far the texture would be from eye-level in the virtual environment. 4K textures (4096 × 4096 px) were used for high-detail textures that would appear at eye level, while 2K textures (2048 × 2048 px) were used for textures that appear well above eye level (for example, the ceiling and roof textures). While many of the textures for this process were adapted from photos taken at the Kaiserdom, supplementary photo-textures were sourced from a creative commons licensed CGI texture database (https://texturehaven.com/). For those materials with more exaggerated relief, such as the building stone and roof tiles, normal maps were also added and accentuated with a parallax displacement effect built with the native UE4 material creation tools.

The 3D models with their corresponding textures were exported into UE4 for placement and real-time visualization (Figure 8A,B). The version of UE4 used in this case was 4.22. Additional elements such as plant meshes, clouds, fog, environmental lighting, and audio were added to heighten the sense of photorealism and immersion. In addition, simple interactions were integrated in order to help the user navigate around the environment. Firstly, a teleportation mechanic was implemented, allowing the user to jump from location to location. This mechanic makes use of a simple ray-tracer, pre-built into UE4, that allows the user to point to any location in the virtual world and check that the location is a valid teleportation point according to a certain set of criteria (these criteria, including the space available and the slope angle at the location, are calculated by UE4 with its "Navigation Mesh" feature). If the location is valid as a teleportation point, the user can teleport there with the click of the trigger button on the controller (Figure 8D). In addition, automatic door-opening animations were added to several doors in the cathedral, allowing users to move between different parts of the building as in the real world. A short trailer of the virtual environment can be viewed online (https://www.youtube.com/watch?v=hmO0JOdlLgw).

Once the real-time environment was built and VR interactions set up, the 360° panoramas could be integrated. A simple mechanism was implemented in the UE4 engine to make each panorama viewable. This mechanism was made up of: (1) a visual clue in the virtual world that indicated where the panorama was located. As an example we used a glowing ring, which stands out well from the rest of the environment (Figure 8C)—a wide variety of other visual clues may be appropriate; (2) A trigger box overlapping with the ring, coupled with a function that fires when a certain button is pressed on the HTC Vive motion controller; (3) A separate, empty map or level in the UE4 editor; and (4) A skybox in the empty level onto which to project the cube-map panorama. Using this mechanism, the player can approach a glowing ring, representing a panorama taken on that spot, press a button on the motion controller, and be transported into the 360° panorama. By toggling the button press on the motion controller, the player can come out of the panorama and be placed back in the virtual world (Figure 9). Certain variations in this mechanic were tested (e.g., projecting the panoramic photo on the inside of a sphere in the virtual world, then using a button on the motion controller to alternately showing and hiding this sphere when the player was in the right area), but the method described above was found to provide the simplest and most robust way of toggling between the panoramic photos in the scene while retaining the original perspective of the photographs.

The finished version of the VR experience was tested with the HTC Vive Pro headset running on a PC with an 8-Core Intel Xeon CPU (3.10 GHz), an NVIDIA GTX 1070i GPU, and 32.0 GB RAM. With this setup, the experience achieved an average frame rate of 40–50 frames per second.

Figure 8. Two views of the Kaiserdom, inside (**A**) and outside (**B**). A third image (**C**) shows an example of the teleportation mechanic in action. Image (**D**) shows an example of a visual clue placed in the virtual world, where a panoramic photo can be viewed.

Figure 9. View from the same position in the virtual world, with the panorama switched off (**A**) and on (**B**).

5. Conclusions and Outlook

This paper presented the interest and workflow in creating a VR visualization with integrated 360° panoramic photography of the Kaiserdom in Königslutter. The combination of these two kinds of media—real-time 3D visualization and HDRI panoramic photography—allows the interactive and immersive potential of the former to complement the high-fidelity and photorealism of the latter. While traditionally these two "kinds" of VR have remained separate, it is important to investigate ways of integrating them in order to build experiences that are able to integrate different kinds of data. This is particularly important for those fields, such as heritage, where documentation can take multiple forms, such as photographs, objects, 3D data, or written documents. The future development of the virtual museum, for example, depends on being able to integrate different kinds of data into a virtual space that can be navigated intuitively in virtual reality [28].

Further applications of the workflow described above can also be envisioned. In another recent project, a recreation of the town of Stade (Lower Saxony) in the year 1620 [12], panoramic photography is implemented so that users can jump between the real-time visualization of the town in 1620 and 360° photos from the modern day. This implementation allows users to directly juxtapose the historic and contemporary city, as an entry point to comparing the historical conditions of the two periods. In particular, this feature could have extra meaning for users who are already familiar with the town, by revealing the perhaps unknown history of certain well-known locations. While real-time 3D visualizations on their own may provide a certain degree of immersion, the integration of different kinds of data in these virtual worlds, such as panoramic photography, can greatly enrich the experience by inviting the user to compare different kinds of visualizations.

In addition, by taking real-time visualisations beyond being simply static virtual worlds through the integration of different kinds of information, VR becomes much more powerful as a tool for education in museums. Cultural heritage monuments such as the Kaiserdom of Königslutter are particularly suited to VR exhibition due a substantial existing audience that may be looking for new ways to extend their visitor experience. By extending real-time visualisations through panoramic photography and other kinds of information, VR can come closer to realising its potential as a tool for cultural heritage education.

Author Contributions: T.P.K. and A.P.W. conceived the main research idea about VR; T.P.K. generated the panorama photography; A.P.W. processed all data and developed the VR application; A.P.W. and T.P.K. wrote the manuscript and generated the illustrations. All authors have read and agreed to the published version of the manuscript.

Funding: This research was supported by HafenCity University Hamburg, Lab for Photogrammetry & Laser Scanning.

Conflicts of Interest: The authors declare no conflicts of interest.

References

1. "Virtual Reality". Merriam-Webster.com Dictionary, Merriam-Webster. Available online: https://www.merriam-webster.com/dictionary/virtual%20reality (accessed on 4 February 2020).
2. Dörner, R.; Broll, W.; Grimm, P.; Jung, B. *Virtual und Augmented Reality (VR/AR): Grundlagen und Methoden der Virtuellen und Augmentierten Realität*; Springer: Berlin, Germany, 2014.
3. Freina, L.; Ott, M. A Literature Review on Immersive Virtual Reality in Education: State of The Art and Perspectives; eLearning & Software for Education. Available online: https://ppm.itd.cnr.it/download/eLSE%202015%20Freina%20Ott%20Paper.pdf (accessed on 18 December 2019).
4. Portman, M.E.; Natapov, A.; Fisher-Gewirtzman, D. To go where no man has gone before: Virtual reality in architecture, landscape architecture and environmental planning. *Comput. Environ. Urban Syst.* **2015**, *54*, 376–384. [CrossRef]
5. Fuchs, P. *Virtual Reality Headsets—A Theoretical and Pragmatic Approach*; CRC Press: London, UK, 2017.
6. Addison Alonzo, C. Emerging Trends in Virtual Heritage. *IEEE MultiMedia* **2000**, *7*, 22–25. [CrossRef]
7. Stone, R.; Ojika, T. Virtual heritage: What next? *IEEE MultiMedia* **2000**, *7*, 73–74. [CrossRef]
8. Affleck, J.; Thomas, K. Reinterpreting Virtual Heritage. In Proceedings of the CAADRIA 2005, New Delhi, India, 28–30 April 2005; Volume 1, pp. 169–178.
9. Kersten, T.; Tschirschwitz, F.; Deggim, S. Development of a Virtual Museum including a 4D Presentation of Building History in Virtual Reality. *Int. Arch. Photogramm. Remote Sens. Spat. Inf. Sci.* **2017**, *42*, 361–367. [CrossRef]
10. Tschirschwitz, F.; Richerzhagen, C.; Przybilla, H.-J.; Kersten, T. Duisburg 1566–Transferring a Historic 3D City Model from Google Earth into a Virtual Reality Application. *PFG J. Photogramm. Remote Sens. Geoinf. Sci.* **2019**, *87*, 1–10. [CrossRef]
11. Deggim, S.; Kersten, T.; Tschirschwitz, F.; Hinrichsen, N. Segeberg 1600–Reconstructing a Historic Town for Virtual Reality Visualisation as an Immersive Experience. *Int. Arch. Photogramm. Remote Sens. Spat. Inf. Sci.* **2017**, *42*, 87–94. [CrossRef]
12. Walmsley, A.; Kersten, T. Low-cost development of an interactive, immersive virtual reality experience of the historic city model Stade 1620. *Int. Arch. Photogramm. Remote Sens. Spat. Inf. Sci.* **2019**, *42*, 405–411. [CrossRef]

13. Kersten, T.; Büyüksalih, G.; Tschirschwitz, F.; Kan, T.; Deggim, S.; Kaya, Y.; Baskaraca, A.P. The Selimiye Mosque of Edirne, Turkey—An Immersive and Interactive Virtual Reality Experience using HTC Vive. *Int. Arch. Photogramm. Remote Sens. Spat. Inf. Sci.* **2017**, *42*, 403–409. [CrossRef]
14. Kersten, T.; Tschirschwitz, F.; Lindstaedt, M.; Deggim, S. The historic wooden model of Solomon's Temple: 3D recording, modelling and immersive virtual reality visualisation. *J. Cult. Herit. Manag. Sustain. Dev.* **2018**, *8*, 448–464. [CrossRef]
15. Westfries Museum. Batavia 1627 in Virtual Reality. Hoorn, Netherlands. Available online: https://wfm.nl/batavia-1627vr (accessed on 17 December 2019).
16. Schofield, G.; Beale, G.; Beale, N.; Fell, M.; Hadley, D.; Hook, J.; Murphy, D.; Richards, J.; Thresh, L. Viking VR: Designing a Virtual Reality Experience for a Museum. In Proceedings of the Designing Interactive Systems Conference, ACM DIS Conference on Designing Interactive Systems 2018, Hong Kong, China, 9–13 June 2018; Association for Computing Machinery (ACM): New York, NY, USA, 2018; pp. 805–816.
17. Fassi, F.; Mandelli, A.; Teruggi, S.; Rechichi, F.; Fiorillo, F.; Achille, C. VR for Cultural Heritage. In *International Conference on Augmented Reality, Virtual Reality and Computer Graphics*; Springer: Cham, Switzerland, 2016; pp. 139–157.
18. Dhanda, A.; Reina Ortiz, M.; Weigert, A.; Paladini, A.; Min, A.; Gyi, M.; Su, S.; Fai, S.; Santana Quintero, M. Recreating cultural heritage environments for VR using photogrammetry. *Int. Arch. Photogramm. Remote Sens. Spat. Inf. Sci.* **2019**, *42*, 305–310. [CrossRef]
19. Skarlatos, D.; Agrafiotis, P.; Balogh, T.; Bruno, F.; Castro, F.; Petriaggi, B.D.; Demesticha, S.; Doulamis, A.; Drap, P.; Georgopoulos, A.; et al. Project iMARECULTURE: Advanced VR, iMmersive serious games and augmented REality as tools to raise awareness and access to European underwater CULTURal heritagE. In *Euro-Mediterranean Conference*; Springer: Cham, Switzerland, 2016; pp. 805–813.
20. See, Z.S.; Santano, D.; Sansom, M.; Fong, C.H.; Thwaites, H. Tomb of a Sultan: A VR Digital Heritage Approach. In Proceedings of the 3rd Digital Heritage International Congress (Digital HERITAGE) Held Jointly with 24th International Conference on Virtual Systems & Multimedia (VSMM 2018), San Francisco, CA, USA, 26–30 October 2018; pp. 1–4.
21. Manghisi, V.M.; Fiorentino, M.; Gattullo, M.; Boccaccio, A.; Bevilacqua, V.; Cascella, G.L.; Dassisti, M.; Uva, A.E. Experiencing the sights, smells, sounds, and climate of southern Italy in VR. *IEEE Comput. Graph. Appl.* **2018**, *37*, 19–25.
22. Ramsey, E. Virtual Wolverhampton: Recreating the historic city in virtual reality. *ArchNet Int. J. Archit. Res.* **2017**, *11*, 42–57. [CrossRef]
23. Bergmann, N.; Dobler, G.; Funke, N. *Kaiserdom Königslutter—Geschichte und Restaurierung*; Michael Imhof Verlag: Petersberg, Germany, 2008.
24. Stiftung Braunschweigischer Kulturbesitz. Königslutter Kaiserdom—The Key Facts. Available online: https://www.koenigslutter-kaiserdom.de/images/cache/Kaiserdom%20Koenigslutter%20THE%20KEY%20FACTS.pdf (accessed on 6 January 2020).
25. Kersten, T.; Lindstaedt, M. Virtual Architectural 3D Model of the Imperial Cathedral (Kaiserdom) of Königslutter, Germany through Terrestrial Laser Scanning. In Proceedings of the EuroMed 2012—International Conference on Cultural Heritage, Limassol, Cyprus, 29 October–3 November 2012; Lecture Notes in Computer Science (LNCS). Ioannides, M., Fritsch, D., Leissner, J., Davies, R., Remondino, F., Caffo, R., Eds.; Springer: Berlin/Heidelberg, Germany, 2012; Volume 7616, pp. 201–210.
26. McCaffrey, M. *Unreal Engine VR Cookbook*; Addison-Wesley Professional: Boston, MA, USA, 2017.
27. Painter, L. Hands on with HTC Vive Virtual Reality Headset. 2015. Available online: http://www.pcadvisor.co.uk/feature/gadget/hands-on-with-htc-vive-virtual-reality-headset-experience-2015-3631768/ (accessed on 18 December 2019).
28. Giangreco, I.; Sauter, L.; Parian, M.A.; Gasser, R.; Heller, S.; Rossetto, L.; Schuldt, H. VIRTUE: A Virtual Reality Museum Experience. In Proceedings of the 24th International Conference on Intelligent User Interfaces: Companion IUI' 19, Los Angeles, CA, USA, 16–20 March 2019; pp. 119–120.

applied sciences

MDPI

Article

Semantic 3D Reconstruction with Learning MVS and 2D Segmentation of Aerial Images

Zizhuang Wei [1,2,†], Yao Wang [1,2,†], Hongwei Yi [1,2], Yisong Chen [1,2,3] and Guoping Wang [1,2,3,*]

1 Graphics & Interaction Lab, School of Electronics Engineering and Computer Sciences, Peking University, Beijing 100871, China; 1801111363@pku.edu.cn (Z.W.); yaowang95@pku.edu.cn (Y.W.); hongweiyi@pku.edu.cn (H.Y.); chenyisong@pku.edu.cn (Y.C.)
2 Key Lab of Machine Perception and Intelligent, MOE, Department of Computer Sciences, Peking University, Beijing 100871, China
3 Beijing Engineering Technology Research Center of Virtual Simulation and Visualization, Peking University, Beijing 100871, China
* Correspondence: wgp@pku.edu.cn
† These authors contributed equally to this work.

Received: 21 December 2019; Accepted: 10 February 2020; Published: 14 February 2020

Abstract: Semantic modeling is a challenging task that has received widespread attention in recent years. With the help of mini Unmanned Aerial Vehicles (UAVs), multi-view high-resolution aerial images of large-scale scenes can be conveniently collected. In this paper, we propose a semantic Multi-View Stereo (MVS) method to reconstruct 3D semantic models from 2D images. Firstly, 2D semantic probability distribution is obtained by Convolutional Neural Network (CNN). Secondly, the calibrated cameras poses are determined by Structure from Motion (SfM), while the depth maps are estimated by learning MVS. Combining 2D segmentation and 3D geometry information, dense point clouds with semantic labels are generated by a probability-based semantic fusion method. In the final stage, the coarse 3D semantic point cloud is optimized by both local and global refinements. By making full use of the multi-view consistency, the proposed method efficiently produces a fine-level 3D semantic point cloud. The experimental result evaluated by re-projection maps achieves 88.4% Pixel Accuracy on the Urban Drone Dataset (UDD). In conclusion, our graph-based semantic fusion procedure and refinement based on local and global information can suppress and reduce the re-projection error.

Keywords: semantic 3D reconstruction; deep learning; multi-view stereo; probabilistic fusion; graph-based refinement

1. Introduction

Semantic 3D reconstruction makes Virtual Reality (VR) and Augmented Reality (AR) much more promising and flexible. In computer vision, 3D reconstruction and scene understanding receive more and more attention these days. 3D models with correct geometrical structures and semantic segmentation are crucial in urban planning, automatic piloting, robot vision, and many other fields. For urban scenes, semantic labels are used to visualize targets such as buildings, terrain, and roads. A 3D point cloud with semantic labels makes the 3D map more simple to understand, thereby propelling the subsequent research and analysis. 3D semantic information also shows potential in automatic piloting. For a self-driving vehicle, one of the most important things is to distinguish whether the road is passable or not. Another essential thing for an autonomous automobile is to localize other vehicles in real-time so that it can adapt to their speed, or exceed it if necessary. In the field of robotics, scene understanding is a standard task for recognizing surrounding objects. The semantics of the surrounding environment play a vital role in applications such as loop closure and route planning.

Although 3D semantic modeling has been widely studied in recent years, the approaches of extracting semantic information through the post-processing of point cloud reconstruction generally lead to inconsistent or incorrect results. Performing semantic segmentation on point cloud data is more difficult than it is on 2D images. One major problem is the lack of 3D training data, since labeling a dataset in 3D is much more laborious than in 2D. Another challenge is the unavoidable noise in 3D point clouds, which makes it difficult to accurately distinguish which category a point belongs to. Thus, it is necessary to develop new semantic 3D reconstruction approaches by simultaneously estimating 3D geometry and semantic information over multiple views. In the past few years, many studies on image semantic segmentation have achieved promising results by deep learning techniques [1–4]. Deep learning methods based on well-trained neural networks can help us do pixel-wise semantic segmentation on various images. Meanwhile, deep-learning-based methods are not only able to extract semantic information, but are also practical for solving Multi-View Stereo (MVS) problems. Recently, learning-based MVS algorithms [5,6] have been proposed to generate high precision 3D point clouds for large-scale scenes. These results inspired us much and gave rise to the research of semantic 3D reconstruction. In this paper, we mainly focus on developing accurate, clear, and complete 3D semantic models of urban scenes.

Once satisfactory depth and semantic maps are acquired, 3D semantic models can be easily generated. 3D laser scanners can detect depth directly but only perform well in short-distance indoor scenes. Compared with 3D laser scanners, the purely RGB-based method to reconstruct 3D models from 2D images is cheaper, faster, and more generalized. Recently, Unmanned Aerial Vehicles (UAV) have become applicable to collecting multi-view, high-resolution aerial images of large-scale outdoor scenes. The calibrated camera poses can be obtained from the images by the traditional Structure from Motion (SfM) technique. After that, 3D point clouds are determined by fusing 2D images according to multi-view geometry.

However, due to the occlusions, the complexity of environments, and the noise of sensors, both 2D segmentation and depth estimation results contain errors. As a result, many inconsistencies may occur when projecting the multi-view 2D semantic labels to the corresponding 3D points. There is still plenty of work to do to obtain accurately-segmented 3D semantic models. With the booming of deep learning methods, 2D segmentation tasks are reaching high performance levels, which makes it possible to acquire a large-scale 3D semantic model easily. Nevertheless, errors within depth maps and semantic maps may lead to inconsistency. This can be alleviated by considering 3D geometry and 2D confidence maps together in an optimization module. Moreover, 3D models with coarse segmentation still need further refinement to filter error points. In a nutshell, the main contributions of our work are three folds:

- We present an end-to-end, learning-based, semantic 3D reconstruction framework, which reaches high Pixel Accuracy on the Urban Drone Dataset (UDD) [7].
- We propose a probability-based semantic MVS method, which combines the 3D geometry consistency and 2D segmentation information to generate better point-wise semantic labels.
- We design a joint local and global refinement method, which is proven effective by computing re-projection errors.

2. Related Work

Right before the renaissance of deep learning, it was a hard task to get a good pixel-wise segmentation map on images. Bao, S.Y. et al. [8] take object-level semantic information to constrain camera extrinsic. Some other methods perform the segmentation directly on the point cloud or meshes, according to their geometric characteristics. Martinovic, A. et al. [9] and Wolf, D. et al. [10] take the random forest classifier to do point segmentation, while Häne, C. et al. [11,12] and Savinov, N. et al. [13] treat it as an energy minimization problem in a Conditional Random Field (CRF). Ray potential (likelihood) is frequently adopted in semantic point cloud generation.

The flourishing CNN-based semantic segmentation methods are quickly outperforming traditional methods in image semantic segmentation tasks; take, for example, the Fully Convolutional Network (FCN) [1] and Deeplab [3]. High-level computer tasks such as scene understanding and semantic 3D reconstruction are now steady and rudimentary processes. The goal of 3D semantic modeling is to assign a semantic label to each 3D point rather than each 2D pixel. Several learning-based approaches follow the end-to-end manner, analyzing the point cloud and giving segmentation results directly in 3D. Voxel-based methods such as ShapeNets [14] and VoxNet [15] were proposed naturally. Some methods learn a spatial encoding of each point and then aggregate all individual point features to a global point cloud signature [16,17]. However, current deep learning-based segmentation pipelines cannot handle noisy, large-scale 3D point clouds. Thus, a feasible method is required to firstly perform pixel-wise semantic segmentation on 2D images and then back-project these labels into 3D space using the calibrated cameras to be fused. The methods above handle the point cloud directly, which means they carry a costly computational burden. In other words, they cannot manage large-scale 3D scenes without first partitioning the scene. More than that, because the morphological gap between point clouds in different scenarios is too large. These algorithms may be poorly generalized.

There are several methods doing semantic segmentation on 2D image and making use of multi-view geometric relationships to project semantic labels into 3D space. For RGBD-based approaches, once good semantic maps of each image are acquired, the semantic point clouds can easily be fused. Vineet, V. et al. [18] took advantage of a random forest to classify 2D features to get semantic information, while Zhao, C. et al. [19] used FCN with CRF-RNN to perform segmentation on images. McCormac, J. et al. [20] and Li, X. et al. [21] proposed incremental semantic label fusion algorithms to fuse 3D semantic maps. For RGB-based approaches, also addressed as Structure from Motion (SfM) and MVS, each point in the generated 3D structure corresponds to pixels on several images. Following the prediction of 2D labels, the final step is to assign each 3D pixel a semantic label [20,22]. The refinement process is as essential as the generation process of the semantic point cloud itself. Chen, Y. et al. [7] and Stathopoulou, E.K. et al. [23] filter the mismatching by semantic labels of feature points. With the motivation of denoising, Zhang, R. et al. [24] proposed a Hough-transform-based algorithm called FC-GHT to detect plane on point cloud for further semantic label optimization. Stathopoulou, E.K. et al. [23] used semantic information as a mask to wipe out the meshes belonging to the semantic class *sky*. These methods have two primary drawbacks. Firstly, they only use the final semantic maps, which means the probabilities of other categories are discarded. Secondly, they contain no global constraints integrated into their algorithms. In response, we propose some ideas for improvement.

3. Method

3.1. Overall Framework

The overall framework of our method is depicted in Figure 1. In the Deeplab v2 [3]-based 2D segmentation branch, we discard the last Argmax layer of the network. We save pixel-wise semantic probability maps for every image instead. With the help of COLMAP-SfM [25], we simultaneously estimate the camera parameters and depth ranges for the source images. In order to acquire 3D geometry for large scale scenes, we utilize learning-based MVS method R-MVSNet [6] to estimate depth maps for multiple images. After 2D segmentation and depth estimation, we obtain a dense semantic point cloud by the semantic fusion method according to multi-view consistency. Finally, we propose a graph-based point cloud refinement algorithm integrating both local and global information as the last step of our pipeline.

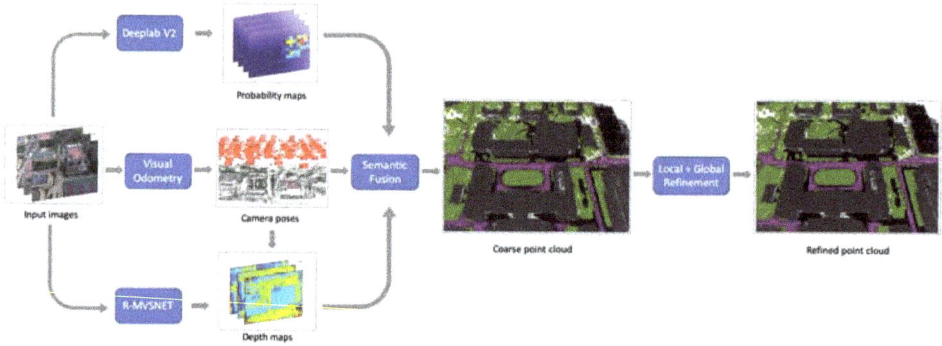

Figure 1. General pipeline of our work. Three branches are implemented to process the reconstruction dataset. The upper branch is the semantic segmentation branch to predict the semantic probability map; the middle branch is SfM to calculate the 3D odometry and camera poses; the lower branch is to estimate the depth map. Then, semantic fusion is applied to fuse them into a coarse point cloud. The last step is to refine the point cloud by local and global methods.

3.2. 2D Segmentation

In this research, Deeplab v2 [3] with Residue Block is adopted as our segmentation network. The pretrained weights of ResNet-101 [26] on Imagenet [27] are used as our initial weights. We adopt the residual block to replace the ordinary 2D convolution layer to improve the training performance. We also modify the softmax layer that classifies the images to fit the label space of the UDD [7] dataset. With the network all set up, the training set of UDD [7] is employed for transfer learning.

The label space of UDD [7] is denoted as $\mathcal{L} = \{l_0, l_1, l_2, l_3, l_4\}$, which contains *Vegetation*, *Building*, *Road*, *Vehicle*, and *Background*. After the transfer learning process, we predict the semantic maps for every image in the reconstruction dataset. Furthermore, we save the weight matrix before the last Argmax layer. This matrix $\boldsymbol{P}(\mathcal{L})$ represents the probability distributions of every pixel in the semantic label space.

3.3. Learning-Based MVS

In order to acquire 3D geometry for large scale scenes, we explore the learning-based MVS method to estimate depth maps for multiple images. R-MVSNet [6], a deep learning architecture with capability to handle multi-scale problem, has advantages in processing high-resolution images and large-scale scenes. Moreover, R-MVSNet utilizes the Gated Recurrent Unit (GRU) to sequentially regularize the 2D cost maps, which reduces the memory consumption and makes the network flexible. Thus, we follow the framework of R-MVSNet to generate corresponding depths of the source images and train it on the DTU [28] dataset. Camera parameters and image pairs are determined by the implementation of COLMAP-SfM [25], while depth samples are chosen within $[d_{min}, d_{max}]$ using the inverse depth setting. The network returns a probability volume $\boldsymbol{P}$ where $P(x, y, d)$ is the probability estimation for the pixel (x, y) at depth d; then, the expectation depth value $d(x, y)$ is calculated by the probability weighted sum over all hypotheses:

$$d(x,y) = \sum_{d=d_{min}}^{d_{max}} P(x,y,d) \cdot d. \tag{1}$$

However, as with most depth estimation methods, the coarse pixel-wise depth data $d(x, y)$ generated by R-MVSNet may contain errors. Therefore, before point cloud fusion by the depth maps, it is necessary to perform a denoising process on the depth data. In this paper, we apply the bilateral filtering method to improve the quality of depth maps with edge preservation; the refined depth data $d'(x,y)$ are obtained by:

$$d'(x,y) = \frac{\sum_{i,j} \omega(x,y,i,j) \cdot d(x,y)}{\sum_{i,j} \omega(x,y,i,j)} \tag{2}$$

where $\omega(x,y,i,j) = exp(-\frac{(x-i)^2+(y-j)^2}{2\sigma_f^2} - \frac{||d(x,y)-d(i,j)||^2}{2\sigma_g^2})$ is the weighted coefficient; σ_f and σ_g are the variance of domain kernel $f(x,y,i,j) = exp(-\frac{(x-i)^2+(y-j)^2}{\sigma_f^2})$ and range kernel $g(x,y,i,j) = exp(-\frac{||d(x,y)-d(i,j)||^2}{\sigma_g^2})$ respectively. As shown in Figure 2, the depth map becomes more smooth with edge preservation after bilateral filtering.

(**a**) Reference image (**b**) Coarse depth (**c**) Refined depth

Figure 2. Visualization of the depth map estimated by the learning-based MVS method. (**a**) The input image. (**b**) Depth estimation by R-MVSNet [6]. (**c**) Refined depth by bilateral filtering.

3.4. Semantic Fusion

With the learning 2D segmentation and depth estimation, pixel-wise 2D semantic labels and depth maps are obtained for each view. However, because of the occlusions, complexities of environments, and the noise of sensors, both image segmentation results and depth maps might have a large number of inconsistencies between different views. Thus, we further cross filter the depth maps by their neighbor views, and then produce the 3D semantic point clouds by combining 2D segmentation and depth maps with multi-view consistency.

Similar to other depth-map-based MVS methods [6,29], we utilize geometric consistency to cross filter the multi-view depth data. Given the pixel (x,y) from image I_i with depth $d(x,y)$, we project (x,y) to the neighbor image I_j through $d(x,y)$ and camera parameters. In turn, we re-project the projected pixel back from the neighbor image I_j to the original image I_i; the re-projected depth on I_i is d_{reproj}. We consider the pixel consistent in the neighbor view I_j when d_{reproj} satisfies:

$$\frac{|d(x,y) - d_{reproj}|}{d(x,y)} < \tau. \tag{3}$$

According to the geometric consistency, we filter the depths which are not consistent in more than k views. In this paper, we take $\tau = 0.01$ and $k = 3$.

After cross filtering, the depths are projected to 3D space to produce 3D point clouds. Since our purpose is to assign point-wise semantic labels for the 3D model, we propose a probabilistic fusion method to aggregate multi-view 2D semantic information. With the fine-tuned CNN, a pixel-wise label probability distribution $\boldsymbol{P}(\mathcal{L})$ has been calculated for each source image. Given a 3D point X which is visible in N views, the corresponding probability on view i for label l_j is $p_i(l_j)$; we accumulate the multi-view probability distribution of each view as follows:

$$P(l_j) = \frac{1}{N} \sum_{i=1}^{N} p_i(l_j), l_j \in \mathcal{L}, \tag{4}$$

where $P(l_j)$ denotes the probability of point X labeling by l_j. In this way, we transfer the probability distribution of multi-view images into 3D space. Generally, the predicted 3D semantic label can be determined by the Argmax operation as:

$$l(X) = \underset{l_j}{Argmax}(P(l_j)), l_j \in \mathcal{L}, \tag{5}$$

where $l(X)$ is the 3D semantic label of X. As depicted in Figure 3, the probabilistic fusion method effectively reduces errors since it integrates information from multiple images.

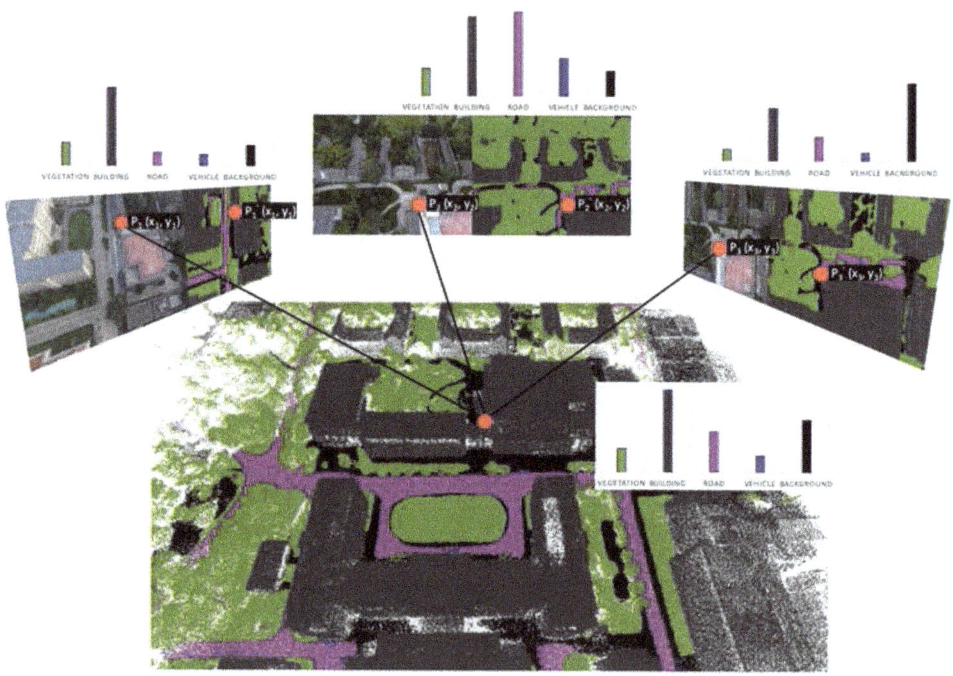

Figure 3. Illustration of our semantic fusion method: the 3D semantic labels are determined by multi-view information; the 3D point's label is decided by the correspondence accumulated probability of 2D pixels in each image.

3.5. Point Cloud Refinement

Through the semantic fusion method, the 3D point cloud is classified into point-wise semantic labels. However, there are still few scattered points with error labels due to incorrect semantics or depths of source images. To remove these unexpected semantic errors, we explore both local and global refinement strategies for point cloud refinement. The *KD-Tree* data structure is employed to accelerate the query speed of the point cloud from O(n) to O(log(n)).

Generally, adjacent point clouds often have some correlation and are more likely to be segmented into the same class. Hence, we utilize the local refinement method for each point by combining the hypothesizes with the neighbor points. Given a 3D point X from the dense semantic model, through the *KD-Tree* structure established by the whole point cloud, the k-nearest neighbor of X could be determined in a short time. $P_i(l_j), i = 1, \cdots, k$ represents the probability for neighbor point i labeling by l_j; the new semantic label $l'(X)$ is updated by:

$$l'(X) = \underset{l_j}{Argmax}\left(\frac{1}{k}\sum_{i=1}^{k} P_i(l_j)\right), l_j \in \mathcal{L}. \tag{6}$$

However, the local refinement method only takes the local adjacency into consideration with the global information ignored. For overall optimization, we further apply a graph-based global refinement method by minimizing an energy function. For every 3D point in the point cloud $\boldsymbol{V}$, a graph $\boldsymbol{G}$ is established by connecting it with its k-nearest neighbor. Then the energy function is defined as:

$$E(\boldsymbol{L}) = \sum_{<X_p, X_q> \in \boldsymbol{D}} B(l(X_p), l(X_q)) + \lambda \cdot \sum_{X \in \boldsymbol{V}} R(l(X)), \tag{7}$$

where $\boldsymbol{L} = \{l(X)|X \in \boldsymbol{V}\}$ are the semantics of $\boldsymbol{V}$ and $\boldsymbol{D}$ is the set of all neighbor pairs. Similarly to [30], $B(l(X_p), l(X_q)) = 1$ and $R(l(X)) = \frac{1}{k}\sum_{i=1}^{k} P_i(l_j)$ are the boundary term and inner region term respectively, while $\lambda \geq 0$ is a constant. Finally, the energy $E(\boldsymbol{L})$ is minimized by a max-flow algorithm, as implemented in [31]. The refined point cloud is illustrated in Figure 4. Compared with the coarse result, our method wipes out semantic outliers and noises.

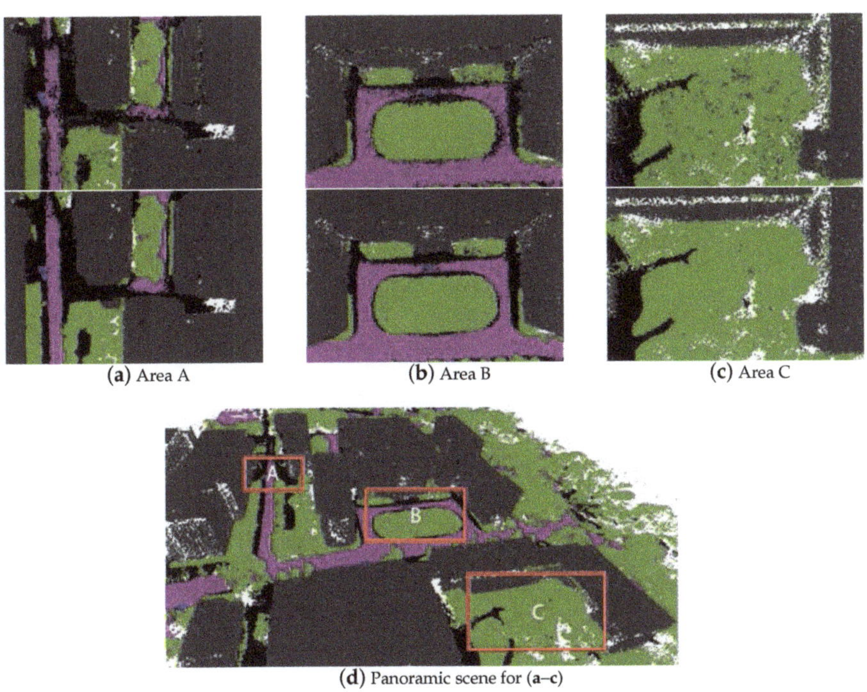

Figure 4. Comparison between the point clouds before and after refinement. (**a**–**c**) Top: coarse result. Bottom: refined result. (**d**): The panoramic scene for (**a**–**c**).

4. Experimental Evaluation

4.1. Experimental Protocol

Dataset: We carry out the training process of semantic segmentation on UDD https://github.com/MarcWong/UDD [7], an UAV collected dataset with five categories, containing 160 and 40 images in the training and validation sets, respectively. The categories are *Building, Vegetation, Road, Vehicle,* and *Background*. The performance is measured on its test set called PKU-M1, which is a reconstruction dataset also collected by a mini-UAV at low altitude. PKU-M1 consists of 288 RGB images at 4000 × 3000 resolution. We down-sample the result to 1000 × 750 to accelerate the prediction speed.

Coloring policy: Cityscapes https://www.cityscapes-dataset.com [32] is the state-of-the-art semantic segmentation dataset for urban scene understanding, which was released in 2016 and received much attention. We borrow the coloring policy of semantic labels from Cityscapes [32].

Training: UDD [7] is trained by Deeplab V2 [3] network structure implemented on TensorFlow [33]. We use the stochastic gradient descending [34] optimizer with weight decaying parameter 5×10^{-5}. Learning rate is initialized to 1×10^{-3} with a momentum of 0.99. The entire apparatus is conducted on a Ubuntu 18.04 server, with an Intel core i7-9700K CPU, 32GB memory, and a single Titan X Pascal GPU.

Measurements recap: Assume the number of non-background classes is k. The confusion matrix **M** for foreground categories can be denoted as below:

$$\mathbf{M} = \begin{pmatrix} c_{11} & c_{12} & \cdots & c_{1k} \\ c_{21} & c_{22} & \cdots & c_{2k} \\ \cdots & \cdots & \cdots & \cdots \\ c_{k1} & c_{k2} & \cdots & c_{kk} \end{pmatrix} \tag{8}$$

For a specific foreground semantic label $l_x \in \mathcal{L}$, the problem can be formulated to a binary classification problem, where:

$$TruePositive(TP) = c_{xx}, \tag{9}$$

$$TrueNegative(TN) = \sum_{i=0}^{k}\sum_{j=0}^{k} c_{ii}, i \neq x, j \neq x, \tag{10}$$

$$FalsePositive(FP) = \sum_{i=0}^{k} c_{xi}, i \neq x, \tag{11}$$

$$FalseNegative(FN) = \sum_{i=0}^{k} c_{ix}, i \neq x. \tag{12}$$

Then, Pixel Accuracy, precision, recall, and F1-score can be deducted as below:

$$PixelAcciracy(PA) = \frac{\sum_{i=0}^{k} c_{ii}}{\sum_{i=0}^{k}\sum_{j=0}^{k} c_{ij}}, \tag{13}$$

$$Precision = \frac{TP}{TP + FP}, \tag{14}$$

$$Recall = \frac{TP}{TP + FN}, \tag{15}$$

$$F1 - score = 2 \times \frac{Precision \times Recall}{Precision + Recall}. \tag{16}$$

4.2. Evaluation Process

We choose proper measurements to quantitatively evaluate the 2D segmentation performance and 3D semantic model. We randomly labeled 16 images in PKU-M1 to test the segmentation performance. An example of PKU-M1 is shown in Figure 5. Table 1 gives class-wise statistics, where the *Building* category is segmented very well, but *Vegetation*, *Road*, and *Vehicle* are segmented relatively poorly. Since hand-crafted 3D semantic labeling is now still a challenging and tedious task, especially for large-scale scenarios, we have to evaluate the 3D semantic model indirectly. Notice that each 3D point is assigned a semantic label during the semantic fusion process; the label can be projected back to each camera coordinate by the geometric relation. We call this step re-projection. Then, we can indirectly evaluate the 3D semantic point cloud by re-projection images in a simpler manner. However, the re-projection map Figure 5d is quite sparse. Only foreground labels, which include *Vegetation*, *Building*, *Vehicle*, and *Road*, are countable for evaluation. So several common measurements for 2D segmentation are not suitable in our cases, such as MIoU (Mean Intersection over Union) and FWIoU (Frequent Weighted Intersection over Union). In our experiment, we choose Pixel Accuracy (Equation (13)) and class-wise F1-score (Equation (16)) for evaluation.

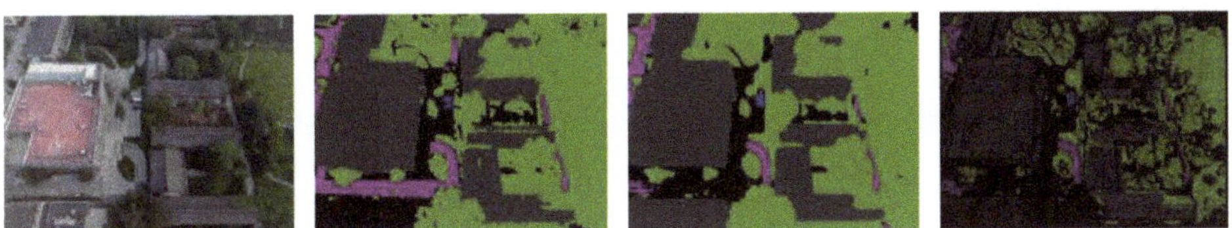

(a) Image **(b)** Ground truth **(c)** Prediction **(d)** 3D re-projection

Figure 5. Visualization of PKU-M1.(**a**): A sample image of PKU-M1, (**b**): ground truth of (**a**), (**c**): prediction of (**a**), and (**d**): 3D re-projection map of (**a**). Since the re-projection map (**d**) is quite sparse, we use Pixel Accuracy to compare the re-projection map and the ground truth map. **Grey**: Building, **Green**: Vegetation, **Blue**: Vehicle, **Pink**: Road, **Black**: Background. Best viewed in color.

Table 1. Evaluation of 2D semantic segmentation.

Category	Accuracy(%)	Precision(%)	Recall(%)	F1 score(%)
Building	95.60	98.25	94.87	96.53
Vegetation	89.85	76.96	71.24	73.99
Vehicle	97.95	67.09	22.02	33.15
Road	87.91	52.58	73.84	61.42

5. Results and Discussion

5.1. Quantitative Results

With the semantic fusion process introduced in Section 3.4, the coarse semantic 3D point cloud was generated. Its quantitative result is denoted as the 3D baseline in Table 2. To be more specific, most points in 3D baseline are correct, yet with outliers and errors. The evaluation result of 3D baseline's re-projection map demonstrates that the 3D baseline is much better than 2D in both PA and F1-score. Figure 6a,b illustrate this fact vividly, where *Vehicle* is segmented badly in 2D segmentation and segmented much better in 3D baseline.

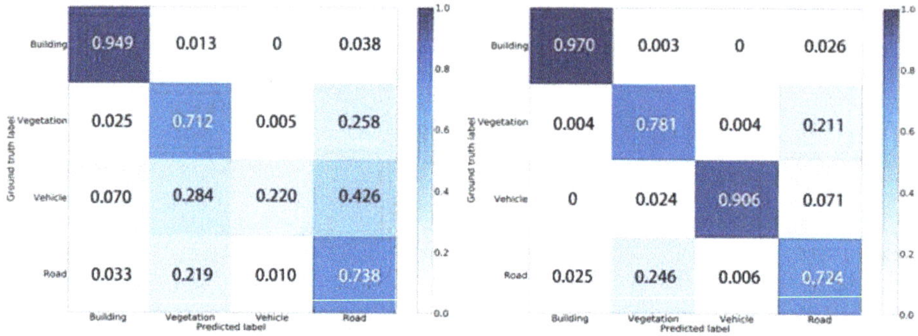

(**a**) 2D segmentation (**b**) 3D re-projection

Figure 6. (**a**) is the Confusion Matrix for 2D segmentation, (**b**) is the Confusion Matrix for re-projection images. Four categories are evaluated, which are *Building*, *Vegetation*, *Road* and *Vehicle*. It shows that the re-projection map from 3D semantic points behaves higher accuracy compared with 2D segmentation, due to considering multi-view information.

Furthermore, as shown in Table 2, the Pixel Accuracy of 3D baseline is 87.76%, and the F1-scores of *Vehicle*, *Vegetation*, and *Road* are relatively low. The refinement methods introduced in Section 3.5 are denoted as Local, Global, and Local+Global in Table 2. Local, Global, and Local+Global methods in Table 2 have been fully tested, and we put the best results under various parameters to this table. With refinement, the F1-score of Vehicle significantly rises, while Building, Vegetation, and Road also have increased scores. In addition, the Local+Global optimization approach is better than the Local or Global approach in each semantic category. It leads to the conclusion that the Local+Global approach outperforms any single Local or Global approach.

Table 2. Quantitative results of different methods for semantic categories.

Pixel Accuracy(%)					
Method	**Building**	**Vegetation**	**Vehicle**	**Road**	**All**
2D prediction	95.60	89.85	97.95	87.91	85.66
3D baseline	97.51	90.06	99.76	75.59	87.76
Local	96.20	91.38	99.74	68.61	88.24
Global	96.16	91.40	99.45	71.44	88.21
Global + Local	96.19	91.40	99.76	68.16	88.40
F1-Score(%)					
Method	**Building**	**Vegetation**	**Vehicle**	**Road**	
2D prediction	96.53	73.99	33.15	61.42	
3D baseline	97.00	74.69	63.66	75.79	
Local	97.13	74.87	62.72	75.63	
Global	97.15	74.69	73.17	75.03	
Global + Local	97.85	76.07	81.40	76.57	

5.2. Discussion

In the following part, the discussion of our semantic fusion method will be arranged in three aspects: the down-sample rate, the parameter chosen for the k-nearest neighbor algorithm, and the decision strategies between soft and hard.

5.2.1. Parameter Selection for K-Nearest Neighbors

There are two criteria for judging neighbor points. As the name k-nearest neighbors itself indicates, the maximum number of neighbors is k. Besides that, the absolute distance in 3D space should also be limited. We down-sample the point could again with a rate of 0.001 to build a small KD-tree. Then we calculate the average distance of these points, setting the value to be the threshold of absolute distance. As indicated in Figure 7, the Pixel Accuracy firstly increases with the growth of k, and reaches its peak with $k = 15$. After crossing the peak, accuracy decreases as k increases. This is because as k increases, the local method negatively optimizes for small areas such as vehicles and narrow roads.

5.2.2. Soft vs. Hard Decision Strategy

The decision strategies based on probability like Bayesian and Markov Decision are soft, while threshold and Argmax layer are hard decision strategies. There is no doubt that hard decision processes discard some information. As demonstrated in Figure 7, Prob outperforms Argmax under the same k in most circumstances. The best result of Prob is also greater than Argmax as well. It reveals that the soft decision strategy leads to better performance.

5.2.3. Down-Sample Rate

Since the dense point cloud's scale of a specific outdoor scene collected by UAV is usually around 20M or bigger, global-wise algorithms cannot handle all points at once. For instance, PKU-M1 contains 27 million points. Table 3 shows a trend that the Pixel Accuracy generally reaches its peak at the down-sample rate of 1, equivalent to which means there are no down sampling process is taken at all. Increasing of down-sample rate makes the filtered point cloud denser, which intends the neighbors of a single point to become closer. The closer points are, the more likely they belong to the same semantic class. So it is sensible that the increasing of the down-sample rate avails the final Pixel Accuracy. If the performance of a method with lower sampling rate is higher than another, it is reasonable to believe that the former method is better.

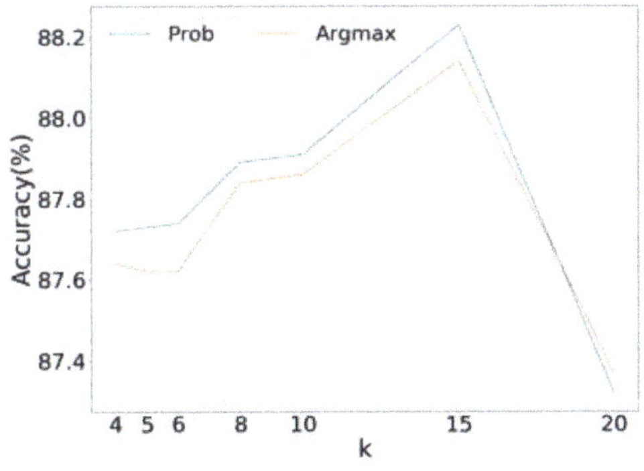

Figure 7. Ablation study on parameter selection for k-nearest neighbor and soft vs. hard decision strategy. For both Prob and Argmax methods, k = 15 is the best parameter. In most circumstances, the soft decision strategy Prob dominates hard decision strategy Argmax.

Table 3. Ablation study on Down-sample rate.

Method	k-Nearest Neighbor	Down-Sample Rate	Pixel Accuracy(%)
2D prediction	0	1	85.66
3D baseline	0	0.1	87.76
Local	15	0.1	88.14
Local	15	0.2	88.02
Local	15	0.5	88.21
Local	15	1	88.24

6. Conclusions

In this paper, we proposed a semantic 3D reconstruction method to reconstruct 3D semantic models by integrating 2D semantic labeling and 3D geometric information. In implementation, we utilize deep learning for both 2D segmentation and depth estimation. Then, the semantic 3D point cloud is obtained by our probability-based semantic fusion method. Finally, we apply the local and global approaches for point cloud refinement. Experimental results show that our semantic fusing procedure with refinement based on local and global information is able to suppress noise and reduce the re-projection error. This work paves the way for realizing finer-grained 3D segmentation and semantic classifications.

Author Contributions: conceptualization, Z.W and Y.W.; methodology, Z.W.; software, Y.W. and Z.W; validation, Y.W. and H.Y.; formal analysis, G.W. and Y.C.; investigation, Y.W and Z.W.; resources, G.W. and Y.C.; data curation, Y.W.; writing—original draft preparation, Z.W. and Y.W.; writing—review and editing, H.Y. and Y.C.; visualization, Y.W. and Z.W.; supervision, G.W.; project administration, Y.W. and H.Y.; funding acquisition, G.W. All authors have read and agreed to the published version of the manuscript.

Funding: This research was supported by The National Key Technology Research and Development Program of China, grant numbers 2017YFB1002705 and 2017YFB1002601; the National Natural Science Foundation of China (NSFC), grant numbers 61632003, 61661146002, and 61872398; and the Equipment Development Project, grant number 315050501.

Conflicts of Interest: The authors declare no conflict of interest.

References

1. Long, J.; Shelhamer, E.; Darrell, T. Fully convolutional networks for semantic segmentation. In Proceedings of the IEEE Conference on Computer Vision and Pattern Recognition, Boston, MA, USA, 7–12 June 2015; pp. 3431–3440.
2. Badrinarayanan, V.; Kendall, A.; Cipolla, R. Segnet: A deep convolutional encoder-decoder architecture for image segmentation. *IEEE Trans. Pattern Anal. Mach. Intell.* **2017**, *39*, 2481–2495. [CrossRef] [PubMed]
3. Chen, L.C.; Papandreou, G.; Kokkinos, I.; Murphy, K.; Yuille, A.L. Deeplab: Semantic image segmentation with deep convolutional nets, atrous convolution, and fully connected crfs. *IEEE Trans. Pattern Anal. Mach. Intell.* **2017**, *40*, 834–848. [CrossRef] [PubMed]
4. Zhao, H.; Shi, J.; Qi, X.; Wang, X.; Jia, J. Pyramid scene parsing network. In Proceedings of the IEEE Conference on Computer Vision and Pattern Recognition, Honolulu, HI, USA, 21–26 July 2017; pp. 2881–2890.
5. Yao, Y.; Luo, Z.; Li, S.; Fang, T.; Quan, L. Mvsnet: Depth inference for unstructured multi-view stereo. In Proceedings of the European Conference on Computer Vision (ECCV), Munich, Germany, 8–14 September 2018; pp. 767–783.
6. Yao, Y.; Luo, Z.; Li, S.; Shen, T.; Fang, T.; Quan, L. Recurrent mvsnet for high-resolution multi-view stereo depth inference. In Proceedings of the IEEE Conference on Computer Vision and Pattern Recognition, Long Beach, CA, USA, 16–20 June 2019; pp. 5525–5534.
7. Chen, Y.; Wang, Y.; Lu, P.; Chen, Y.; Wang, G. Large-scale structure from motion with semantic constraints of aerial images. In Proceedings of the Chinese Conference on Pattern Recognition and Computer Vision (PRCV), Guangzhou, China, 23–26 November 2018; pp. 347–359.
8. Bao, S.Y.; Savarese, S. Semantic structure from motion. In Proceedings of the CVPR 2011, Providence, RI, USA, 20–25 June 2011; pp. 2025–2032.

9. Martinovic, A.; Knopp, J.; Riemenschneider, H.; Van Gool, L. 3D all the way: Semantic segmentation of urban scenes from start to end in 3d. In Proceedings of the IEEE Conference on Computer Vision and Pattern Recognition, Santiago, Chile, 7–13 December 2015; pp. 4456–4465.
10. Wolf, D.; Prankl, J.; Vincze, M. Fast semantic segmentation of 3D point clouds using a dense CRF with learned parameters. In Proceedings of the 2015 IEEE International Conference on Robotics and Automation (ICRA), Seattle, WA, USA, 26–30 May 2015; pp. 4867–4873.
11. Häne, C.; Zach, C.; Cohen, A.; Pollefeys, M. Dense semantic 3d reconstruction. *IEEE Trans. Pattern Anal. Mach. Intell.* **2016**, *39*, 1730–1743. [CrossRef] [PubMed]
12. Hane, C.; Zach, C.; Cohen, A.; Angst, R.; Pollefeys, M. Joint 3D scene reconstruction and class segmentation. In Proceedings of the IEEE Conference on Computer Vision and Pattern Recognition, Portland, OR, USA, 23–28 June 2013; pp. 97–104.
13. Savinov, N.; Ladicky, L.; Hane, C.; Pollefeys, M. Discrete optimization of ray potentials for semantic 3d reconstruction. In Proceedings of the IEEE Conference on Computer Vision and Pattern Recognition, Santiago, Chile, 7–13 December 2015; pp. 5511–5518.
14. Wu, Z.; Song, S.; Khosla, A.; Yu, F.; Zhang, L.; Tang, X.; Xiao, J. 3d shapenets: A deep representation for volumetric shapes. In Proceedings of the IEEE Conference on Computer Vision and Pattern Recognition, Santiago, Chile, 7–13 December 2015; pp. 1912–1920.
15. Maturana, D.; Scherer, S. Voxnet: A 3d convolutional neural network for real-time object recognition. In Proceedings of the 2015 IEEE/RSJ International Conference on Intelligent Robots and Systems (IROS), Hamburg, Germany, 28 September–3 October 2015; pp. 922–928.
16. Qi, C.R.; Su, H.; Mo, K.; Guibas, L.J. Pointnet: Deep learning on point sets for 3d classification and segmentation. In Proceedings of the IEEE Conference on Computer Vision and Pattern Recognition, Honolulu, HI, USA, 21–26 July 2017; pp. 652–660.
17. Qi, C.R.; Yi, L.; Su, H.; Guibas, L.J. Pointnet++: Deep hierarchical feature learning on point sets in a metric space. In Proceedings of the Advances in Neural Information Processing Systems, Long Beach, CA, USA, 4–9 December 2017; pp. 5099–5108.
18. Vineet, V.; Miksik, O.; Lidegaard, M.; Nießner, M.; Golodetz, S.; Prisacariu, V.A.; Kähler, O.; Murray, D.W.; Izadi, S.; Pérez, P.; et al. Incremental dense semantic stereo fusion for large-scale semantic scene reconstruction. In Proceedings of the 2015 IEEE International Conference on Robotics and Automation (ICRA), Seattle, WA, USA, 26–30 May 2015; pp. 75–82.
19. Zhao, C.; Sun, L.; Stolkin, R. A fully end-to-end deep learning approach for real-time simultaneous 3D reconstruction and material recognition. In Proceedings of the 2017 18th International Conference on Advanced Robotics (ICAR), Hong Kong, China, 10–12 July 2017; pp. 75–82.
20. McCormac, J.; Handa, A.; Davison, A.; Leutenegger, S. Semanticfusion: Dense 3d semantic mapping with convolutional neural networks. In Proceedings of the 2017 IEEE International Conference on Robotics and Automation (ICRA), Singapore, Singapore, 29 May–3 June 2017; pp. 4628–4635.
21. Li, X.; Wang, D.; Ao, H.; Belaroussi, R.; Gruyer, D. Fast 3D Semantic Mapping in Road Scenes. *Appl. Sci.* **2019**, *9*, 631. [CrossRef]
22. Zhou, Y.; Shen, S.; Hu, Z. Fine-level semantic labeling of large-scale 3d model by active learning. In Proceedings of the 2018 International Conference on 3D Vision (3DV), Verona, Italy, 5–8 September 2018; pp. 523–532.
23. Stathopoulou, E.; Remondino, F. Semantic photogrammetry: boosting image-based 3D reconstruction with semantic labeling. *Int. Arch. Photogramm. Remote Sens. Spat. Inf. Sci.* **2019**, *42*, 2/W9. [CrossRef]
24. Zhang, R.; Li, G.; Li, M.; Wang, L. Fusion of images and point clouds for the semantic segmentation of large-scale 3D scenes based on deep learning. *ISPRS J. Photogramm. Remote Sens.* **2018**, *143*, 85–96. [CrossRef]
25. Schonberger, J.L.; Frahm, J.M. Structure-from-motion revisited. In Proceedings of the IEEE Conference on Computer Vision and Pattern Recognition, Las Vegas, NV, USA, 27–30 June 2016; pp. 4104–4113.
26. He, K.; Zhang, X.; Ren, S.; Sun, J. Deep residual learning for image recognition. In Proceedings of the IEEE Conference on Computer Vision and Pattern Recognition, Las Vegas, NV, USA, 27–30 June 2016; pp. 770–778.
27. Krizhevsky, A.; Sutskever, I.; Hinton, G.E. Imagenet classification with deep convolutional neural networks. In Proceedings of the Advances in Neural Information Processing Systems, Lake Tahoe, NV, USA, 3–8 December 2012; pp. 1097–1105.

28. Jensen, R.; Dahl, A.; Vogiatzis, G.; Tola, E.; Aanæs, H. Large scale multi-view stereopsis evaluation. In Proceedings of the IEEE Conference on Computer Vision and Pattern Recognition, Columbus, OH, USA, 23–28 June 2014; pp. 406–413.
29. Galliani, S.; Lasinger, K.; Schindler, K. Massively parallel multiview stereopsis by surface normal diffusion. In Proceedings of the IEEE International Conference on Computer Vision, Santiago, Chile, 7–13 December 2015; pp. 873–881.
30. Sedlacek, D.; Zara, J. Graph cut based point-cloud segmentation for polygonal reconstruction. In Proceedings of the International Symposium on Visual Computing, Las Vegas, NV, USA, 30 November–2 December 2009; pp. 218–227.
31. Boykov, Y.; Kolmogorov, V. An experimental comparison of min-cut/max-flow algorithms for energy minimization in vision. *IEEE Trans. Pattern Anal. Mach. Intell.* **2004**, *26*, 1124–1137. [CrossRef] [PubMed]
32. Cordts, M.; Omran, M.; Ramos, S.; Rehfeld, T.; Enzweiler, M.; Benenson, R.; Franke, U.; Roth, S.; Schiele, B. The cityscapes dataset for semantic urban scene understanding. In Proceedings of the IEEE Conference on Computer Vision and Pattern Recognition, Las Vegas, NV, USA, 27–30 June 2016; pp. 3213–3223.
33. Abadi, M.; Agarwal, A.; Barham, P.; Brevdo, E.; Chen, Z.; Citro, C.; Corrado, G.S.; Davis, A.; Dean, J.; Devin, M.; et al. TensorFlow: Large-scale machine learning on heterogeneous systems. *arXiv* **2015**, arXiv:1603.04467.
34. Bottou, L. Large-scale machine learning with stochastic gradient descent. In *Proceedings of COMPSTAT'2010*; Springer: Paris, France, 2010; pp. 177–186.

applied sciences

Article

Semantic 3D Reconstruction for Robotic Manipulators with an Eye-In-Hand Vision System

Fusheng Zha [1], Yu Fu [1], Pengfei Wang [1], Wei Guo [1], Mantian Li [1,2,*], Xin Wang [2,*] and Hegao Cai [1]

1 State Key Laboratory of Robotics and System, Harbin Institute of Technology, Harbin 150080, China; zhafusheng@hit.edu.cn (F.Z.); 6120810528@hit.edu.cn (Y.F.); wangpengfei@hit.edu.cn (P.W.); wguo01@hit.edu.cn (W.G.); zfsh751228@163.com (H.C.)

2 Shenzhen Academy of Aerospace Technology, Shenzhen 518057, China

* Correspondence: limt@hit.edu.cn (M.L.); xin.wang@chinasaat.com (X.W.)

Received: 16 December 2019; Accepted: 3 February 2020; Published: 10 February 2020

Abstract: Three-dimensional reconstruction and semantic understandings have attracted extensive attention in recent years. However, current reconstruction techniques mainly target large-scale scenes, such as an indoor environment or automatic self-driving cars. There are few studies on small-scale and high-precision scene reconstruction for manipulator operation, which plays an essential role in the decision-making and intelligent control system. In this paper, a group of images captured from an eye-in-hand vision system carried on a robotic manipulator are segmented by deep learning and geometric features and create a semantic 3D reconstruction using a map stitching method. The results demonstrate that the quality of segmented images and the precision of semantic 3D reconstruction are effectively improved by our method.

Keywords: semantic 3D reconstruction; eye-in-hand vision system; robotic manipulator

1. Introduction

In an unstructured environment, the type and shape of the objects are unpredictable. While, in order to achieve autonomous operations, the robot must be able to use visual sensors, such as lasers or cameras, to get the information about the scene [1–3]. Therefore, the robot can obtain features and identify relevant objects in the surrounding environment and then plan the motion accordingly. In the process, besides providing the location information of objects, a semantic 3D map can facilitate its decision-making based on actual world processes, such as judging the stability of the scene objects [4–6], grasping and placing objects by imitating human beings [7], and generating relevant action sequences [8–10].

Environmental information is usually collected by different sensors, such as lasers [11], a monocular camera [12], or a depth camera [13], and is then processed through a series of algorithms, such as height estimation [14,15], target detection, image segmentation, visual odometer, and image stitching to generate an environmental map, which is called simultaneous localization and mapping (SLAM) or structure from motion (SFM). The visual odometer-based method seriously affects the accuracy of the mapping due to the position error caused by the sensors. However, the eye-in-hand vision system is more accurate than the visual odometer. Therefore, it is necessary to make full use of the high accuracy of the robotic manipulator to improve the quality of the 3D reconstruction of the scene [16,17]. Another problem is that the precision of semantic segmentation is still insufficient, even by the latest method, so it is necessary to find a way to improve the quality of semantic segmentation.

Therefore, we explore to establish an integrated 3D object semantic reconstruction framework for eye-in-hand manipulators, including RGBD image segmentation, camera pose optimization, and map stitching. This enables us to achieve the following: (1) combine deep learning with geometric feature methods to perform the semantic segmentation; (2) employ the object point cloud segmentation-based

Segment Iterative Closest Point (SICP) method to optimize the camera pose and position; and (3) stitch together a semantic 3D map by data association.

In summary, the main contributions of this work are:

- The accuracy of image segmentation and the quality of object modeling are improved with an eye-in-hand manipulator through combining deep learning with geometric methods.
- A high-precision semantic 3D map is established by applying the SICP method to optimize the camera position.

The paper is organized as follows: related works and the present work are described in Sections 2 and 3, respectively. In Section 4, the experimental results are detailed and presented. The discussion and conclusion are given in Section 5.

2. Related Works

As previous 3D reconstruction using an eye-in-hand camera rarely contains semantic information and, currently, a large number of semantic 3D reconstruction is based on hand-held cameras, we discuss the following two parts: semantic 3D reconstruction based on an eye-in-hand camera and a hand-held camera.

2.1. Semantic 3D Reconstruction Based on an Eye-in-Hand Camera

Since the position of the object in the 3D space is necessary for robotic manipulators to operate objects, the eye-in-hand camera is usually applied to get this information and make 3D scene reconstruction. Fuchs et al. [18] used Time of Flight (ToF) cameras to acquire images and optimize the images through the Iterative Closest Point (ICP) algorithm. Barth et al. [19] used the LSD-SLAM method to create sparse scene maps, using object edge information to identify objects. Chang et al. [20] used a monocular eye-in-hand camera and a laser radar to obtain the point cloud of the scene and combined it with the Computer Closer Point (CCP) and ICP methods to improve the matching accuracy. The above methods can only build 3D maps without semantic information, causing them have to use all the point clouds to perform ICP matching. That induced a low calculation speed and low matching precision due to the background interference. Moreover, since there is no semantic segmentation of the scene, the object-level 3D reconstruction cannot be achieved.

2.2. Semantic 3D Reconstruction Based on a Hand-Held Camera

After years of development, 3D scene reconstruction based on vision has been relatively mature and has produced a large number of excellent algorithms [21–23]. With the improvement of target detection and image segmentation algorithms, semantic 3D scene reconstruction has become a research hotspot in recent years [24–29]. Its essence is the effective combination of semantic information with the SLAM system to generate 3D maps with semantic labels. Single Shot Detectors (SSDs) are introduced to handle geometric feather-based point cloud segmentation on the foundation of the orb-slam and processed map fusion through data association and the ICP [30]. Based on probabilistic methods, lots of previous works conduct 2D image segmentation through Random Decision Forests (RDF) and integrate 2D image labels into a 3D reconstruction map with a conditional random field and Bayesian updating model [31]. McCormac et al. [32] used the Convolutional Neural Network (CNN) to obtain the probability distribution of Classification for each 2D pixel, and then the Bayesian updating model would track the classification probability distribution of each curved surface, which would be updated based on the information regarding the data association provided by the SLAM system. In the subsequent work, they created a SLAM system with 6 degrees of freedom by merging 2D object detection with a 3D voxel foreground [33]. Bowman et al. [34] proposed a theoretical framework for the fusion of scale and semantic information, realizing the dynamic tracking of objects through ICP and RGB error and achieving the real-time object 3D reconstruction by asynchronous frame updating. Although the above works have established an environmental semantic map, the map scale is usually too large to reach a high accuracy, which limits its application in elaborate 3D modeling, such as desktop objects. The aforementioned 3D reconstruction method commonly use a hand-held camera and need a visual odometer, while in the eye-in-hand vision system in robotic manipulators, the 3D reconstruction method can be simplified through a forward kinematics analysis of robotic manipulators.

3. Overview of the Proposed Method

Our algorithm includes fusion segmentation, combining deep learning with geometric feature methods, camera pose optimization, and map stitching. The algorithm flow is shown in Figure 1. The deep learning adopts the R-50-FPN structure of mask R-CNN, and the geometric feature method adopts supervoxel segmentation and the Locally Colored Convex Connected Patches (LCCCP) clustering method with color information. The fusion segmentation uses neural network segmentation results to further cluster LCCCP segmentation mass to generate a high-precision segmented point cloud with semantic information and then apply the split point cloud of two adjacent frames for ICP matching to get the real camera position. The segmented point cloud is transformed to the world coordinate system through the current real camera position, and the data association method based on the gravity center distance is adopted to judge whether the segmented point cloud is a false recognition. If there is no false recognition, the segmented point cloud is spliced in the map. A 3D model reconstruction of each object is realized by splicing the point cloud at different positions from multiple angles.

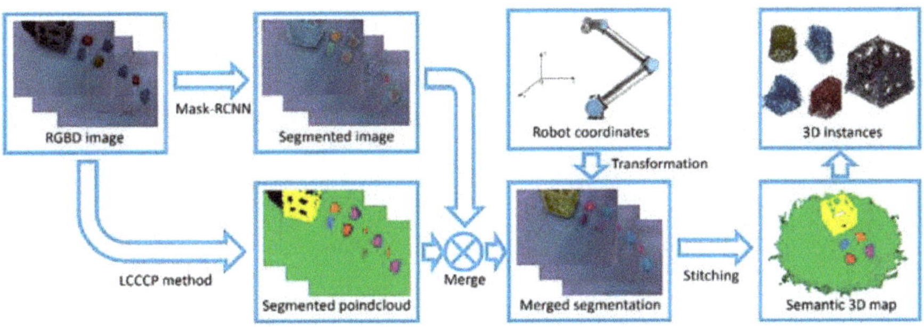

Figure 1. Overview of our method. This process is mainly divided into two parts: image segmentation and map stitching.

3.1. *Object Recognition and Fusion Segmentation*

The semantic segmentation algorithm is the basis of map stitching. Pictures and point clouds are segmented by neural networks and geometric features, respectively, and finally the two parts are fused together to generate semantic information. Therefore, this algorithm includes three parts: 2D semantic segmentation, point cloud segmentation, and semantic fusion.

3.1.1. Target Detection and Instance Segmentation Based on 2D Images

Among numerous methods for object detection and instance segmentation based on 2D images, see, e.g., [35–39], mask R-CNN is one of the most pragmatic instance segmentation frameworks at present, which can effectively detect objects in images and simultaneously generate a high-quality segmentation mask for each instance. Based on previous classification and regression branches in Faster-CNN, it adds another branch, which segment and output each region of interest (ROI) to achieve semantic segmentation [40]. The object recognition and 2D image segmentation in our work are constructed according to mask R-CNN framework.

3.1.2. Point Cloud Segmentation Based on the Geometric Feature Method

Although mask R-CNN has a relatively high recognition accuracy, the image segmentation accuracy is still insufficient, so it is difficult to achieve high-precision 3D reconstruction by merely adopting 2D image segmentation. In order to improve the accuracy of segmentation, we also take advantage of the 3D point cloud segmentation method. Firstly, the point clouds have been decomposing into many small patches by way of supervoxel segmentation to implement over-segmentation and then perform clustering analysis using the locally convex connected patches (LCCP) method [41].

The aforementioned LCCP method merely utilizes position and normal as not relying on the point cloud color. Suppose $\vec{p_i}$ and $\vec{p_j}$ represent two adjacent supervoxels, $couv(\vec{p_i}, \vec{p_j})$ represents whether the connection between two supervoxels is convex. Extended Convexity Criterion and Sanity Criterion can be expressed with $\mathrm{CC}_e\left(\vec{p_i}, \vec{p_j}\right)$ and $SC(\vec{p_i}, \vec{p_j})$, respectively [41].

Since the conventional LCCP method is not able to recognize two objects when the surface of different objects is tangential, it is necessary to differentiate objects by means of color information. In consideration of this problem, we improve the LCCP method by adding a parameter named the Point Color Criterion (PCC). We define γ as the maximum value of color-difference between two adjacent supervoxels, that is:

$$\gamma(\vec{p_i}, \vec{p_j}) = \max\left(\left|R_{\vec{p_i}} - R_{\vec{p_j}}\right|, \left|G_{\vec{p_i}} - G_{\vec{p_j}}\right|, \left|B_{\vec{p_i}} - B_{\vec{p_j}}\right|\right) \tag{1}$$

where $\gamma(\vec{p_i}, \vec{p_j})$ is larger than the threshold value γ_{thresh}, the two supervoxels are recognized as two different objects. γ_{thresh} is an important parameter, which depends on the color difference between the objects. It is generally set to be a small value. Therefore, even if the color differences between the objects are small, the algorithm can also distinguish between them. However, too small a γ_{thresh} will cause over-segmentation. The color criterion of point cloud can be defined as:

$$\mathrm{PCC}(\vec{p_i}, \vec{p_j}) := \begin{cases} \text{true} & \gamma(\vec{p_i}, \vec{p_j}) < \gamma_{\text{thresh}} \\ \text{false} & \text{otherwise} \end{cases} \tag{2}$$

As a result, the LCCCP method is judged by the criteria:

$$\mathrm{conv}(\vec{p_i}, \vec{p_j}) = \mathrm{CC}_e(\vec{p_i}, \vec{p_j}) \wedge SC(\vec{p_i}, \vec{p_j}) \wedge \mathrm{PCC}(\vec{p_i}, \vec{p_j}) \tag{3}$$

3.1.3. Fusion Segmentation

As described above, the 2D image segmentation method relying on the neural network can segment multiple objects simultaneously with poor accuracy, while the geometric feature segmentation method is characterized by high edge accuracy but a tendency towards over-segmentation and a lack of semantic information in the segmented block. So, it is indispensable to combine the two methods to achieve a high-precision semantic instance segmentation. Assuming that 50% of the segmented patches generated by the LCCCP method are in the segmented image produced by mask R-CNN, the segmented block is marked as the object. Count all the segmented patches belonging to the object and merge them into the point cloud P_0^c of the current frame object in the camera coordinate system.

3.2. Camera Pose Optimization

Due to the motion error of the manipulator, the position of the eye-in-hand camera will deviate from the target position. If the point cloud of the current frame is directly spliced into the map, it will lead to point cloud model misalignment, so the registration method is necessary to be employed to optimize the camera pose and the SICP method is applied to calculate the camera pose deviation.

Supposing that the point cloud of the current frame in the camera coordinate system is P_0^c, the point cloud in the world coordinate system is P^w, the transform matrix at the end of the manipulator of the current frame relative to the world coordinate system is $T_{w,t}$, the transform matrix of the camera relative to the end of manipulator is $T_{t,c}$, the transform matrix of the current frame in the world coordinate system is $T_{w,c} = T_{w,t}T_{t,c}$. After being transformed by $T_{w,c}$, the current frame object point cloud matches with the map point cloud by the SICP algorithm P^w to obtain the optimization transformation T_{ICP}, and the point cloud P_0^w of current frame object after compensation in the world coordinate system is:

$$P_0^w = T_{ICP}T_{w,t}T_{t,c}P_0^c \tag{4}$$

3.3. Data Association and Map Stitching

After transforming the point cloud of the current frame to the world coordinate system, it is essential to judge whether the transformed point cloud label is correct. Based on the previously reported method, the point cloud of the instance object in the world coordinate system is P_0^w. Assuming that there are m objects of the same category in the current map, we calculate the point cloud gravity center C_0 of each object point cloud $\left\{P_1^w, \cdots P_m^w\right\}$. The object point cloud P_t^w is:

$$P_t^w = arg\min_p ||C_i - C_0|| \tag{5}$$

Using this, we are able to calculate the Euclidean distance between all point pairs, which is from the current object point cloud P_0^w to the target object point cloud P_t^w. The value of ζ depends on the similarity between two sequential images. Parameter ζ usually takes a small value. The algorithm can identify semantic errors and avoid wrong splicing. However, we cannot set ζ too low, because when the similarity between two sequential images is poor, many segmentation results will be discarded. If more than 50% of the distance between point pairs is less than ζ, then the matching is considered successful, otherwise it is classified as a misidentification. Generally, this process takes $\zeta = 2$ mm. After successful matching, the object point cloud is merged into the point cloud map with voxel filtering.

4. Experimental Results

To verify the precision and reliability of our algorithm, we completed a series of experiments on image segmentation and 3D reconstruction by a robotic manipulator with the eye-in-hand vision system. Each experiment has been repeated 10 times.

4.1. Experimental Conditions

We assembled a RealSense D435 camera at the end of UR10 robotic manipulator to take photos at 400 mm away from the desktop with resolution at 640 × 480 pixels. We controlled the robotic manipulator with an eye-in-hand camera system to take 16 pictures every 360 degrees around the object.

We validate our algorithm by employing two different datasets. Our dataset contains five types of toys, namely cylinder, half-cylinder, L-block, cuboid, and box, as shown in Figure 2a. The dataset has a total of 1200 images shot at different angles. The other dataset comes from the Yale-CMU-Berkeley (YCB) Benchmarks [42], which contain objects of different sizes and shapes in daily life. We chose lemon, pear, orange, plum, bowl, and mug for a total of six objects, as shown in Figure 2b.

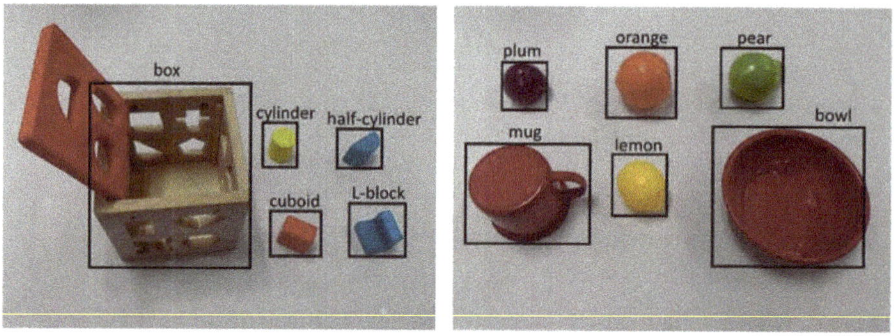

(a) (b)

Figure 2. Datasets in this paper. (**a**) The objects in our dataset, including the cylinder, half-cylinder, L-block, cuboid, and box, and (**b**) the objects in the Yale-CMU-Berkeley (YCB) Benchmarks, including lemon, pear, orange, plum, bowl, and mug.

4.2. Image Segmentation Results

The mask R-CNN adopts an R-50-FPN structure and is trained by 1200 manually labeled images with 5 types of objecting in the training set. The images are processed with instance segmentation according to the above method, and the segmentation result, which is shown in Figure 3, is compared with the mask R-CNN method. Figure 3a is the qualitative segmentation result of mask R-CNN. Each color represents one type of object. The edge of the segmented image is far from the edge of the actual object, and a hole may be generated in the segmentation area. Figure 3b is the segmentation result of our method. Because the geometric features at the edge of the object change drastically, while the geometric features of the object are stable, the image segmentation method based on geometric features makes the segmentation on the edge of the object more delicate with the segmentation edge closer to the real value and the segmentation region more complete.

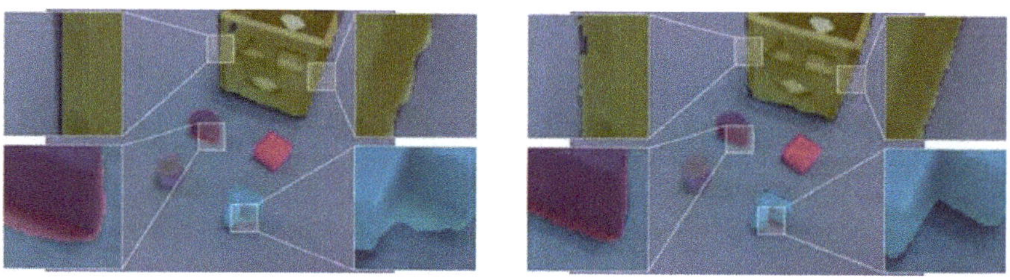

(a) (b)

Figure 3. Comparison of the segmented image. (**a**) The segmentation results of mask R-CNN, and (**b**) the segmentation results of our method.

The quantitative comparison criteria is referred to in [43]. The Intersection-over-Union (IoU) is a standard metric used to evaluate the accuracy of image segmentation, which calculates the ratio of the intersection and union between the true value and the predicted segmentation. For each type of object, we respectively calculate the results of true positives (TP), false positive (FP), and false negative (FN), and then acquire the IoU of each object using the following formula:

$$\mathrm{IoU} = \frac{TP}{TP + FP + FN} \tag{6}$$

The experimental results of our dataset and YCB Benchmarks are shown in Tables 1 and 2, respectively. Since the mask R-CNN method is not sensitive to image boundaries, the geometric method can clearly discriminate image boundaries, so we combine the deep learning and geometric feature methods to merge and segment. Since it can compensate for the edge and internal defects of mask R-CNN, our method is more accurate than the Mask R-CNN method. The MIoU (Mean Intersection over Union) increased by 2.18% and 5.70% on our dataset and YCB Benchmarks, respectively. Whether the object is large or small, square or round, our method performed better than the Mask R-CNN method in all results, which proved that our algorithm can be suitable for a variety of objects. In extreme cases, like lemon and orange, our method did not perform as good as usual. This is mainly caused by the bad quality of the point cloud. The precision of our method is influenced by the quality of point clouds. When an RGBD camera shoots spherical object, the point clouds of the edges are distorted, which has great effects on the image segmentation. Even so, our method is still more accurate than the Mask R-CNN method. Thus, the applicability and accuracy of our method is better than the Mask R-CNN method. The performance of the two above methods on the two datasets is quite different, because the background on the YCB Benchmarks is not exactly the same as our background, and each image in the datasets contains only one object, while our captured image contains several objects. The Mask R-CNN method preformed quite good in our dataset, it is difficult to improve the precision of segmentation. However, while the Mask R-CNN method preformed poor in the YCB dataset, our method made a greater improvement.

Table 1. Our dataset results on instance segmentation Intersection-over-Union (IoU) (%).

Method	Mean	Cylinder	Half-Cylinder	L-Block	Cuboid	Box
Mask R-CNN [40]	90.782	92.128	90.168	92.788	86.498	92.327
Our method	92.958	92.174	91.330	92.991	92.443	95.852

Table 2. YCB dataset results on instance segmentation IoU (%).

Method	Mean	Lemon	Pear	Orange	Plum	Bowl	Mug
Mask-RCNN [40]	85.221	84.364	82.427	85.868	81.811	87.500	89.356
Our method	90.919	88.194	88.782	91.486	91.440	92.525	93.085

4.3. Three-Dimensional Reconstruction Results

In order to prove the accuracy of the algorithm, we tested the following four methods:

1. Mask-only: mask R-CNN for image segmentation and the forward kinematics for camera position calculation;
2. Mask+ICP: mask R-CNN for image segmentation, the forward kinematics, and ICP registration for camera position calculation;
3. SLIC+ICP: Simple Linear Iterative Cluster (SLIC), the forward kinematics, and ICP registration for camera position calculation; and
4. Our method: mask R-CNN is combined with the LCCCP method for image segmentation, the forward kinematics, and SICP registration for camera position calculation.

After building the 3D model of five types of objects with these methods, we made the ICP match with the ground that was true of the object to calculate the object reconstruction accuracy and the Cloud to Cloud (C2C) absolute distance. The results are shown in Figure 4. Figure 4a shows the original image, and Figure 4b–e represent the 3D reconstruction results of mask-only, mask + ICP, SLIC + ICP, and our method, respectively, with different color points representing different types of objects. Table 3 shows the Cloud to Cloud (C2C) absolute distance between object models and 3D reconstruction by four methods. The higher value of C2C absolute distance means the lower precision of the 3D reconstruction. The comparison results show that as the camera position is inaccurate due to the robotic manipulator motion error, the mask-only method has the lowest modeling accuracy, and the image of each frame does not overlap well. Since the mask + ICP and the SLIC + ICP method optimizes the camera position, the image coincides well, and the model accuracy is greatly improved compared to the non-optimization method. Our method improves the accuracy of 3D reconstruction based on the mask + ICP method because it improves the segmentation quality of each frame of image.

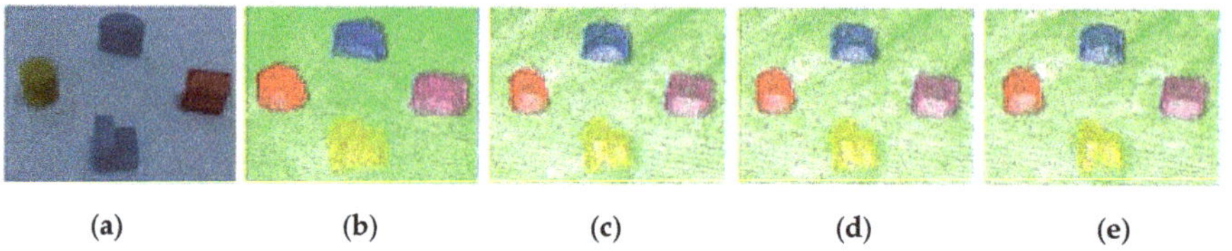

Figure 4. Comparison of the semantic map with four methods. (**a**) The original image, (**b**) the reconstruction results of the mask-only method, (**c**) the reconstruction results of the mask + ICP method, (**d**) the reconstruction results of the SLIC + ICP method, and (**e**) the reconstruction results of our method.

Table 3. Cloud to Cloud (C2C) absolute distances between our dataset models and 3D reconstruction (mm).

Method	Cylinder	Half-Cylinder	L-Block	Cuboid	Box
Mask-only [40]	4.786	5.851	5.622	4.806	3.442
Mask + ICP [34]	3.534	4.597	4.535	3.250	3.083
SLIC + ICP [44]	3.504	4.250	4.380	3.262	3.012
Our method	3.449	4.074	4.142	2.992	2.973

We evaluated the accuracy of 3D reconstruction by the method introduced in [43]. The 3D reconstruction accuracy of the four methods on our dataset is shown in Figure 5. Due to the poor quality of the image segmentation boundary of mask R-CNN, the reconstructed model has a large number of misidentification points. The motion error of the robotic manipulator and the camera calibration error result in an inaccurate position of the camera in the world coordinate system, so, when directly using the mask R-CNN method, the 3D modeling accuracy is low, with only 28.18% of the points in 1 mm distance to the model. Since the mask + ICP method optimizes the camera position, the 3D modeling accuracy is improved compared to the non-optimization method, but as the quality of image segmentation is still poor, only 48.23% of the points are within 1 mm of the model. Our method employs fusion segmentation to improve the quality of image segmentation and also uses the segmentation SICP method to finely correct the image, so it has the highest modeling accuracy among the four methods, and the average precision reaches 53.16%, which is improved by 25.49% over the mask-only method, 4.93% over the mask + ICP method, and 3.50% over the SLIC+ICP method.

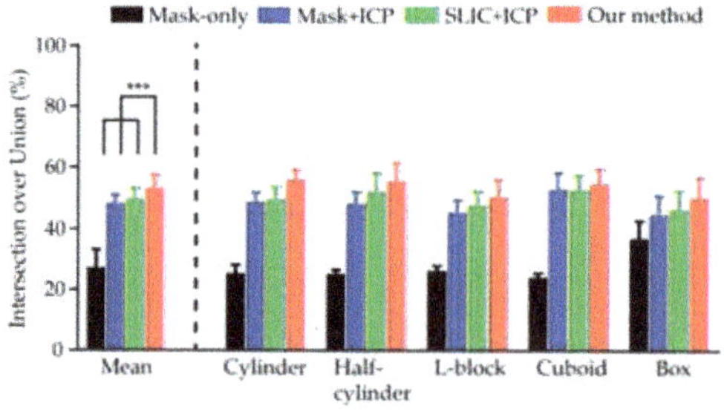

Figure 5. Our dataset results on 3D reconstruction IoU (%). Our method significantly improved the 3D reconstruction IoU compared with the mask-only method, the mask + ICP method, and the SLIC + ICP method by two-way Analysis of Variance (ANOVA) repeat measures with Tukey's multiple comparison test (*** $p < 0.001$).

Similarly, we validate our algorithm on the YCB Benchmarks. The C2C absolute distances between YCB models and 3D reconstruction by four methods are shown in Table 4. The C2C absolute distance can be used to evaluate the similarity between 3D reconstruction results and object models. The lower the value, the higher the accuracy of 3D reconstruction and the more significant the similarity is between the object models. The results in Table 4 indicate that the C2C absolute distances of each object decrease successively in the four methods of mask-only, mask + ICP, SLIC + ICP, and our method, which suggests that the 3D reconstruction results of our method are closer to the object models with highest accuracy. Compared with the other three methods, our method improves the accuracy of image segmentation and reduces the number of outlier points, so the 3D reconstruction results of our method is more accurate. The YCB dataset results on 3D reconstruction are shown in Figure 6. Since the point cloud model of the YCB Benchmarks is obtained by the depth camera in multi-angle shooting, it is closer to the actual situation than the point cloud model generated by Computer Aided Design (CAD), so the four methods perform better on the YCB Benchmarks. As shown in Figure 6, the average accuracy of our method on the YCB Benchmarks is 13.65% over the mask-only method, 4.01% over the mask + ICP method, and 3.27% over the SLIC + ICP method.

Table 4. C2C absolute distances between YCB models and 3D reconstruction (mm).

Method	Lemon	Pear	Orange	Plum	Bowl	Mug
Mask-only [40]	3.727	8.717	4.802	4.918	4.079	3.601
Mask + ICP [34]	3.646	6.244	4.240	4.574	3.341	3.131
SLIC + ICP [44]	3.121	5.849	4.106	4.421	3.251	3.110
Our method	2.593	5.406	3.586	3.693	3.067	2.968

We counted the average CPU time of each methods, as shown in Table 5. Due to the missing geometric feature segmentation and camera pose optimization, the mask-only method ran fastest with lowest precision. Without geometric feature segmentation, the mask + ICP method saved time in segmentation, but the precision of the 3D reconstruction was still poor. The SLIC + ICP method balanced performance and CPU time. Our method took a little more time in segmentation than the SLIC + ICP method, but we saved much more time in the 3D mapping. Because we utilized the SICP method to remove unrelated objects and accelerate point clouds matching.

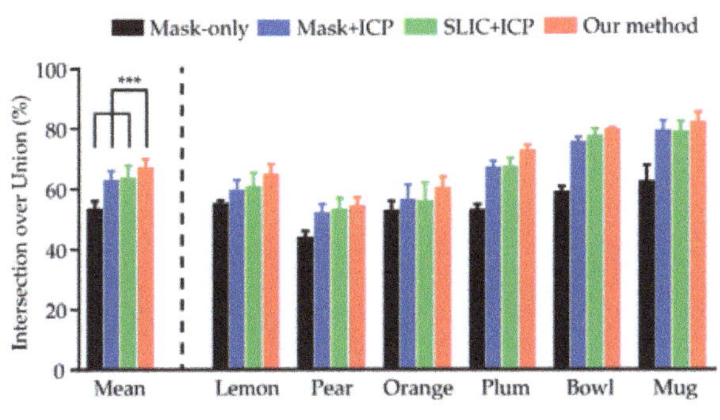

Figure 6. YCB dataset results on 3D reconstruction IoU (%). Our method significantly improved the 3D reconstruction IoU compared with the mask-only method, the mask + ICP method, and the SLIC + ICP method by two-way ANOVA repeat measures with Tukey's multiple comparison test (*** $p < 0.001$).

Table 5. Average CPU Time of each method (ms).

Step	Mask-Only	Mask + ICP	SLIC + ICP	Our Method
Segmentation	81	81	784	802
3D Mapping	12	330	334	271

5. Discussion and Conclusions

This paper proposes an algorithm framework for semantic 3D reconstruction using a robotic manipulator with an eye-in-hand camera. Unlike SLAM, SFM, and other multi-angle modeling methods, our approach adds semantic information into the 3D reconstruction process. We have improved the precision of image segmentation by combining deep learning and geometric feature analysis, and we have increased the accuracy of the 3D reconstruction model through the SICP algorithm to optimize camera pose. The semantic information plays two important roles in 3D reconstruction, one of which is providing the foundation for voxel block merging in image segmentation works, and the other is to remove background during the point cloud matching process and improve the accuracy of the ICP algorithm.

We evaluated the four methods on the YCB Benchmarks and the dataset created by ourselves. The experimental results show that, compared with the deep learning methods, our algorithm is more accurate in the edge segmentation of objects, leading to an improvement of 3D reconstruction. Moreover, the accuracy of the 3D reconstruction of objects is remarkably improved due to the removal of the background interference. Compared with the mask-only, mask + ICP, and SLIC + ICP methods, our method improved the accuracy of the 3D reconstruction on the YCB Benchmarks by 13.65%, 4.01%, and 3.27%, respectively. The same trend was showed on our dataset, with the increasing of the accuracy by 25.49%, 4.93%, and 3.50%, respectively.

In the future, we will apply this method to more scenarios and objects. Based on semantic 3D reconstruction, we will use the object point cloud model to analyze the spatial topological relationship between objects to obtain the decision of the corresponding capture strategy and then make the autonomous robot planning perform a variety of tasks in a semantic map.

Author Contributions: Conceptualization, F.Z. and Y.F. methodology, F.Z. software, Y.F. validation, Y.F., P.W. and W.G. formal analysis, Y.F. writing-original draft preparation, F.Z. writing-review and editing, Y.F.; visualization, Y.F.; supervision, M.L. and H.C. project administration, M.L. funding acquisition, M.L. and X.W. All authors have read and agreed to the published version of the manuscript.

Funding: This research was funded by [Natural Science Foundation of China] grant number [61773139], [Shenzhen Science and Technology Program] grant number [KQTD2016112515134654], and [Shenzhen Special Fund for Future Industrial Development] grant number [JCYJ20160425150757025].

Acknowledgments: This work was supported by Natural Science Foundation of China (No.61773139), Shenzhen Science and Technology Program (No.KQTD2016112515134654) and Shenzhen Special Fund for Future Industrial Development (No.JCYJ20160425150757025).

Conflicts of Interest: The funders had no role in the design of the study; in the collection, analyses, or interpretation of data; in the writing of the manuscript, or in the decision to publish the results.

References

1. Foumani, M.; Razeghi, A.; Smith-Miles, K. Stochastic optimization of two-Machine flow shop robotic cells with controllable inspection times: From theory toward practice. *Robot. Comput. Integr. Manuf.* **2020**, *61*, 101822. [CrossRef]
2. Foumani, M.; Smith-Miles, K.; Gunawan, I. Scheduling of two-Machine robotic rework cells: In-Process, post-Process and in-Line inspection scenarios. *Robot. Auton. Syst.* **2017**, *91*, 210–225. [CrossRef]
3. Foumani, M.; Smith-Miles, K.; Gunawan, I.; Moeini, A. A framework for stochastic scheduling of two-Machine robotic rework cells with in-Process inspection system. *Comput. Ind. Eng.* **2017**, *112*, 492–502. [CrossRef]
4. Jia, Z.; Gallagher, A.; Saxena, A.; Chen, T. 3d-Based reasoning with blocks, support, and stability. In Proceedings of the 2013 IEEE Conference on Computer Vision and Pattern Recognition (CVPR), Portland, OR, USA, 23–28 June 2013; pp. 1–8.
5. Jia, Z.; Gallagher, A.C.; Saxena, A.; Chen, T. 3d reasoning from blocks to stability. *IEEE Trans. Pattern Anal. Mach. Intell.* **2014**, *37*, 905–918. [CrossRef] [PubMed]
6. Zheng, B.; Zhao, Y.; Yu, J.; Ikeuchi, K.; Zhu, S.-C. Scene understanding by reasoning stability and safety. *Int. J. Comput. Vis.* **2015**, *112*, 221–238. [CrossRef]

7. Tremblay, J.; To, T.; Molchanov, A.; Tyree, S.; Kautz, J.; Birchfield, S. Synthetically trained neural networks for learning human-Readable plans from real-World demonstrations. In Proceedings of the 2018 IEEE International Conference on Robotics and Automation (ICRA), Brisbane, QLD, Australia, 21–25 May 2018; pp. 1–5.
8. Blodow, N.; Goron, L.C.; Marton, Z.-C.; Pangercic, D.; Rühr, T.; Tenorth, M.; Beetz, M. Autonomous semantic mapping for robots performing everyday manipulation tasks in kitchen environments. In Proceedings of the 2011 IEEE/RSJ International Conference on Intelligent Robots and Systems, San Francisco, CA, USA, 25–30 September 2011; pp. 4263–4270.
9. Galindo, C.; Fernández-Madrigal, J.-A.; González, J.; Saffiotti, A. Robot task planning using semantic maps. *Robot. Auton. Syst.* **2008**, *56*, 955–966. [CrossRef]
10. Silberman, N.; Hoiem, D.; Kohli, P.; Fergus, R. Indoor segmentation and support inference from rgbd images. In Proceedings of the 2012 European Conference on Computer Vision (ECCV), Florence, Italy, 7–13 October 2012; pp. 746–760.
11. Zhang, J.; Singh, S. Visual-Lidar odometry and mapping: Low-Drift, robust, and fast. In Proceedings of the 2015 IEEE International Conference on Robotics and Automation (ICRA), Seattle, WA, USA, 26–30 May 2015; pp. 2174–2181.
12. Paxton, C.; Barnoy, Y.; Katyal, K.; Arora, R.; Hager, G.D. Visual robot task planning. In Proceedings of the 2019 International Conference on Robotics and Automation (ICRA), Montreal, QC, Canada, 20–24 May 2019; pp. 8832–8838.
13. Endres, F.; Hess, J.; Sturm, J.; Cremers, D.; Burgard, W. 3-D mapping with an RGB-D camera. *IEEE Trans. Robot.* **2013**, *30*, 177–187. [CrossRef]
14. Smith, W.A.; Ramamoorthi, R.; Tozza, S. Height-From-Polarisation with unknown lighting or albedo. *IEEE Trans. Pattern Anal. Mach. Intell.* **2018**, *41*, 2875–2888. [CrossRef] [PubMed]
15. Tozza, S.; Smith, W.A.; Zhu, D.; Ramamoorthi, R.; Hancock, E.R. Linear differential constraints for photo-Polarimetric height estimation. In Proceedings of the 2017 IEEE International Conference on Computer Vision (ICCV), Venice, Italy, 22–29 October 2017; pp. 2279–2287.
16. Walck, G.; Drouin, M. Progressive 3D reconstruction of unknown objects using one eye-In-Hand camera. In Proceedings of the 2009 IEEE International Conference on Robotics and Biomimetics (ROBIO), Guilin, China, 19–23 December 2009; pp. 971–976.
17. Tozza, S.; Falcone, M. Analysis and approximation of some shape-From-Shading models for non-Lambertian surfaces. *J. Math. Imaging Vis.* **2016**, *55*, 153–178. [CrossRef]
18. Fuchs, S.; May, S. Calibration and registration for precise surface reconstruction with Time-Of-Flight cameras. *Int. J. Intell. Syst. Technol. Appl.* **2008**, *5*, 274–284. [CrossRef]
19. Barth, R.; Hemming, J.; van Henten, E.J. Design of an eye-In-Hand sensing and servo control framework for harvesting robotics in dense vegetation. *Biosyst. Eng.* **2016**, *146*, 71–84. [CrossRef]
20. Chang, W.-C.; Wu, C.-H. Eye-In-Hand vision-Based robotic bin-Picking with active laser projection. *Int. J. Adv. Manuf. Technol.* **2016**, *85*, 2873–2885. [CrossRef]
21. Bescos, B.; Fácil, J.M.; Civera, J.; Neira, J. DynaSLAM: Tracking, mapping, and inpainting in dynamic scenes. *IEEE Robot. Autom. Lett.* **2018**, *3*, 4076–4083. [CrossRef]
22. Brachmann, E.; Rother, C. Learning less is more-6d camera localization via 3d surface regression. In Proceedings of the 2018 IEEE/CVF Conference on Computer Vision and Pattern Recognition (CVPR), Salt Lake City, UT, USA, 18–23 June 2018; pp. 4654–4662.
23. Mur-Artal, R.; Tardós, J.D. Orb-slam2: An open-Source slam system for monocular, stereo, and rgb-d cameras. *IEEE Trans. Robot.* **2017**, *33*, 1255–1262. [CrossRef]
24. Fan, H.; Su, H.; Guibas, L.J. A point set generation network for 3d object reconstruction from a single image. In Proceedings of the 2017 IEEE Conference on Computer Vision and Pattern Recognition (CVPR), Honolulu, HI, USA, 21–26 July 2017; pp. 605–613.

25. Fehr, M.; Furrer, F.; Dryanovski, I.; Sturm, J.; Gilitschenski, I.; Siegwart, R.; Cadena, C. TSDF-Based change detection for consistent long-Term dense reconstruction and dynamic object discovery. In Proceedings of the 2017 IEEE International Conference on Robotics and Automation (ICRA), Singapore, 29 May–3 June 2017; pp. 5237–5244.
26. Karpathy, A.; Miller, S.; Fei-Fei, L. Object discovery in 3d scenes via shape analysis. In Proceedings of the 2013 IEEE International Conference on Robotics and Automation (ICRA), Karlsruhe, Germany, 6–10 May 2013; pp. 2088–2095.
27. Koppula, H.S.; Anand, A.; Joachims, T.; Saxena, A. Semantic labeling of 3d point clouds for indoor scenes. In Proceedings of the 2011 Advances in Neural Information Processing Systems (NIPS), Granada, Spain, 12–14 December 2011; pp. 244–252.
28. Salas-Moreno, R.F.; Newcombe, R.A.; Strasdat, H.; Kelly, P.H.; Davison, A.J. Slam++: Simultaneous localisation and mapping at the level of objects. In Proceedings of the 2013 IEEE Conference on Computer Vision and Pattern Recognition (CVPR), Portland, OR, USA, 23–28 June 2013; pp. 1352–1359.
29. Tateno, K.; Tombari, F.; Navab, N. Real-Time and scalable incremental segmentation on dense slam. In Proceedings of the 2015 IEEE/RSJ International Conference on Intelligent Robots and Systems (IROS), Hamburg, Germany, 28 September–2 October 2015; pp. 4465–4472.
30. Sünderhauf, N.; Pham, T.T.; Latif, Y.; Milford, M.; Reid, I. Meaningful maps with object-Oriented semantic mapping. In Proceedings of the 2017 IEEE/RSJ International Conference on Intelligent Robots and Systems (IROS), Vancouver, BC, Canada, 24–28 September 2017; pp. 5079–5085.
31. Hermans, A.; Floros, G.; Leibe, B. Dense 3d semantic mapping of indoor scenes from rgb-d images. In Proceedings of the 2014 IEEE International Conference on Robotics and Automation (ICRA), Hong Kong, China, 31 May–7 June 2014; pp. 2631–2638.
32. McCormac, J.; Handa, A.; Davison, A.; Leutenegger, S. Semanticfusion: Dense 3d semantic mapping with convolutional neural networks. In Proceedings of the 2017 IEEE International Conference on Robotics and automation (ICRA), Singapore, 29 May–3 June 2017; pp. 4628–4635.
33. McCormac, J.; Clark, R.; Bloesch, M.; Davison, A.; Leutenegger, S. Fusion++: Volumetric object-level slam. In Proceedings of the 2018 International Conference on 3D Vision (3DV), Verona, Italy, 5–8 September 2018; pp. 32–41.
34. Bowman, S.L.; Atanasov, N.; Daniilidis, K.; Pappas, G.J. Probabilistic data association for semantic slam. In Proceedings of the 2017 IEEE International Conference on Robotics and Automation (ICRA), Singapore, 29 May–3 June 2017; pp. 1722–1729.
35. Girshick, R. Fast r-cnn. In Proceedings of the 2015 IEEE international conference on computer vision (ICCV), Santiago, Chile, 7–13 December 2015; pp. 1440–1448.
36. Girshick, R.; Donahue, J.; Darrell, T.; Malik, J. Rich feature hierarchies for accurate object detection and semantic segmentation. In Proceedings of the 2014 IEEE Conference on Computer Vision and Pattern Recognition (CVPR), Columbus, OH, USA, 23–28 June 2014; pp. 580–587.
37. He, K.; Zhang, X.; Ren, S.; Sun, J. Spatial pyramid pooling in deep convolutional networks for visual recognition. *IEEE Trans. Pattern Anal. Mach. Intell.* **2015**, *37*, 1904–1916. [CrossRef] [PubMed]
38. Li, Y.; Qi, H.; Dai, J.; Ji, X.; Wei, Y. Fully convolutional instance-aware semantic segmentation. In Proceedings of the 2017 IEEE Conference on Computer Vision and Pattern Recognition (CVPR), Honolulu, HI, USA, 21–26 July 2017; pp. 2359–2367.
39. Lin, G.; Milan, A.; Shen, C.; Reid, I. Refinenet: Multi-Path refinement networks for high-Resolution semantic segmentation. In Proceedings of the 2017 IEEE Conference on Computer Vision and Pattern Recognition (CVPR), Honolulu, HI, USA, 21–26 July 2017; pp. 1925–1934.
40. He, K.; Gkioxari, G.; Dollár, P.; Girshick, R. Mask r-cnn. In Proceedings of the 2017 IEEE International Conference on Computer Vision (ICCV), Venice, Italy, 22–29 October 2017; pp. 2961–2969.
41. Christoph Stein, S.; Schoeler, M.; Papon, J.; Worgotter, F. Object partitioning using local convexity. In Proceedings of the 2014 IEEE Conference on Computer Vision and Pattern Recognition (CVPR), Columbus, OH, USA, 23–28 June 2014; pp. 304–311.
42. Calli, B.; Walsman, A.; Singh, A.; Srinivasa, S.; Abbeel, P.; Dollar, A.M. Benchmarking in manipulation research: Using the Yale-CMU-Berkeley object and model set. *IEEE Robot. Autom. Mag.* **2015**, *22*, 36–52. [CrossRef]

43. Garcia-Garcia, A.; Orts-Escolano, S.; Oprea, S.; Villena-Martinez, V.; Garcia-Rodriguez, J. A review on deep learning techniques applied to semantic segmentation. *arXiv* **2017**, arXiv:1704.06857.
44. Runz, M.; Buffier, M.; Agapito, L. Maskfusion: Real-Time recognition, tracking and reconstruction of multiple moving objects. In Proceedings of the 2018 IEEE International Symposium on Mixed and Augmented Reality (ISMAR), Munich, Germany, 16–20 October 2018; pp. 10–20.

applied sciences

MDPI

Article

3D Face Model Super-Resolution Based on Radial Curve Estimation

Fan Zhang [1], Junli Zhao [2], Liang Wang [3] and Fuqing Duan [1,4,*]

1 College of Artificial Intelligence, Beijing Normal University, Beijing 100875, China; fzhang@mail.bnu.edu.cn
2 School of Data Science and Software Engineering, Qingdao University, Qingdao 266071, China; zhaojl@yeah.net
3 Faculty of Information Technology, Beijing University of Technology, Beijing 100124, China; wangliang@bjut.edu.cn
4 Engineering Research Center of Virtual Reality and Applications, Ministry of Education, Beijing 100875, China
* Correspondence: fqduan@bnu.edu.cn

Received: 9 December 2019; Accepted: 31 January 2020; Published: 5 February 2020

Abstract: Consumer depth cameras bring about cheap and fast acquisition of 3D models. However, the precision and resolution of these consumer depth cameras cannot satisfy the requirements of some 3D face applications. In this paper, we present a super-resolution method for reconstructing a high resolution 3D face model from a low resolution 3D face model acquired from a consumer depth camera. We used a group of radial curves to represent a 3D face. For a given low resolution 3D face model, we first extracted radial curves on it, and then estimated their corresponding high resolution ones by radial curve matching, for which Dynamic Time Warping (DTW) was used. Finally, a reference high resolution 3D face model was deformed to generate a high resolution face model by using the radial curves as the constraining feature. We evaluated our method both qualitatively and quantitatively, and the experimental results validated our method.

Keywords: 3D face model; super-resolution; radial curve; Dynamic Time Warping

1. Introduction

In recent years, 3D face modeling has received extensive attention due to its widespread applications in face recognition, animation and 3D video games. The usual way to obtain 3D face models is 3D scanning by some high resolution 3D scanners, such as Artec Eva and Minolta Vivid. However, these professional 3D scanners are expensive and have a high computational cost. For this reason, some consumer depth cameras, such as Microsoft Kinect and Intel RealSense, have drawn wide attention because of their low cost and easy integration. A depth camera is able to acquire the depth information of objects in a scene; i.e., the distances between the camera and the surfaces of objects. Depth information can be transformed into 3D information; i.e., corresponding 3D models of objects can be constructed from depth images. The emergence of consumer depth cameras makes a cheap and fast acquisition of 3D face model possible. However, the precision and resolution of these consumer depth cameras cannot satisfy the requirements of some 3D face applications. How to acquire high precision and high resolution face models fast and cheaply still is a challenging task. Much research about 3D reconstruction based on depth cameras has been done [1–3], but the resolution is not high enough when involving the human face. Improving the precision and resolution of 3D face models acquired by consumer cameras, i.e., 3D face model super-resolution, is a valuable study.

In this work, we built a database including 111 sets of 3D face models, where each set contains a low resolution 3D face model and the corresponding high resolution one of the same participant. The low resolution model was acquired by Kinect with the Kinect Fusion method [2], while the high resolution one was acquired by Artec Eva. With this database, we propose a 3D face model super-resolution method based on radial curves. In the method, we estimate the radial curves of the high resolution 3D face model from the corresponding low resolution one, and generate the high resolution 3D face model by deforming a high resolution face reference with the landmarks on the radial curves being control points. Experiments validated the proposed method.

The rest of the paper is organized as follows: The related work is described in Section 2; radial curves' extraction and the 3D face super-resolution method are described in Section 3; experimental results are reported and discussed in Section 4; conclusions are given in Section 5.

2. Related Work

Formerly, super-resolution [4] was introduced for 2D images, and its aim was to obtain a high-resolution image from one or more possibly contaminated low resolution observations. Super-resolution methods in the 3D space can be divided into two categories: methods based on multi-view fusion [2,5–8] and methods based on learning [9–11].

Methods based on multi-view fusion obtain a high-resolution depth map or 3D model by fusing depth scans from multiple viewpoints. In order to solve the problem of low resolution and high noise, Sebastian et al. [5] proposed a depth image super-resolution method named Lidar Boost to deal with depth images acquired by ToF cameras. Its main idea is to minimize an energy function, which consists of a data fidelity term and a geometry prior regularization term. The data fidelity term is to ensure the similarity between the super-resolution image and the low-resolution image. The regularization term is to ensure the smoothness of super-resolution image edges, and it is defined as the sum of L2 norm of the gradient in each pixel. Based on the Lidar Boost method, Cui et al. [12] proposed a shape scanning algorithm. They tried three non-linear regularization terms to replace the linear regularization term in Lidar Boost for improving the super-resolution accuracy, and fused the high resolution depth data to generate a high resolution 3D model by a probabilistic scan alignment approach. Methods in [2,6–8] reduce the noise in depth data by aligning multiple scans to construct a high resolution 3D model, where Kinect Fusion [2] is a classical real-time 3D reconstruction method designed for general objects using Microsoft Kinect, while the methods in [6–8] are designed for 3D face models. These methods intend to approach the noise problem in depth data by fusing multiple scans using geometry priors, and still cannot well solve the low accuracy problem of the depth sensors.

Learning based methods [10] obtain high resolution models by teaching the mapping from the low resolution models to high resolution models. Methods in [9,10] use mesh simplification and down-sampling to high-resolution 3D face models to produce low-resolution models. The super-resolution is realized by building the mapping in the regular representation domain. However, the way they generate the low resolution models cannot well simulate the imaging conditions of the sensors so that the methods are not necessarily applicable for the observed real low resolution models. Liang et al. [11] proposed a super-resolution method for 3D face models from a single depth frame. They divide the input depth frame into several regions, eyes, nose, etc., and search the best matching shape per region from a database they built, which includes 3D face models from 1204 distinct individuals. Then the matched database shapes are combined with the input depth frame to generate a high resolution face model. This method relies on the similarity measure between the low resolution face regions and the corresponding high-resolution regions. But this similarity measure between the heterogeneous data is unreliable. Unlike the methods mentioned above, our method is to estimate a high resolution 3D face model from a low resolution face model based on radial curve estimation. We use radial curves to represent 3D face models, and estimate the radial curves on the high resolution model through the radial curve matching of the low resolution models. The high resolution 3D face model is recovered using the estimated radial curves.

3. Method

In this section, we introduce the proposed 3D face model super-resolution method, where radial curves are used to represent the face model. For a low resolution face model, we first extract a set of radial curves from it, and then estimate the radial curves on its corresponding high resolution face model. Finally, we recover the high resolution face model using the estimated radial curves.

3.1. The Radial Curves' Extraction

As shown in Figure 1, radial curves on a face model are a set of curves passing the nose tip, and they can be defined by the intersections between the face model and a set of planes obtained by rotating the facial symmetrical plane around the normal direction of the nose tip. Thus, we have to locate the nose tip point and the symmetrical plane of a 3D face model for radial curve extraction. Assuming 3D face models are triangle mesh models, we first perform principal component analysis for the vertices of the face model to establish an initial coordinate system, i.e., three principal orientations as the coordinate axes with Y axis throughout the front and back of the face, and then fit a cylinder to the 3D face model [13] and adjust the Z axis to be parallel to the cylinder's axis. The nose tip is the most protruding point on the face's surface, so the point of the maximal value in Y direction is chosen as the nose tip. In order to establish a uniform coordinate system, we adjust the Y axis to be parallel to the normal vector of the nose tip with the nose tip being the coordinate center (see Figure 2). In order to get rid of the influence caused by the head size on the following radial curve estimation, we normalize all high resolution face models in size by a scale transform determined by several landmarks, such as nose tip and the corner points of the eyes and mouth, which can be labeled manually. With the nose tip, we estimate the symmetrical plane of the 3D face model using the method proposed in [14].

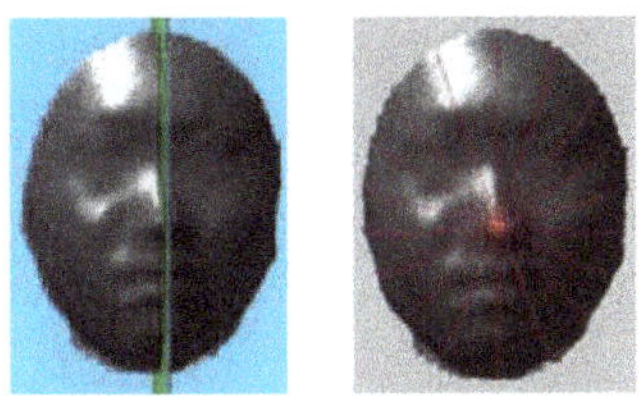

(a) (b)

Figure 1. (**a**) Symmetrical plane (shown in green). (**b**) Radial curves (shown in red) on 3D face model.

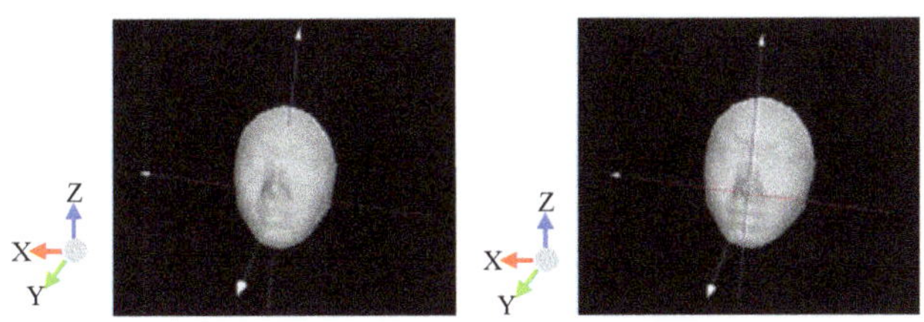

(a) (b)

Figure 2. (**a**) Initial coordinate system. (**b**) Standard face model coordinate system. X, Y and Z axes are shown in red, green and blue respectively.

As illustrated in Figure 1, we get the first radial curve by calculating the intersection curve of the symmetrical plane and face mesh, and then rotate the symmetrical plane around the normal direction of the nose tip by a fixed angle β gradually to extract the other radial curves. A radial curve is initially represented in a group of intersection points of the plane and edges of the triangle patches in the face mesh. For the following radial curve registration, we uniformly sample the radial curves with a fixed sampling internal Δ (here, 0.01 by experience). Then each radial curve is represented in a point sequence.

3.2. Radial Curves Database

As radial curves are used as the main feature of face models, we establish a radial curve database for the subsequent processing. As described in Section 1, the face database we used contains 111 sets of face models. Each set consists of a low resolution 3D face model and a high resolution 3D face model. Considering N sets of low resolution and high resolution face models in face database D, for low resolution face models, we extract K curves $C = \{c_1, c_2, ..., c_K\}$ from each model as described in Section 3.1 using β as rotation angle. Then we can get the low resolution face model radial curve database $D_{lowC} = \{D_{lowC1}, D_{lowC2}, ..., D_{lowCK}\}$, where $D_{lowCi}(i \in [1, K])$ represents the set of the i-th radial curve from all the low resolution face models. Similarly, for high resolution face models, we extract corresponding radial curves using the same β. These curves form the high resolution face model radial curve database $D_{highC} = \{D_{highC1}, D_{highC2}, ..., D_{highCK}\}$. Figure 3 shows radial curves and numbering when $\beta = 10°$; i.e., $K = 18$.

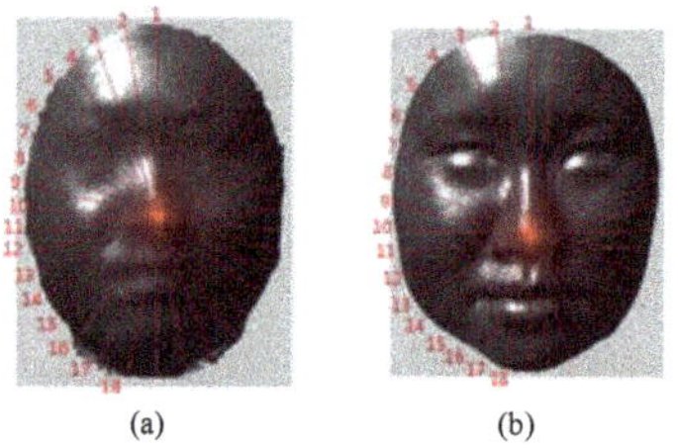

Figure 3. The correspondence of radial curves from (**a**) the low-resolution model and (**b**) the high-resolution model.

3.3. Face Model Super-Resolution

3.3.1. Registration of Radial Curves

Registration of radial curves is to establish a point correspondence between two radial curves. Dynamic Time Warping (DTW) [15] is one method for finding an optimal match and measuring similarity between two temporal sequences in time series analysis. It has been applied to analyze any data that can be transferred into a linear sequence, such as temporal sequences of video, audio and graphics data. Here radial curve registration and radial curve match are realized by using DTW of two point sequences of two radial curves. In order to keep the central feature area of human faces for radial curve match, we eliminate the boundary area by cropping all the radial curves in the following way: keep 40 points in both directions centered on the nose tip point. After cropping, each radial curve has 81 sampling points at equal intervals. The point number 40 is set by the distribution of face features. Of course, it can vary in a wide range; for example, 38 or 42 are also good selections, since they will have little influence on the radial curve match as long as the sampling points cover the central face feature area.

For point sequences of two radial curves, DTW is used to align them and measure their similarity. In order to eliminate the pose effect, we use the curvature difference of the two curve points to calculate their local match cost in DTW. For example, the curvature of the point a_n can be calculated as follows:

$$(a_n)_\kappa = \frac{1}{\rho(a_{n-1}, a_n, a_{n+1})}, 1 < n < N \tag{1}$$

where ρ is radius of curvature; i.e., $\rho(d,e,f)$ is radius of circle determined by $d, e,$ and f. Assume the matching point pair of two curves X and Y is $(x_i, y_i), i = 1, ..., N$; then the match cost of the two curves is:

$$\text{DTW}(X,Y) = \sum_{i=1}^{N} |(x_i)_\kappa - (y_i)_\kappa| \tag{2}$$

3.3.2. Radial Curve Estimation

For a given low resolution model M_{low}, we want to estimate radial curves $\hat{B}_{high}$ of its unknown high resolution model M_{high}. Here we assume that if two radial curves from two low resolution models are similar, the corresponding curves from their high resolution models are similar too. The assumption is rational since the data acquisition conditions (and thus noise models) are consistent for low resolution models. Thus, for a low resolution 3D face model, we first extract its radial curves $A = \{a_1, a_2, ..., a_K\}$. Then, for each radial curve a_i, we search for the best matching curve l_i in the radial curve database D_{lowCi} by the match cost defined in Equation (2); i.e., the curve with the least match cost. Finally, the radial curve in the database D_{highCi} corresponding to the best matching curve l_i is considered as the estimated high resolution radial curve $\hat{B}^i_{high}$. Then, $\hat{B}_{high} = \{\hat{B}^1_{high}, \hat{B}^2_{high}, ..., \hat{B}^K_{high}\}$. Considering the symmetry of the human face, the i-th and $(K+2-i)$-th radial curve should be symmetric about the symmetry plane of the human face with a even number K (see Figure 3). Thus, assume the i-th radial curve comes from the high resolution face model labeled G^i; that is,

$$\hat{B}^i_{high} = D^{G^i}_{highCi}, i \in (1, K/2) \tag{3}$$

Correspondingly, the $(K+2-i)$-th radial curve should come from the same face model; that is,

$$\hat{B}^{K+2-i}_{high} = D^{G^i}_{highC(K+2-i)}, i \in (1, K/2) \tag{4}$$

3.3.3. High Resolution Face Model Estimation

At this point, although we obtained a radial curve representation of the estimated high resolution face model, its mesh model was not yet constructed. For constructing the mesh model, the mean high resolution face model was taken as the reference and radial curves were extracted from it. Some landmarks were labeled on each radial curve manually (see Figure 4). Then, the corresponding landmarks on each radial curve we estimated by DTW registration of corresponding radial curves could be obtained. Therefore, we obtained a group of landmark correspondences between the estimated high resolution face model and the reference face model. Using these landmark correspondences as control points, we deformed the reference face to obtain the high resolution face models by using Thin Plate Splines (TPS) deformation [13].

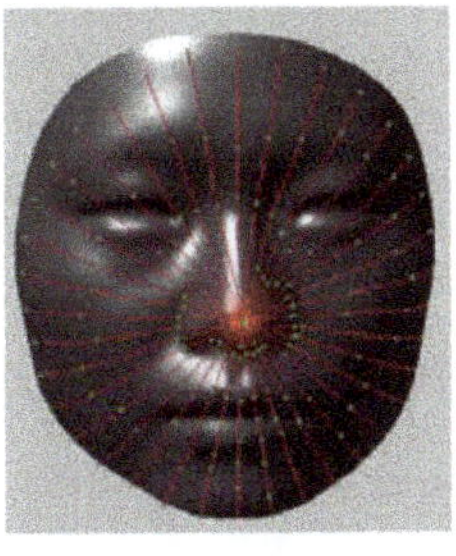

Figure 4. Landmarks (green) on reference model.

4. Experiment

4.1. Experiment Setting

4.1.1. Dataset

In this research, we used face models from 111 people. For each person, one low resolution face model was acquired using Kinect and one high resolution face model was acquired using Artec Eva, with a neutral expression. The high resolution face model was considered ground truth. In the experiment, we performed leave-one-out cross validation; i.e., one low resolution face model M_{low} from the face database was chosen as a probe; meanwhile, we removed its corresponding high resolution face model M_{high} from the face database. The remaining 110 sets of face models formed the training database D. We chose a high resolution model from D randomly as the reference model. When labeling landmarks on the reference model, it is better to choose inflection points or points that have large curvature. On the reference model, 158 landmarks are labeled.

4.1.2. Error Metric

To evaluate experimental results quantitatively, we used the attribute deviation metric [16] to measure geometric error between the high resolution estimation model and the ground truth model. Given a 3D surface S and a 3D point p, the distance between p and S based on attribute i is defined as:

$$d_i(p, S) = \|f_i(p) - f_i(N_S(p))\| \tag{5}$$

where $p' = N_S(p)$ is the nearest point from point p to surface S, attribute deviation distance $d_i(p, S)$ is the difference between p and p' by attribute i, and $f_i(p)$ denotes the attribute i of the point p. Here we use two attributes; i.e., the point coordinate and the normal vector [17]. We calculate mean error for all vertexes of the estimation model to the ground truth model.

4.2. Results and Analysis

Figure 5 shows 10 super-resolution results we randomly chose, and their error metrics are shown in Figure 6. In Figure 5, low resolution models are shown in the first column; the high resolution models estimated using our method are shown in the second column; the ground truth models are shown in the third column; and the comparison diagrams of the estimation models and the ground truths by normal vector are shown in the fourth column. We can see from Figure 5 that each estimation model and its ground truth are similar overall; the error in most regions of the human face is small (i.e., under 15 degrees); and higher error mainly concentrates on boundaries. That is because the calculation of the normal vector is not stable in boundary region. From Figure 6 we can see that the average error in normal direction for each model is less than 15 degrees, and the average error in Euclidean distance is less than 4 mm, but is less than 2 mm for eight models of the total 10 models. By comparing the Figures 5 and 6, we can see that the error metrics of the models of higher error metrics, such as model 4, 7 and 10, are mainly affected by the higher errors in the boundary region. That is, the overall similarity is high for each estimation model. The experiment demonstrates the proposed method is effective.

4.3. Discussion

In the proposed method, radial curves are used to represent the face model. The advantage is that we can estimate the radial curves on the high resolution face models by curve matching among low resolution face models, and then use the estimated radial curves to construct the high resolution face model. In fact, if the point correspondence among 3D face models is established, we also can teach the mapping from low resolution models to high resolution models such as the learning based methods [9,10]. However, it is difficult to establish point correspondence among low resolution 3D face models acquired form a consumer depth camera due to the high noise. The methods in [9,10] use mesh simplification and down-sampling of high-resolution 3D face models to produce low-resolution models. They only need to perform data registration among high-resolution 3D face models, which can be realized easily by many existing methods [13]. We also tried to use the registration method [13] to establish the point correspondence among low resolution 3D face models. As a result, the registration accuracy was very bad. This is the reason why we do not use a learning based method like [9,10]. On the other hand, a larger training dataset is necessary for learning based methods. Our dataset only contains 111 samples; it is not enough in a statistical sense for machine learning. In our work, to improve the statistical sense, we performed leave-one-out validation in the testing phase. The visual results show high similarity against the ground truth. This validates the effectiveness of our proposed method.

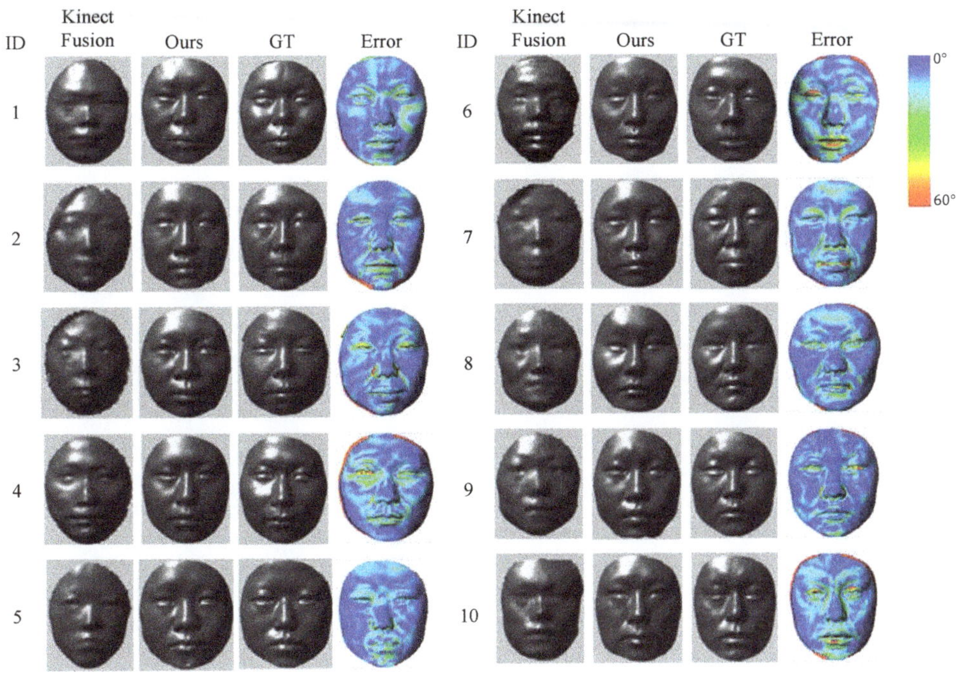

Figure 5. Experimental results of 10 face models. For each model, we show the low resolution model, made by the Kinect Fusion [2] method, the high resolution model derived by our method, the ground truth model (short for GT) and the normal direction error displayed with color map from left to right. The color scale of error map is shown in the right.

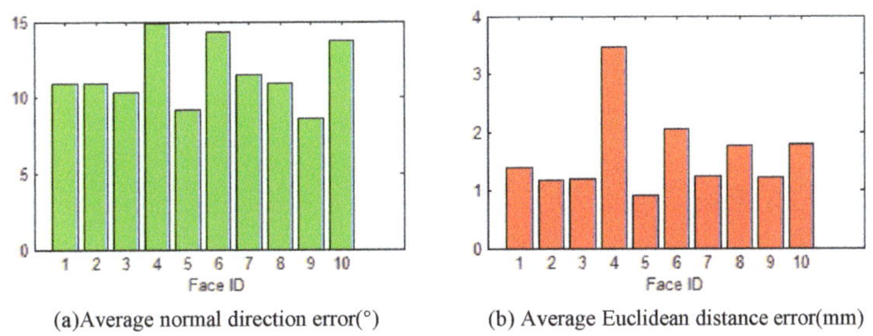

Figure 6. Errors of the estimated models are shown in histograms.

5. Conclusions

In this paper, we presented a super-resolution method for 3D face models, aiming to improve the resolution and resolution of 3D face model acquired by consumer depth cameras. We established a face model database which contains low resolution and high resolution face models of 111 participants acquired respectively using Kinect and Artec Eva. Based on this database, we estimated a radial curve database which includes low resolution radial curves and their corresponding high resolution ones. For a given high resolution 3D face model, we first extracted a set of radial curves on it, and then estimated their corresponding high resolution ones by utilizing the radial curve database. Finally, we deformed a reference high resolution 3D face model to generate a high-resolution face model by using radial curves as the main feature. We evaluated the method both quantitatively and qualitatively. The evaluation results show that the proposed method is effective. Our method has practical implications for improving the quality of 3D face models and promoting applications such as 3D face recognition and 3D games. However, our assumption in the radial curve estimation phase, i.e., if two radial curves from two low resolution models are similar, the corresponding curves from their high resolution models are similar too, is too strict. In the future, we will improve the radial curve estimation method to relax this assumption.

Author Contributions: Conceptualization, J.Z. and F.D.; methodology, F.Z., J.Z. and L.W.; software, F.Z.; validation, F.Z.; data curation, F.Z.; writing—original draft preparation, J.Z., L.W. and F.Z.; writing—review and editing, F.Z., J.Z., L.W. and F.D.; visualization, F.Z. and J.Z.; supervision, F.D.; project administration, F.D.; funding acquisition, F.D. All authors have read and agreed to the published version of the manuscript.

Funding: This work was funded by the Natural Science Foundation of China, grant numbers 61572078, 61702293 and 61772050, and the Chinese Postdoctoral Science Foundation, 2017M622137.

Conflicts of Interest: The authors declare no conflict of interest.

References

1. Berretti, S.; Del Bimbo, A.; Pala, P. Superfaces: A super-resolution model for 3D faces. In Proceedings of the European Conference on Computer Vision, Florence, Italy, 7–13 October 2012; pp. 73–82.
2. Newcombe, R.A.; Izadi, S.; Hilliges, O.; Molyneaux, D.; Kim, D.; Davison, A.J.; Kohli, P.; Shotton, J.; Hodges, S.; Fitzgibbon, A.W. Kinectfusion: Real-time dense surface mapping and tracking. In Proceedings of the 2011 10th IEEE International Symposium on Mixed and Augmented Reality, Basel, Switzerland, 26–29 October 2011; Volume 11, pp. 127–136.
3. Drira, H.; Amor, B.B.; Daoudi, M.; Srivastava, A. Pose and expression-invariant 3d face recognition using elastic radial curves. In Proceedings of the British Machine Vision Conference, Aberystwyth, UK, 31 August–3 September 2010; pp. 1–11.
4. Nasrollahi, K.; Moeslund, T.B. Super-resolution: A comprehensive survey. *Mach. Vis. Appl.* **2014**, *25*, 1423–1468. [CrossRef]
5. Schuon, S.; Theobalt, C.; Davis, J.; Thrun, S. Lidarboost: Depth superresolution for tof 3d shape scanning. In Proceedings of the 2009 IEEE Conference on Computer Vision and Pattern Recognition, Miami, FL, USA, 20–25 June 2009; pp. 343–350.
6. Berretti, S.; Pala, P.; Del Bimbo, A. Face recognition by super-resolved 3D models from consumer depth cameras. *IEEE Trans. Inf. Forensics Secur.* **2014**, *9*, 1436–1449. [CrossRef]
7. Hernandez, M.; Choi, J.; Medioni, G. Laser scan quality 3-d face modeling using a low-cost depth camera. In Proceedings of the 2012 Proceedings of the 20th European Signal Processing Conference (EUSIPCO), Bucharest, Romania, 27–31 August 2012; pp. 1995–1999.
8. Bondi, E.; Pala, P.; Berretti, S.; Del Bimbo, A. Reconstructing high-resolution face models from kinect depth sequences. *IEEE Trans. Inf. Forensics Secur.* **2016**, *11*, 2843–2853. [CrossRef]
9. Pan, G.; Han, S.; Wu, Z.; Wang, Y. Super-resolution of 3d face. In Proceedings of the European Conference on Computer Vision, Graz, Austria, 7–13 May 2006; pp. 389–401.
10. Peng, S.; Pan, G.; Wu, Z. Learning-based super-resolution of 3D face model. In Proceedings of the IEEE International Conference on Image Processing, Genoa, Italy, 1–14 September 2005; Volume 2, p. II-382.

11. Liang, S.; Kemelmacher-Shlizerman, I.; Shapiro, L.G. 3d face hallucination from a single depth frame. In Proceedings of the 2014 2nd International Conference on 3D Vision, Tokyo, Japan, 8–11 December 2014; Volume 1, pp. 31–38.
12. Cui, Y.; Schuon, S.; Thrun, S.; Stricker, D.; Theobalt, C. Algorithms for 3d shape scanning with a depth camera. *IEEE Trans. Pattern Anal. Mach. Intell.* **2012**, *35*, 1039–1050.
13. Chen, Y.; Zhao, J.; Deng, Q.; Duan, F. 3D craniofacial registration using thin-plate spline transform and cylindrical surface projection. *PLoS ONE* **2017**, *12*, e0185567. [CrossRef] [PubMed]
14. Wang, Y.; Liu, J.; Tang, X. Robust 3D face recognition by local shape difference boosting. *IEEE Trans. Pattern Anal. Mach. Intell.* **2010**, *32*, 1858–1870. [CrossRef] [PubMed]
15. Zhao, J.; Itti, L. Shapedtw: Shape dynamic time warping. *Pattern Recognit.* **2018**, *74*, 171–184. [CrossRef]
16. Roy, M.; Foufou, S.; Truchetet, F. Mesh comparison using attribute deviation metric. *Int. J. Image Graph.* **2004**, *4*, 127–140. [CrossRef]
17. Fouhey, D.F.; Gupta, A.; Hebert, M. Data-driven 3D primitives for single image understanding. In Proceedings of the IEEE International Conference on Computer Vision, Sydney, Australia, 1–8 December 2013; pp. 3392–3399.

applied sciences

MDPI

Article

Projection-Based Augmented Reality Assistance for Manual Electronic Component Assembly Processes

Marco Ojer [1,*], Hugo Alvarez [1], Ismael Serrano [1], Fátima A. Saiz [1], Iñigo Barandiaran [1], Daniel Aguinaga [2], Leire Querejeta [2] and David Alejandro [3]

1 Vicomtech Foundation, Basque Research and Technology Alliance (BRTA), 20009 Donostia, Spain; halvarez@vicomtech.org (H.A.); iserragr@everis.com (I.S.); fsaiz@vicomtech.org (F.A.S.); ibarandiaran@vicomtech.org (I.B.)
2 Ikor Technology Centre, 20018 Donostia, Spain; daguinaga@ikor.es (D.A.); lquerejeta@ikor.es (L.Q.)
3 IKOR Sistemas Electrónicos, 20018 Donostia, Spain; david.alejandro@ikor.es
* Correspondence: mojer@vicomtech.org

Received: 19 December 2019 ; Accepted: 16 January 2020; Published: 22 January 2020

Abstract: Personalized production is moving the progress of industrial automation forward, and demanding new tools for improving the decision-making of the operators. This paper presents a new, projection-based augmented reality system for assisting operators during electronic component assembly processes. The paper describes both the hardware and software solutions, and depicts the results obtained during a usability test with the new system.

Keywords: computer vision; augmented reality; projection mapping

1. Introduction

It is evident that initiatives such as the German paradigm of "Industry 4.0" or some other similar ones all around the world are having a deep impact on the manufacturing sector, and thus are reshaping the industry. The development of such paradigms is accelerating the development and deployment of advanced ITC-related technologies [1], transforming many aspects, such as the industrial workforce and the way they develop their tasks. Even though customer-centric and demand-driven production is moving forward through the progress of industrial automation, the need for a better and more empowered human workforce is more demanding than ever. The next human workforce should have new and more powerful tools that allow them to improve their decision-making processes, to more easily adapt to changing production conditions and to adopt strategies for continuous training. Along with the development of the Industry 4.0 paradigm appears the concept of Operator 4.0 [2]. This concept is driven by several objectives, such as to simplify the day-to-day work, while improving efficiency and autonomy by focusing on added value tasks, all in a comfortable and healthy working environment. This paper proposes a new system based on augmented reality (AR) for assisting operators during manual assembly of electronic components. As mentioned before, a customer-centric oriented and personalized production requires continuous changes in production lines. The electronics sector is not an exception in this regard. This industry has many automated processes for the assembly of electronic components for electronic boards, also known as printed circuit boards (PCB), but there are also many manual assembly stages along the production lines. Operators perform the monotonous task of board assembly over considerable periods of time; therefore, they are likely to experience fatigue and distractions. Furthermore, the low profile needed for this task favors rotation of personnel, which is undesirable because new employees take a certain amount of time to adapt. As a consequence, manual processes have the highest error ratio of the production process; electronic manufacturers have identified the necessity of improving these processes as a key point. Therefore, This paper proposes a system which aims to reduce assembly errors and adaptation times for new employees while increasing operator comfort, confidence and assembling speed by means of AR.

This paper is structured as follows: Section 2 describes the current state of the art related works with the application of augmented reality to the manufacturing sector. In Section 3, we show our approach to assist the operators during the manual assembly of electronic components. Section 4 outlines the results of a usability test we carried out with several operators using the proposed approach. Section 5 discusses the proposed approach and shows how a significant and positive impact has been achieved in the production line evaluated. Finally, Section 6 gives some conclusive remarks and also mentions some future research directions for improving the next generation of the system.

2. Related Work

Visual Computing technologies (including augmented reality) will be key enabling technologies for the smart factories of the future [1]. These technologies have demonstrated good capacities for empowering human operators when performing industrial tasks by providing tools that assist them and improve their comfort and performance [3]. Consequently, the research community has focused on these technologies and several related approaches have been proposed [4]. Next, we mention a few AR works applied to the manufacturing sector.

Augmented reality has been extensively used in many industrial processes, such as maintenance operations [5]. Some of these solutions [6–11] are oriented toward assembly tasks, in which an AR technology provides virtual instructions in order to guide the operators. In those solutions, the virtual content is shown in a screen, forcing the operators to constantly change the attention between the physical workspace and the screen. As stated by [12], switching attention between two sources during a maintenance task (for example, between the documentation and the workspace when using a traditional paper based instructions, or between a screen and the workspace) might cause a high cognitive load, which translates into greater probability of errors and an increase of the task completion time. On the contrary, projection based augmented reality (also cited as spatial augmented reality (SAR) [13] in a broader meaning, or just projection mapping) projects the virtual data directly in the physical space. This approach allows the operator to have their hands free and is considered an enabling technology to face the challenge of supporting operators performing tasks [14]. Attracted by these advantages, several SAR works have been developed for industrial environments [15–19]. Most of these works are focused on providing guidance to the operators, without verifying if the task is correct or not. To face that, [20] proposes an AR system that also verifies the operator task by comparing the status of every step along the maintenance procedure, represented by a captured image, with a reference virtual 3D representation of the expected status, which is converted to an image as well by rendering the virtual 3D data using the tracked real camera location.

Moreover, as more and more visual computing solutions are integrated into industrial shop floors, the complexity of communication and interaction across different peripherals and industrial devices increases. Nonetheless, [21] has recently proposed a middleware architecture that enables communication and interaction across different technologies without manual configuration or calibration.

From the works cited above, only [6,11] deal with PCBs and are focused on a similar domain to our work. However, they only address the part of offering augmented instructions on the screen (without projection). Additionally, compared to all the works cited, our work combines the best characteristics of each of them. Thus, our work has the following strong points:

- The proposed system verifies if the operator has performed the operation correctly.
- Instructions are simple, so there is no need to create the multimedia content that is projected. The authoring effort is minimized to only set the position of each component in the reference board.
- The projection is done on a flat surface, so the calibration step has been simplified to be easy, fast and automatic (the user only has to put the calibration pattern in the workspace).

- The proposed system uses advanced visualization techniques (flickering) to deal with reflections when projecting on PCBs.
- The proposed system supports dynamic projection; i.e., the projection is updated in real time when the PCB is moved.
- A normal RGB camera is used; no depth information is required.

3. Proposed Method

This paper proposes a SAR system to guide and assist operators during the process of assembling electronic components. This system performs real-time checking of the state of a PCB; i.e., checks presence or absence of electronic components, by means of computer vision techniques. It offers visual information about which component should be assembled and whether previous assemblies have been correctly done. This work is based on [22], but with the improvement that the virtual content is directly projected on the PCB using projection mapping techniques. In the following sections we provide a brief description of the SAR system that we rely on and give a detailed explanation of the components newly-added to the aforementioned system. The system has two work modes, one consists of the model generation (during an offline phase) and the other consists of the real-time board inspection and operator guiding (during an online phase); see Figure 1. We explain each component in the following subsections.

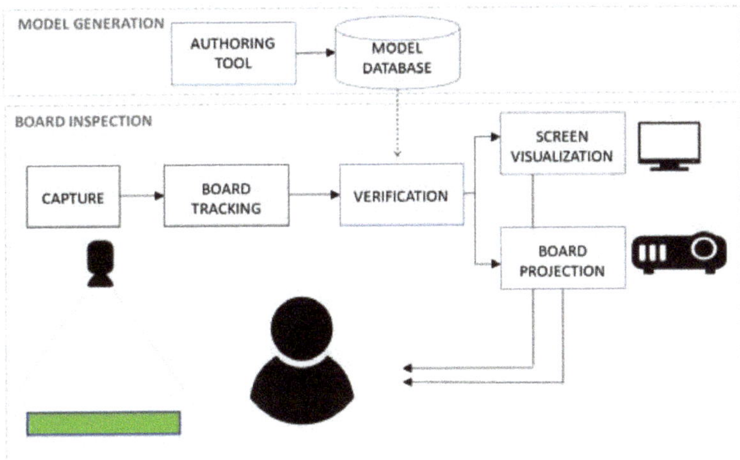

Figure 1. Pipeline of the system.

3.1. Setup

The proposed system consists of four different parts: an illumination system, a 2D high-resolution image acquisition setup, a screen and a projector (see Figure 2). The illumination system, the camera and the projector must be located at sufficient height in order to not disturb the operator during manual operation. Given user experiences and comments, the minimum ergonomic height settled on was 600 mm. A 12 mega-pixel camera is at the center of the illumination system, at a height of 700 mm. This positioning, combined with the optical lens, offers a field of view of 500 × 420 mm. A PCB's maximum size was established to 320 × 400 mm, which is covered by the proposed setup.

The projector model used is conventional, more specifically, an Optoma ML750e, which uses LED technology and has a light output of 700 lumens. It is not a very powerful projector, but it has proven to be sufficient (Section 3.6.2), and, in return, thanks to its small dimensions, it has allowed us to achieve a fairly compact setup. It is positioned next to the camera, covering all the field of view of the camera.

Figure 2. Hardware setup of the proposed system. The camera and projector are highlighted with light-blue and dark-red rectangles, respectively.

The screen is in front of the operator, hopefully at the most ergonomic position. The screen shows the outputs and feedback of the proposed system. It is a complementary visualization, since this output is also shown directly on the board using the projector.

3.2. Authoring Tool

The main goal of this tool is to generate models which are able to distinguish between the presence and absence of electronic components in the board. This tool is intended to be used before board inspection in case there are any components unknown to the system. In this case, an operator with correct access rights will use this tool to generate the model for this specific component.

The component catalog is immense, of the order of 10,000 different components, which it is being constantly updated. Furthermore, these components present huge variations in their characteristics such as size, shape, texture and color. In order to tackle this problem, [22] proposed a one-classifier-per-component approach and the definition of a training phase that only needs a single image of a minimum number of components to generate a model. This training phase can be divided into different stages: segmentation, image generation and training.

- Segmentation: In this stage the operator takes an image of the new referenced component, selecting a foamy material with chromatic contrast to the background. The operator has to place a set of components with the same reference almost covering all the camera field of view. Experiments show that five well distributed components are enough to capture the prospective distortion of the camera. When the capture is ready, the segmentation process starts. The first step consists of applying a rough or an approximate segmentation. After this process, a more accurate segmentation is carried out using GrabCut algorithm [23] for improving component segmentation result.

- Image generation: To get a high performance classifier, a substantial number of image samples that include as much component variability as possible, is necessary. In [22], the authors propose generating synthetic images of the components and different backgrounds by applying geometric and photo-metric transformations. This step ensures the robustness of trained classifiers during operation.
- Training: In order to generate the classification model from the generated set of images, the first part is to extract the relevant features from these images. The images of this dataset have a huge variety in terms of background; some of them are totally uniform, while others have numerous pinholes and tracks. For this reason, global features obtained from the whole image should be used instead of focusing on local keypoints. Once features are extracted, a classifier is trained with them, in order to to discriminate between components and background, and it is saved in a database.

In [22], a study is conducted which compares the accuracy of different combinations of features and classifiers. Training and validation were performed with artificially generated images, whereas testing was performed with real images taken with the proposed setup ensuring performance in real environments. This study was conducted using 21 different components chosen in order to cover a big spectrum of components, ranging from multi-colored big components to uniform small components. In conclusion, a combination of color histograms, histogram Of gradients (HOG) and local binary patterns (LBPs) were chosen as features. Along with a radial-basis function support vector machine (RBF-SVM) as the classifier, this combination achieved more than 90% accuracy in validation and testing. Furthermore, this combination was assured to have low computation time; that is enough for a real-time application.

3.3. Board Tracking

As the proposed system uses the image captured by the camera to recognize components, it is essential to avoid distortions in the image due to the camera lens. It is therefore necessary to calibrate the camera, i.e., to know the intrinsic camera parameters, before or prior to using the system. In our system, we propose to use the well known Zhang's camera calibration algorithm [24]. This calibration process only needs to be done once, and it allows us to calculate the compensation that has to be applied to each image captured by the camera to avoid distortions.

During the component assembly phase, the boards have a non-fixed position, having one degree of freedom for horizontal displacement. They have also different sizes, shapes and colors due to the mounting boards and used materials. Owing to a component's position being referred to via the bottom-left corner of the board, the use of some markers is proposed with the final purpose of tracking the board position. In this system, the ArUco markers are used [25].

Two ArUco markers are placed to locate the vertical board position, an other two ArUco markers are placed to locate the horizontal board position. During the assembly, the operator might occlude the horizontal markers, but if it happens, the system assumes the previously captured horizontal marks positions as current positions (temporal coherence assumption). The corner of the board is calculated by intersecting the vertical line and the horizontal line referenced to the markers; see Figure 3. This corner is necessary to obtain the reference system of the PCB, and therefore, to locate component positions. If vertical line calculation is not possible, the component inspection stops. Thus, visible vertical markers are necessary to track correctly the board.

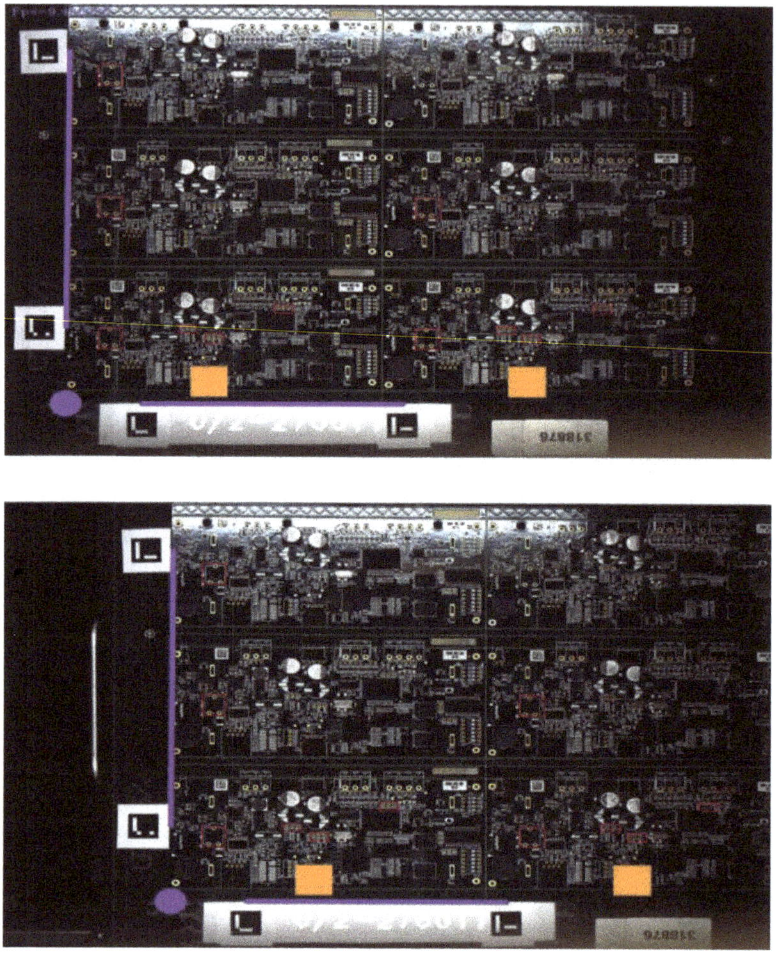

Figure 3. Images where the printed circuit boards (PCB) is in different positions. The purple lines are located thanks to the ArUco markers; the corner (purple circle) is the intersection between them and denotes the PCB reference system.

3.4. Verification

In this step, the main goal is to verify the presence of the components on the board.

First, the assembly region of each components should be located. A list of component relative coordinates with respect to the board corner is feed to the system, and because the board corner is already located, the assembly regions can be situated in the image. This coordinate list is created by the quality engineer during the design process of the board using the manufacturing execution system (MES) of the company.

A further step is to calculate the detection probability of each component, using the cropped image of the assembly region. The classification models of the board components are loaded from the model database. Then, for each cropped image, the selected combination of features is extracted and feed to the classification model, an RBF-SVM in this case.

The output of the model is a probability for the analyzed image crop of the component. A high value of this probability represents component presence, whereas low probability means absence. Note that a larger region usually provides a stronger response than smaller region because it has more borders, texture, colors, etc. To adjust this response, a threshold calculated proportionally using the region size is given. This operation minimizes false positives.

When these values are obtained, the output is visualized on the screen and on the board. The visualization strategy is explained in the next section.

3.5. Screen Visualization

With the verification output, the region location is highlighted in the screen by a rectangle; if the component is mounted, the rectangle is green, whereas if it is not mounted, the color is red. The current group of components to be mounted is highlighted with a blinking orange solid rectangle in the visualization. On the right side of the screen, the reference and image of the component to be mounted are shown; see Figure 4.

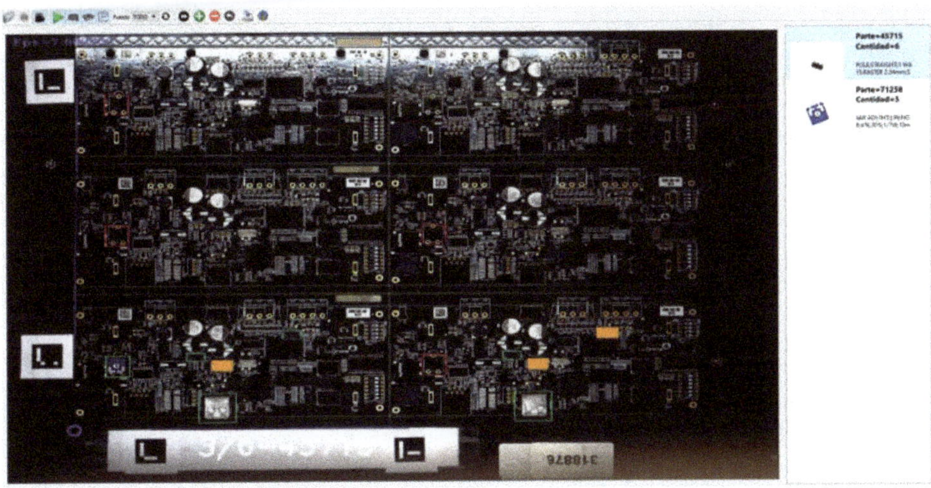

Figure 4. Screen visualization of the current state of the PCB.

3.6. Projection Mapping

The main problem of screen based visualization is that the operator has to constantly check the state of the assembly on the screen, switching attention between the board and screen. A more ergonomic solution is obtained when the projector is used to visualize this output directly onto the PCB. This improves the posture of the worker and increases the assembly speed and quality, since the operator does not have to look up to receive work-instructions.

Apart from offering assistance in a conventional screen, the proposed system also provides guidance by projecting relevant virtual content directly onto the PCB. However, to project content in the desired place and with an adequate degree of immersion, it is first necessary to calibrate the camera–projector pair.

3.6.1. Camera–Projector Calibration

To project virtual content adequately in a physical space, we must calibrate the setup; i.e., find a geometric transformation that adapts the virtual data to the shape of the projection surface. This transformation can be fixed manually by modifying the position or shape of the virtual content until the projection gives the desired results, which is a laborious and expensive process that requires technical skills. However, in those cases where there is also a camera in the setup, the camera–projector calibration, i.e., finding the correct geometric transformation, can be calculated automatically. The projector can emit a pattern that is captured and recognized by the camera and which can be used to estimate the transformation that moves content from the camera's coordinate system to the projector's coordinate system. Additionally, when an object is recognized in the camera image and the camera pose is known, i.e., the position and orientation respect to the object is known (Section 3.3), we have the transformation that relates the object and camera coordinate systems. Thus, since the virtual content is defined in the same coordinate system as the object, its projection can be calculated using the chain rule. In this work, we have followed this methodology to calibrate the camera–projector pair. We propose to place a planar checkerboard in the physical space, and the projector projects a complete gray code sequence. This structured-light sequence can be decoded, so that each pixel of the camera is associated with a projector row and column. Therefore, since the 3D coordinates of the checkerboard corners and their 2D positions (pixels) in the camera and projectors images are known, a traditional stereo calibration method can be applied to solve the three-dimensional camera–projector relationship (see [26]). Nonetheless, in our setup, the projection surface is a plane (a PCB), and it is always parallel to the camera image plane, so we have simplified the camera–projector relationship to 2D. We have modified the [26] implementation to estimate a 2D homography that represents the camera–projector relationship. Although this simplification can be inaccurate for more complex projection surfaces, it offers good results for planar surfaces and simplifies the calibration process. In the original calibration version [26], a structured-light sequence must be captured from several points of view, but in our simplified version, only one point of view is required. Therefore, our simplified and not optimized version only takes approximately 85 s to do the calibration (50 s to project and capture the gray code sequence and 35 s to decode the patterns and to estimate the homography). Nevertheless, this time is not usually critical, since the calibration process is only executed once when the setup is built. Likewise, the setup must be recalibrated when there is a change in the camera, the projector or the projection surface.

3.6.2. Virtual Content Projection

In the proposed projection mapping pipeline (Figure 5), as stated in the previous subsection, first, the virtual content is transferred to the camera image using the camera tracking data (T_{track}, Section 3.3), which creates the view that is displayed in the screen. Then, this content, which is already referenced with respect to the camera image coordinate system, is again transformed using the camera–projector calibration (H_{calib}, Section 3.6.1) to the projector image area that is subsequently projected. Thus, to project any content, we define its location in the reference 2D coordinate system of the board and then we apply the chain rule, which can be represented conceptually by $T_{track} * H_{calib}$.

In our application, we decided to project the following information (Figure 6), which answers three simple questions that are very useful for operators:

- "Where?": The place where the operator has to assemble the current electronic component, which is highlighted with the projection of a white flicking rectangle.
- "What?": The reference number of the electronic components that must be assembled in the current step.
- "How many?": The number of the current electronic components that have already been assembled regarding the total number to be assembled. A fraction "i/j" is projected, where i is the number of current components already assembled from the total of j.

The projection of "What?" and "How many?" is located at the border frame (Figure 6), outside the electronic board, as this area is not used for anything and it offers good visibility. The projection of "Where?" on the other hand, is superimposed on the real position that corresponds to the inside the electronic board (Figure 6). This was not an appropriate area to get good contrast due to the material of the PCB and the limited power of the projector that was used, so we opted to flick the projection to capture the operator's visual attention, and, consequently, improve its visibility. This has been proven as a good solution, since the result of the usability test was positive (Section 4). A sample of the system performance can be seen in the Supplementary Video S1.

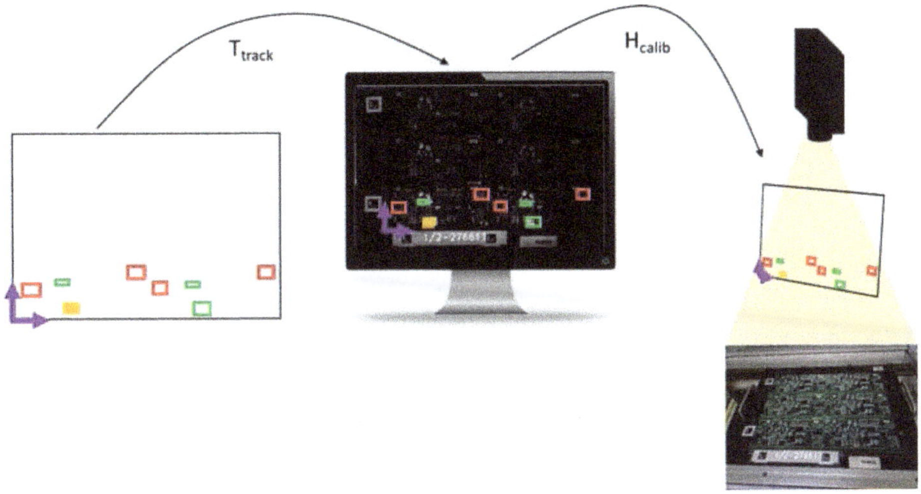

Figure 5. Conceptual scheme of the projection mapping pipeline. Virtual content (left) is transferred to the camera image (T_{track}), and then, this content is transformed again (H_{calib}) to the projector image area that is subsequently projected. See text for details.

Figure 6. Example of virtual content that is projected (highlighted in white) in the printed circuit board during the component assembly process. The projection is more clearly seen live, so we provided the bottom row that has zoomed-in versions of the top images to see the projections in these images with more quality.

4. Usability Test

With the aim of evaluating the benefits of the AR extension compared to the previous system, a system usability scale (SUS) survey was made, which compares the usability between the two systems. On one hand, the original system presented in [22], where instructions are only displayed on the screen. On the other hand, the proposed system, where instructions are displayed both in the screen and on the board directly via the projector. SUS survey is a ten-item scale test giving a global view of subjective assessments of usability [27], which is used as a standard survey for usability tests.

The proposed test consists of mounting the same PCB with the aid of both solutions, the original and the proposed ones, wherein every mounted process is timed and the number of mounting errors is measured. Finally, the SUS test was completed.

A total of 21 people were surveyed. They were between 20–45 years old; there were 15 men and six women, one of them color-blind. They did not have any experience in PCB assembly. This was done in order to emulate a newcomer to the production line, since rotation of personnel is common. The test was performed in a laboratory where a replica of the manufacturing production workspace was located.

They were divided into three groups: seven participants for each group. Group 1 used the original system in the first place and later the proposed one. Contrarily, Group 2 used the proposed system first and original system second. Groups 1 and 2 did not have any experience mounting the electronic board; thus, it was fair to assume that the first mounting would take longer than the second, as the users had more experience for the second mounting. For this reason, Group 3 was created. This group had already mounted the PCB using a different solution, so they already had some knowledge of the PCB when using both processes. This grouping was done in order to measure time efficiency among processes, but it did not have any impact from the usability point of view.

Figure 7 displays the SUS scores. The higher the score, the more usable the systems is. The systems achieved average values of 80 and 90 out of 100, respectively. Although a SUS score interpretation is not straightforward, Bangor et al. [28] concluded that any system above 68 can be considered usable; he also proposed an adjective scale, where a mean SUS score of around 70 is considered good, one around 85.5 is considered excellent and one around 90.9 is referred as the best imaginable. Thus, both systems are highly usable, but the use of augmented reality is preferable.

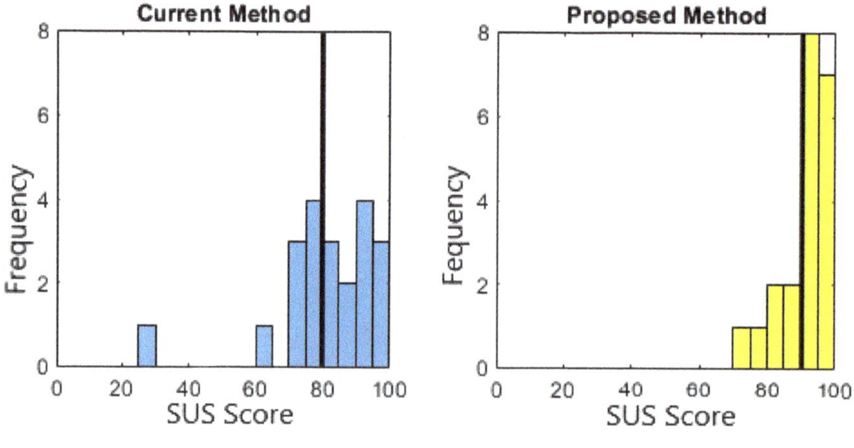

Figure 7. Distributions of SUS scores. Blue represents the original system and Yellow the proposed one. Black lines mark the average value of both distributions.

As mentioned, mounting times were measured in order to get some objective insights about system efficiency; see Figure 8. As predicted, for Groups 1 and 2, the first mounting was usually the more time consuming one. However, for Group 3, where participants started both mountings with the same experience, the proposed solution yielded lower mounting times for all participants. In addition, the feedback provided by the two systems prevented the users from making any errors.

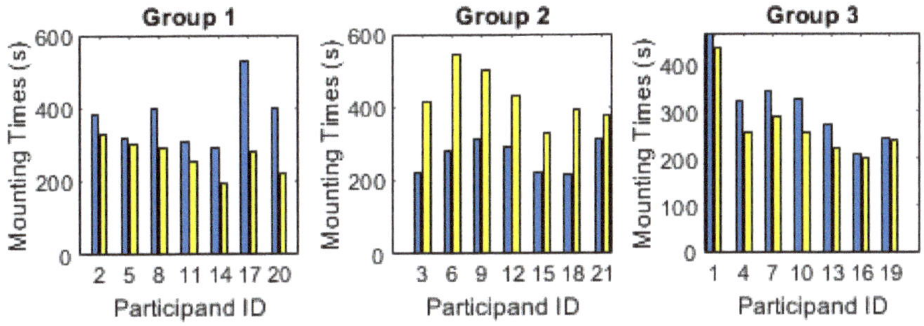

Figure 8. Mounting times of each group. Blue and yellow bars represent the original and proposed system, respectively. Group 1 started with the original, Group 2 started with the proposed and Group 3 already had experience.

These results show that the AR system is even faster and more comfortable than the previous system. From the users' comments, it can be deduced that both velocity and comfort are increased because the user only needs to look and focus on the board, instead of changing their focus between screen and board, thereby helping the operator to maintain the same posture. Moreover, the direct projection onto the board allows the operator to find placing location easier, saving operational time and reducing placement errors. The system was also validated by experienced workers of the manufacturing company, who also pointed out the enhancement provided by the projection mapping. In [22], the usability of the only-screen system is compared with the traditional system used by the manufacturer; the system proposed achieved a much higher satisfaction levels than the traditional system. Therefore, the AR extension is also much more usable than the traditional system.

5. Discussion

We propose to use direct projection in the workspace for improving user satisfaction and at the same time reducing assembly errors. The previous section shows that operators actually find the system more usable, feel more secure with it and require less time to do their tasks. A further advantage is that operators requires less training time, as the system gives assistance throughout the assembly. Moreover, this system allows the production managers to have traceability of the most complex components or PCBs to be assembled. This enables them to take further measures for ensuring operator satisfaction while also optimizing production because of the reduction of potential errors.

To guarantee that the projection-based solution is effective, the illumination conditions of the workspace have to be considered. The ambient light cannot be strong, so that the light emitted by the projector is predominant and the projected content is shown with contrast and sharpness. A balance must be achieved between a valid ambient light for object detection (electronic components in our case) and light that does not defeat the visibility of the projector. Similarly, it is preferable to work on non-specular surfaces, so that no brightness is generated that hinders the visibility of the projection. In our scenario, we had to deal with this difficulty, since PCBs are specular, and therefore, we had to use more sophisticated visualization techniques to capture the operator's attention (flickering).

In the use case presented in this paper (assembly small electronic components in a PCB) we have not had problems with hidden areas of projection. These areas appear when an object that is in the workspace and in front of the projector has large dimensions and occludes the area behind it. Thus, the rays emitted by the projector cannot reach this area, and therefore, it is not possible to project content in this zone. To solve this limitation, a multiprojector configuration should be used.

6. Conclusions and Future Work

Despite the improvements in the last few decades, the use of augmented reality in industry has not been extended yet due to several reasons, including ergonomics, visual fatigue, content creation, the lack of IT infrastructure, etc. [29]. In fact, ergonomics is the main obstacle for AR glasses; thus, projection based AR systems have been positioned as the alternative because they project data directly in the workspace, leaving the operator's hands free and avoiding discomfort due to motion sickness or vergence-accommodation conflicts [14].

The fast adoption of new, advanced ITC-related technologies such as cloud computing and augmented reality by the manufacturing sector is having a real positive impact in several terms, such as increasing flexibility, productivity and efficiency. In this paper, we propose integrating an AR system to support operators during the manual assembly of electronic components for improving workers' ability to adapt to very variable production conditions. Our results show that, compared with the old procedure, with the new system the operators generate less errors, especially when they face a new PCB they have not assembled before. In addition, they feel more comfortable because they know that there is an additional system that ensures that their work is being done correctly. In the future, we plan to implement some additional features, such as one to verify the polarity; i.e., the orientations of some components. Also, we plan to evaluate the impact of using deep learning approach for recognizing components in order to increase robustness against severe illumination changes.

Supplementary Materials: The following are available online at http://www.mdpi.com/2076-3417/10/3/796/s1, Video S1: title: Automatic System To Assist Operators in The Assembly of Electronic Components.

Author Contributions: Conceptualization, I.S. and I.B.; Formal analysis, M.O.; Funding acquisition, D.A. and L.Q.; Investigation, M.O. and H.A.; Methodology, I.S.; Resources, D.A.; Software, M.O., H.A. and I.S.; Validation, F.A.S.; Writing–original draft, H.A.; Writing–review and editing, M.O. and F.a.S. All authors have read and agreed to the published version of the manuscript.

Funding: We would like also to thanks to SPRI agency for founding the SIRA applied research project under the Hazitek 2018 calls, where the research described in this paper was carried on.

Acknowledgments: We would like to thank the expert operators of Ikor for doing the user evaluation tests. We also thank Sara Garcia for her help in generating multimedia content.

Conflicts of Interest: The authors declare no conflicts of interest.

References

1. Posada, J.; Toro, C.; Barandiaran, I.; Oyarzun, D.; Stricker, D.; De Amicis, R.; Pinto, E.B.; Eisert, P.; Döllner, J.; Vallarino, I. Visual Computing as a Key Enabling Technology for Industrie 4.0 and Industrial Internet. *IEEE Comput. Graph. Appl.* **2015**, *35*, 26–40. [CrossRef] [PubMed]
2. Romero, D.; Stahre, J.; Wuest, T.; Noran, O.; Bernus, P.; Fast-Berglund, Å.; Gorecky, D. Towards an operator 4.0 typology: A human-centric perspective on the fourth industrial revolution technologies. In Proceedings of the International Conference on Computers and Industrial Engineering (CIE46), Tianjin, China, 29–31 October 2016.
3. Segura, A.; Diez, H.; Barandiaran, I.; Arbelaiz, A.; Álvarez, H.; Simões, B.; Posada, J.; García-Alonso, A.; Ugarte, R. Visual computing technologies to support the Operator 4.0. *Comput. Ind. Eng.* **2018**. [CrossRef]
4. De Lacalle, L.N.L.; Posada, J. Special Issue on New Industry 4.0 Advances in Industrial IoT and Visual Computing for Manufacturing Processes. *Appl. Sci.* **2019**, *9*, 4323. [CrossRef]
5. Palmarini, R.; Erkoyuncu, J.; Roy, R.; Torabmostaedi, H. A systematic review of augmented reality applications in maintenance. *Robot. Comput.-Integr. Manuf.* **2018**, *49*, 215–228. [CrossRef]
6. Hahn, J.; Ludwig, B.; Wolff, C. Augmented reality-based training of the PCB assembly process. In Proceedings of the 14th International Conference on Mobile and Ubiquitous Multimedia, Linz, Austria, 30 November–2 December 2015; Volume 46, pp. 395–399.
7. Sanna, A.; Manuri, F.; Lamberti, F.; Member, S.; Paravati, G.; Pezzolla, P. Using handheld devices to support augmented reality-based maintenance and assembly tasks. In Proceedings of the IEEE International Conference on Consumer Electronics, Las Vegas, NV, USA, 9–12 January 2015; pp. 178–179.

8. Wang, X.; Ong, S.K.; Nee, A. Real-virtual components interaction for assembly simulation and planning. *Robot. Comput.-Integr. Manuf.* **2016**, *41*, 102–114. [CrossRef]
9. Webel, S.; Engelke, T.; Peveri, M.; Olbrich, M.; Preusche, C. Augmented Reality Training for Assembly and Maintenance Skills. *BIO Web Conf.* **2011**, *1*, 00097. [CrossRef]
10. Yuan, M.; Ong, S.K.; Nee, A. Augmented reality for assembly guidance using a virtual interactive tool. *Int. J. Prod. Res.* **2008**, *46*, 1745–1767. [CrossRef]
11. InspectAR. Available online: https://www.inspectar.com (accessed December 10, 2019).
12. Neumann, U.; Majoros, A. Cognitive, performance, and systems issues for augmented reality applications in manufacturing and maintenance. In Proceedings of the IEEE 1998 Virtual Reality Annual International Symposium, Atlanta, GA, USA, 14–18 March 1998; pp. 4–11.
13. Bimber, O.; Raskar, R. *Spatial Augmented Reality Merging Real and Virtual Worlds*; A.K. Peters: Natick, MA, USA, 2005.
14. Posada, J.; Zorrilla, M.; Dominguez, A.; Simões, B.; Eisert, P.; Stricker, D.; Rambach, J.; Dollner, J.; Guevara, M. Graphics and Media Technologies for Operators in Industry 4.0. *IEEE Comput. Graph. Appl.* **2018**, *38*, 119–132. [CrossRef] [PubMed]
15. Alvarez, H.; Lajas, I.; Larrañaga, A.; Amozarrain, L.; Barandiaran, I. Augmented reality system to guide operators in the setup of die cutters. *Int. J. Adv. Manuf. Technol.* **2019**, *103*, 1543–1553. [CrossRef]
16. Kern, J.; Weinmann, M.; Wursthorn, S. Projector-based Augmented Reality for quality inspection of scanned objects. *ISPRS Ann. Photogramm. Remote Sens. Spat. Inf. Sci.* **2017**, *IV-2/W4*, 83–90. [CrossRef]
17. Korn, O.; Funk, M.; Schmidt, A. Design Approaches for the Gamification of Production Environments: A Study Focusing on Acceptance. In Proceedings of the 8th ACM International Conference on Pervasive Technologies Related to Assistive Environments, Corfu, Greece, 1–3 July 2015; pp. 1–7.
18. Rodriguez, L.; Quint, F.; Gorecky, D.; Romero, D.; Siller, H.R. Developing a Mixed Reality Assistance System Based on Projection Mapping Technology for Manual Operations at Assembly Workstations. *Procedia Comput. Sci.* **2015**, *75*, 327–333. [CrossRef]
19. Sand, O.; Büttner, S.; Paelke, V.; Röcker, C. smARt. Assembly—Projection-Based Augmented Reality for Supporting Assembly Workers. In Proceedings of the 8th International Conference Virtual, Augmented and Mixed Reality, Toronto, ON, Canada, 17–22 July 2016; pp. 643–652.
20. Manuri, F.; Pizzigalli, A.; Sanna, A. A State Validation System for Augmented Reality Based Maintenance Procedures. *Appl. Sci.* **2019**, *9*, 2115. [CrossRef]
21. Simões, B.; De Amicis, R.; Barandiaran, I.; Posada, J. X-Reality System Architecture for Industry 4.0 Processes. *Multimodal Technol. Interact.* **2018**, *2*, 72. [CrossRef]
22. Ojer, M.; Serrano, I.; Saiz, F.; Barandiaran, I.; Gil, I.; Aguinaga, D.; Alejandro, D. Real-Time Automatic Optical System to Assit Operator in the Assembling of Electronic Components. *Int. J. Adv. Manuf. Technol.* **2019**, in press.
23. Rother, C.; Kolmogorov, V.; Blake, A. "GrabCut": Interactive Foreground Extraction Using Iterated Graph Cuts. *ACM Trans. Graph.* **2004**, *6*, 309–314. [CrossRef]
24. Zhang, Z. A flexible new technique for camera calibration.
25. Romero Ramirez, J.; Muñoz Salinas, R.; Medina Carnicer, R. Speeded up detection of squared fiducial markers. *Image Vis. Comput.* **2018**, *76*, 38–47. [CrossRef]
26. Moreno, D.; Taubin, G. Simple, Accurate, and Robust Projector-Camera Calibration. In Proceedings of the Second International Conference on 3D Imaging, Modeling, Processing, Visualization and Transmission, Zurich, Switzerland, 13–15 October 2012; pp. 464–471.
27. Brooke, John. SUS: A quick and dirty usability scale. *Usability Eval. Ind.* **1996**, *189*, 4–7.
28. Bangor, A.; Kortum, P.; Miller, J. Determining What Individual SUS Scores Mean: Adding an Adjective Rating Scale. *J. Usabitily Stud.* **2009**, *4*, 114–123.
29. Martinetti, A.; Marques, H.; Singh, S.; Dongen, L. Reflections on the Limited Pervasiveness of Augmented Reality in Industrial Sectors. *Appl. Sci.* **2019**, *9*, 3382. [CrossRef]

applied sciences

MDPI

Article

Automatic Lip Reading System Based on a Fusion Lightweight Neural Network with Raspberry Pi

Jing Wen and Yuanyao Lu *

School of Information Science and Technology, North China University of Technology, Beijing 100144, China; wwwinj@163.com

* Correspondence: luyy@ncut.edu.cn

Received: 11 November 2019; Accepted: 10 December 2019; Published: 11 December 2019

Abstract: Virtual Reality (VR) is a kind of interactive experience technology. Human vision, hearing, expression, voice and even touch can be added to the interaction between humans and machine. Lip reading recognition is a new technology in the field of human-computer interaction, which has a broad development prospect. It is particularly important in a noisy environment and within the hearing- impaired population and is obtained by means of visual information from a video to make up for the deficiency of voice information. This information is a visual language that benefits from Augmented Reality (AR). The purpose is to establish an efficient and convenient way of communication. However, the traditional lip reading recognition system has high requirements of running speed and performance of the equipment because of its long recognition process and large number of parameters, so it is difficult to meet the requirements of practical application. In this paper, the mobile end lip-reading recognition system based on Raspberry Pi is implemented for the first time, and the recognition application has reached the latest level of our research. Our mobile lip-reading recognition system can be divided into three stages: First, we extract key frames from our own independent database, and then use a multi-task cascade convolution network (MTCNN) to correct the face, so as to improve the accuracy of lip extraction. In the second stage, we use MobileNets to extract lip image features and long short-term memory (LSTM) to extract sequence information between key frames. Finally, we compare three lip reading models: (1) The fusion model of Bi-LSTM and AlexNet. (2) A fusion model with attention mechanism. (3) The LSTM and MobileNets hybrid network model proposed by us. The results show that our model has fewer parameters and lower complexity. The accuracy of the model in the test dataset is 86.5%. Therefore, our mobile lip reading system is simpler and smaller than other PC platforms and saves computing resources and memory space.

Keywords: mobile lip reading system; lightweight neural network; face correction; virtual reality (VR)

1. Introduction

Lip reading refers to recognition of what people are saying by catching the speaker's lip motion. Especially in a noisy environment of voice superposition, or people with hearing impairment, the system will automatically detect lip area and identify the information [1]. Lip reading technology can supplement speech information by visual perception based on enhanced learning. Meanwhile, automatic lip reading technology can be widely used in Virtual Reality (VR) systems [2], information security [3], speech recognition [4] and auxiliary driving systems [5]. The lip reading system is mainly divided into the lip reading system based on traditional methods and the lip reading system based on in-depth learning. Traditional lip reading systems usually include two aspects: feature extraction and classification. For feature extraction, there are two kinds of methods: pixel-based and model- based. Pixel-based feature extraction uses the pixel values extracted from the interested mouth Region of Interest (ROI) as visual information. Then, the abstract image features are extracted by Discrete Cosine Transform (DCT) [6], Discrete Wavelet Transform (DWT) [7], and Principal Component Analysis (PCA) [8]. The method based model is to express the lips by a mathematical model, approximate the lip contour infinitely with curves and special features, and obtain the lip geometric features. For classification, the extracted features are sent to the classifier for classification. The commonly used classifiers are Artificial Neural Network (ANN) [9], Support Vector Machine (SVM) [10], and Hidden Markov Models (HMM) [11]. The breakthrough of in-depth learning also affects the development of lip reading technology. It has changed from the research direction of combining artificial design-based features with traditional classification model to an end-to-end complete system based on a deep-level neural network [12].

In recent years, researchers of the Google team proposed the MobileNets. MobileNets model to be an efficient model for mobile and embedded visual applications; it can combine depth separable convolution to construct a lightweight depth neural network [13]. This type of network offers an extremely efficient network architecture that can easily be matched to the requirements for mobile and embedded applications [14]. Considering that lip feature extraction has voice information and visual perception, we propose a hybrid neural network which combines MobileNets and LSTM to build a mobile lip reading system based on Raspberry Pi. Raspberry Pi is a credit card sized microcomputer which can do everything a normal PC can do and is widely supported by a large number of users. For example, it can be embedded in VR wearable devices. The whole lip reading recognition system runs on Raspberry Pi which is based on the Linux system; we deployed our project to the destination folder (/home/pi/lip-recognition) of our Pi to realize the self-startup. In this article, we realized self-startup by adding a script file. Also, we compared the Raspberry Pi with android smartphones and computers. Smartphones are limited by the space of PCBA (Printed Circuit Board Assembly), therefore they cannot allow the corresponding USB, HDMI and other interfaces. The low hardware cost of smartphones leads to low software adaptability. Although the computer has powerful process capability, it is inconvenient to move and cannot be deployed in simple devices. In contrast, Raspberry Pi has the advantages of small size, easy to carry, and low cost.

Our lip reading recognition system on mobile devices can be divided into the following stages: First, a lip reading video is obtained by a camera connected to the Raspberry Pi, and frames are extracted by using our own design rules to reduce the complexity of redundant information [15]. In the second stage, the multi-task cascade convolution network (MTCNN is used to correct the face and extract the key points of lip region [16]. Then MobileNets are used to extract lip features. After this, the attention-based LSTM network is used to learn the sequence information and attention weight between key frame features of the video. Finally, the final recognition results are predicted by two full connection layers and softmax. The softmax function converts the prediction results into probability [17]. The advantages of this mobile lip reading system are: (1) Face correction and lip key point detection using the MTCNN network can improve the accuracy of feature extraction. (2) Compared with PC-based mobile devices, Raspberry Pi has the advantages of small size, low power consumption and low cost. It can also accomplish some PC tasks and applications as usual. (3) Hybrid neural networks based on MobileNets and LSTM can reduce the number of parameters, the model complexity and the interference of invalid information.

The rest of this paper is organized as follows: In Section 2, we introduce the preparation and architecture of mobile lip reading system. Section 3 contains the analysis and experimental results of our proposed method. Section 4 provides conclusions and suggestions for future research directions.

2. Proposed Model

In this section, we propose the research framework and main steps. The framework we designed is a video recognition system based on mobile devices. Considering the performance limitations of mobile devices, we propose a framework as shown in Figure 1. First, we need to handle the dynamic video. We design an efficient method to extract the fixed frame. Second, we implement face location and face correction. Then we segment the mouth image region using MobileNets to extract features. Finally, we learn the temporal features and predict recognition results from LSTM.

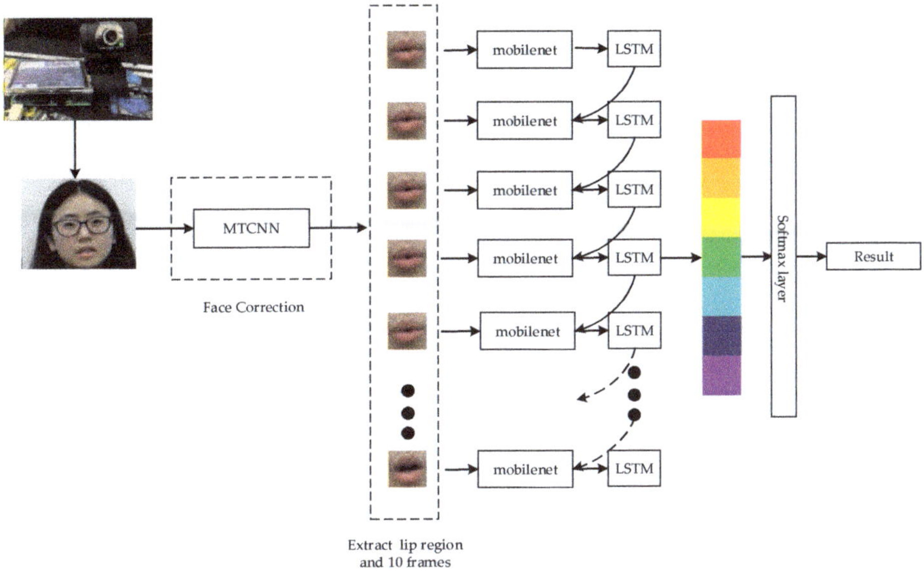

Figure 1. The architecture of our proposed Lip Reading System.

2.1. Extraction of Fixed Frames, MTCNN Detection and Correction of Lips

Lip detection is an essential part of lip reading recognition. However, previous studies were based on Dlib to locate the lips of a face [18]. Then the lip is segmented for feature extraction. In this paper, we independently design a frame extraction rule and propose a multi-task cascade convolution network (MTCNN) to extract the lip region as training data and locate key lip points to correct lip areas.

The quality of extracting fixed frames directly determines the quality of the recognition results. Therefore, we design a frame extraction scheme for lip recognition. In order to increase the robustness of the model, we design and implement a partition-based random selection method. If the total number of frames in a video segment is V, we first divide the video V into x blocks (x = 1, 2, 3, 4, 5 ... n). F represents the sequence number of each frame, because there may be situations where it cannot be divisible, we reduce the total number of frames. As shown in Formula (1).

$$x = v - \frac{v}{n} * n \tag{1}$$

Among them, $\lfloor \ \rfloor$ is the downward integer operator, the first X blocks the increase of the number of frames by, for each block, two frames are extracted as fixed frames. As shown in Formula (2).

$$F = A^{i}_{block_n} \tag{2}$$

Among them, $A^{i}_{block_n}$ represents selecting i frames in $block_n$ orderly.

MTCNN has an absolute advantage in the performance and recognition speed of face detection [19]. It is based on a cascade framework and can be divided into three layers: P-Net, R-Net, and O- Net [20]. The specific network structure is as follows:

- Proposal Network (P-Net): The network structure mainly obtains the regression vectors of the candidate windows, and the boundary areas of the face. The boundary areas are used for regression analysis to calibrate the candidate windows, and then merge the highly overlapping candidate windows by Non-maximum Suppression (NMS). (See Figure 2a).

- Refine Network (R-Net): The network structure also removes the false regions by boundary area regression analysis and NMS. However, due to the difference between the network structure and the P-Net network structure, there is an additional full-connection layer, so it can achieve a better effect of restraining the misjudgment rate. (See Figure 2b).
- Output Network (O-Net): This layer has one more convolution layer than the R-Net layer, so the processing results will be better. It works the same as the R-Net layer, but the layer monitors more of the face area and outputs five landmarks. (See Figure 2c).

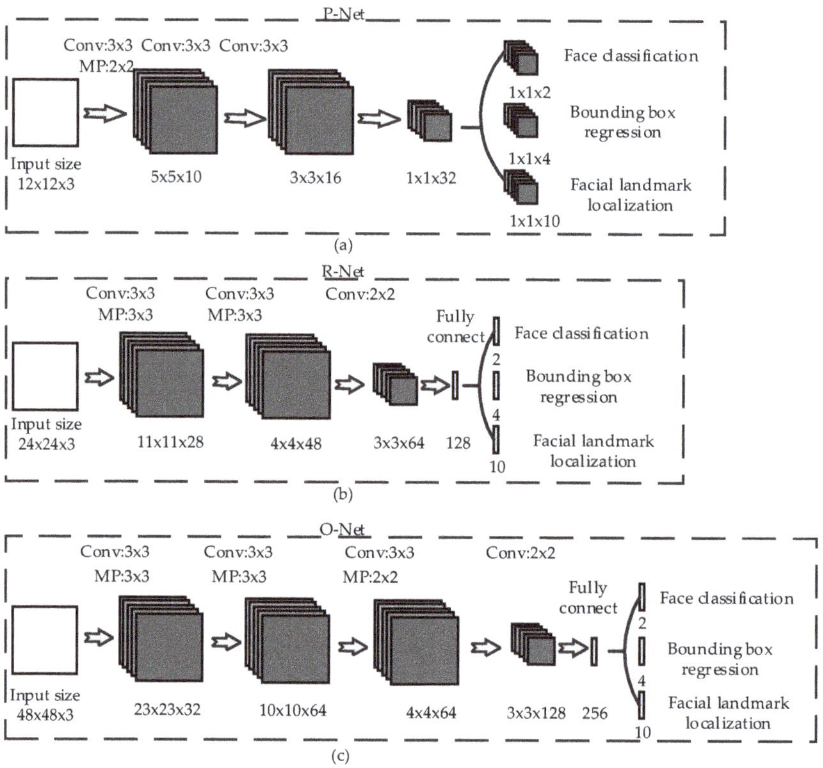

Figure 2. Multi-task cascade convolution network (MTCNN) architecture.

MTCNN model can detect the face area and face landmarks concurrently, and realize the calibration of feature landmarks. In this process, the model uses the method of Non-maximum Suppression. Based on this, we can achieve the goal of correcting the face. We achieve an effect by using MTCNN as shown in Figure 3 to improve the accuracy of the following recognition.

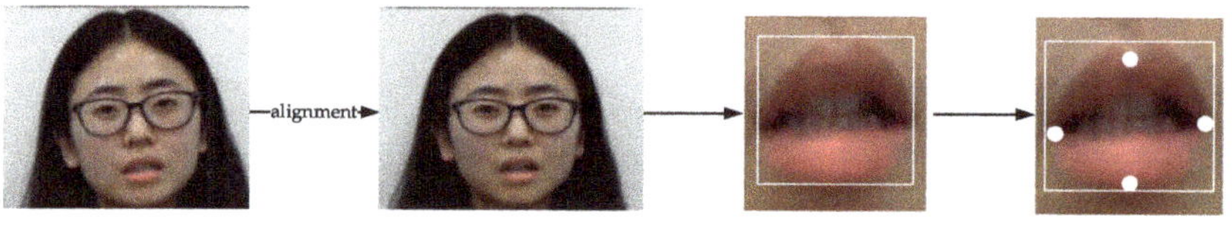

Figure 3. MTCNN face correction and lip extraction.

2.2. MobileNets Architecture

MobileNets based on a Streamlined Architecture uses Depthwise Separable Convolutions to construct a lightweight deep neural network. We introduce two simple global hyper-parameters. These hyper-parameters allow the model generator to select the appropriate size model for its application according to the constraints of the problem, thus reducing the complexity of the model [21].

The main work of MobileNets is using Depthwise Separable Convolutions instead of Standard Convolutions to solve the problems of computing efficiency and the parameters of the convolutional network [22–24]. The Standard Convolutions are shown in Figure 4a. It decomposes the standard convolution into Depthwise Convolutions and Pointwise Convolution. It is a key component of many effective neural network structures. The basic idea is to use a decomposition version instead of a complete convolution operator to decompose the convolution into two separate layers. The first layer is shown in Figure 4b, called Depthwise Convolution, which performs lightweight filtering by applying a convolution filter to each input channel. The second layer is Figure 4c, which is a 1×1 convolution called Pointwise Convolution. It is responsible for building new features by calculating the linear combination of the input channels.

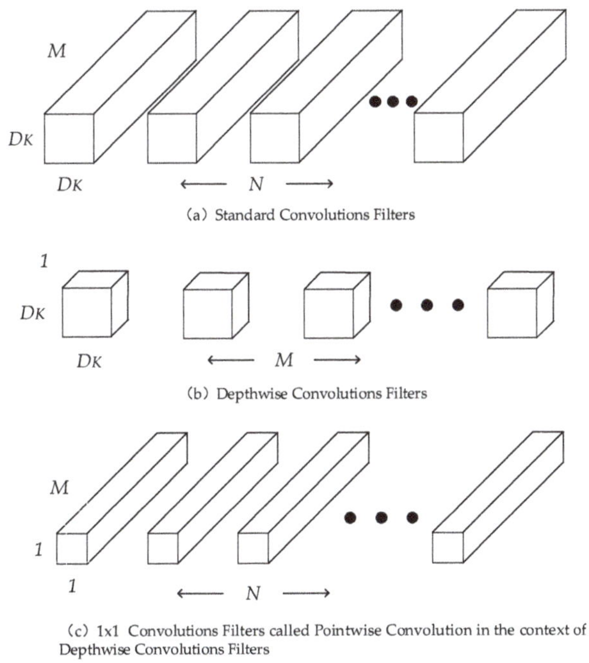

Figure 4. MoblieNets model structure.

In addition to the Depthwise Separable Convolutions, which is the basic component of MobileNets, the ReLU activation function is used in the model. Therefore the basic structure of Depthwise Separable Convolutions is shown in Figure 5. BN and ReLU are used to speed up the training speed and improve the recognition precision of the model [25].

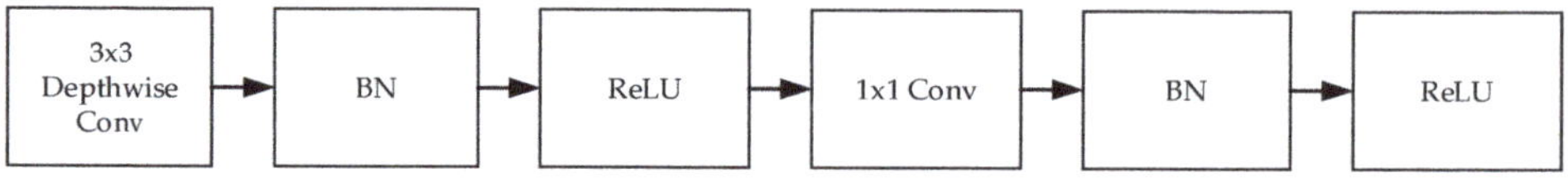

Figure 5. Depthwise separable convolutions basic structure.

2.3. LSTM (Long Short-Term Memory)

In order to solve the problem of gradient disappearance and gradient explosion when RNN processes long-sequence data, Hochreiter [26] proposed an improved form of RNN, called long short-term memory (LSTM), which is specially used to deal with information missing in long-term dependent sequences [27]. LSTM stores historical information by introducing memory units. By introducing three control gate structures, including the input gate, forget gate, and output gate, LSTM controls the increase and removal of information flow in the network. To better discover and utilize long-term dependencies from sequence data (such as video, audio, and text), memory cell remembers the associated information that needs to be remembered in a long sequence and forgets some of the useless information. Figure 6 shows the operations performed within a single LSTM cell. Among them, x_t represents the input vector of the network node at t time, h_t represents the output vector of the network node at t time, i_t, f_t, o_t, and c_t represent the input gate, forget gate, output gate and memory unit at t time respectively.

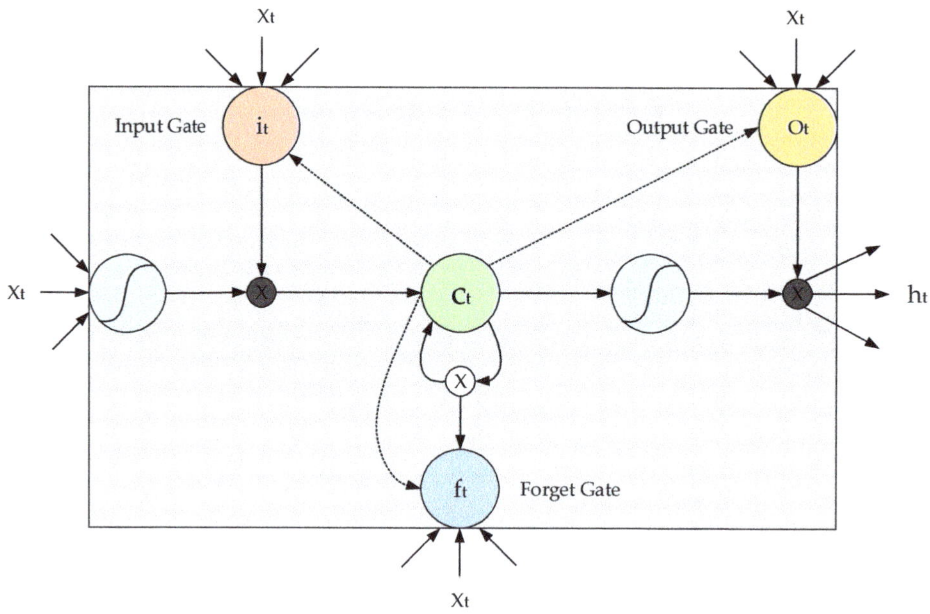

Figure 6. Long Short-Term Memory basic unit diagram.

The calculation steps of the input gate, forget gate, memory unit, and output gate in the LSTM unit are as follows:

1. Input gate: This gate is used to control the input node information. The mathematical expressions of the input gate output and candidate information are as follows:

$$i_t = \sigma(U_i x_t + W_i h_{t-1} + b_i) \tag{3}$$

$$g_t = \tan h\left(U_g x_t + W_g h_{t-1} + b_g\right) \quad (4)$$

Among them, U_i, W_i, and b_i represent the weights and biases of input gates, U_g, W_g, and b_g represent the weights and biases of candidate states, σ represents the sigmoid activation function, and $\tan h$ is the activation function.

2. Forget gate: This gate is used to control which information is discarded by the current LSTM unit. The mathematical expression of the forget gate is as follows:

$$f_t = \sigma\left(U_f x_t + W_f h_{t-1} + b_f\right) \quad (5)$$

Among them, U_f, W_f, and b_f denote the weights and biases of the forget gates respectively, and σ represents the sigmoid activation function.

3. The memory unit (memory cell): is used to save the state information and update the state. The mathematical expression of the memory unit c is as follows:

$$c_t = f_t \odot c_{t-1} + i_t \odot g_t \quad (6)$$

Among them, $\odot$ represents the Hadamar product.

4. Output gate: The gate is used to output the control of node information. The mathematical expression of the initial output value and the output of the LSTM unit is:

$$o_t = \sigma(U_o x_t + W_o h_{t-1} + b_o) \quad (7)$$

$$h_t = o_t \odot \tan h(c_t) \quad (8)$$

Among them, U_o W_o, and b_o represent the weight and bias of the output gate, respectively.

We input the pre-processed images into Mobilenets, extract the high-dimensional features of the images in fully connected layers, and input the features into LSTM model for learning the past and future information of the sequence features. In the memory unit of LSTM, putting in all the data passes through only one cell unit in different timing states. Also, it can reduce the number of parameters by keeping updating the weights. (See Figure 7) Among them, W(f), W(i), W(j), W(o) are weight parameters in the cell unit of LSTM. We aim to train these four weight parameters to optimize the LSTM network and reduce the input parameters.

The Dropout technique is used to mitigate the over-fit problems that have occurred during the training process. The Dropout technique reduces the complexity of the model by randomly dropping part of the neurons during each training process, thus improving the generalization ability of the model. In particular, it is assumed that a neural network with n nodes, in each training procedure, randomly discards the neurons in the network hidden layer at a probability p, and the probability of the retention of the neurons is 1–p. In general, this probability value p is set to 0.5 (referred to as the Dropout rate), since the randomly generated network structure is the most, that is, a set corresponding to 2n models. In addition, the joint action between the various neurons can be reduced, so that the appearance of a certain feature does not depend on the characteristic of the fixed relation, and can be used for weakening the interaction between the various features, so that the model is not too dependent on some local characteristics. Thus, the generalization ability of the model is enhanced.

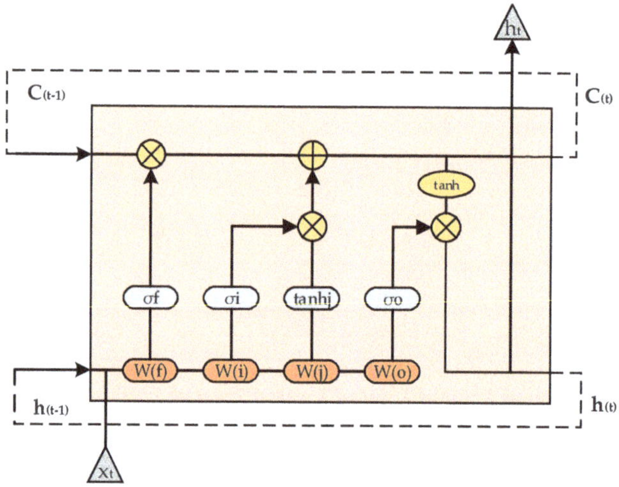

Figure 7. Neural network model based on weight analysis.

3. Experimental Dataset and Results

3.1. Dataset

The dataset in this paper is our self-made lip-language video database, recorded by six different producers (three men and three women) in a single constant environment. At this stage, it is worth emphasizing that privacy restrictions on datasets cause most of the data that we get to be Asian. The system has a good performance with Asian features. During recording, the head and the camera remain relatively static. The recorded content is the independent pronunciation of ten English words 0–9. Each person makes 100 sounds and divides them into different video clips. Then the sample size of the database is 6000. At the same time as we made the data enhancement to the original data, the dataset was expanded to 12,000 samples by increasing the light-and-dark, the image, the rotation, the Gaussian noise, the pepper and salt noise, etc. The original image has a resolution of 1920 × 1020, approximately 25 frames per second.

3.2. Results and Discussion

In this section, we evaluate the designed mobile lip reading recognition system, and analyze and compare the results on our dataset. We randomly disrupt the dataset and divide the training set and the test set according to 90% and 10%. We built MTCNN and LSTM networks with PyTorch. The random gradient drop method is used to train the network. The training model is inputted in 64 units. The learning rate of the first 100 iterations is 0.1, and then changed to 0.001 (in order to speed up the convergence rate).

We choose Raspberry Pi (Raspberry Pi 4, 4GB of LPDDR4 SDRAM, Dual monitor support, at resolutions up to 4K) based on the Linux system to realize dynamic lip reading recognition on the mobile end, as shown in Figure 8. Compared with the general PC computer platform, Raspberry Pi has the advantages of small size, low power consumption, and low cost. It can complete some tasks and applications that a PC platform can normally realize.

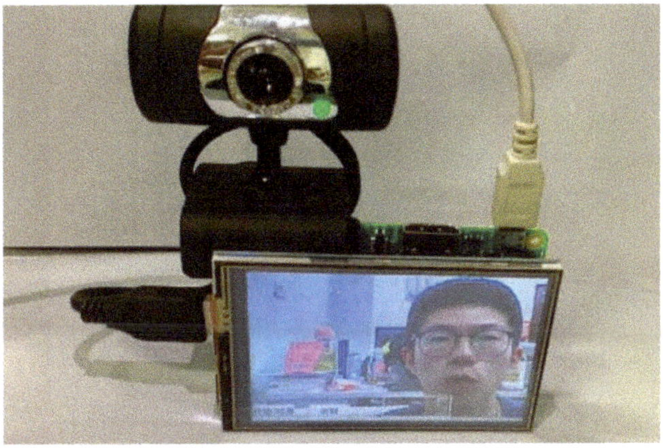

Figure 8. Physical photo of Lip Reading System on Raspberry Pi.

In order to evaluate the performance of the mobile lip reading system, we compared the mainstream research methods [28,29] through a large number of experiments, and the results are shown in Table 1. The proposed method can reduce a large number of parameters and reduce the complexity of the model and does not significantly degrade the performance of the model. We propose that the recognition time of the model is the time of video recognition, including decision- making. It can be seen that the recognition accuracy of the lightweight model proposed by us is smaller than that of the deep convolution hybrid network, and the recognition speed is greatly improved, which can meet the deployment and application of the mobile terminal.

Table 1. Performance comparison of the mainstream research methods.

Network	Accuracy	Time	Model Parameter (Million)
BiLSTM + AlexNet (No data expanded)	85.7%	10.0 s	61
AttentionLSTM+VGG16 (No data expanded)	88.2%	16.3 s	139
LSTM + MobileNets (Data expanded)	86.5%	7.3 s	5.2

The proposed system can be adapted well to the real environment without excessive degradation of model performance.

The training dataset and the test dataset are input into two MobileNets respectively, and then the sequence features of 4096 × 10 are extracted with the same LSTM model. Loss, accuracy, and recall of each period are shown in Figures 9 and 10.

In Figure 9, when the period (epochs) is about 19, the loss tends to be stable, indicating that the optimal solution has been reached at this time. The accuracy of the proposed network model in the test dataset is 86.5%. The test set performs very well, and the accuracy and loss eventually tend to balance, which shows that the spatial and temporal characteristics have been learnt.

As we want to identify the results as accurately as possible and reduce the situation of confusion recognition, we therefore pay more attention to the recall evaluation index of the model. Figure 10 below is the recall of our proposed model and the recall of the comparative model. It can be seen that our model performs well in pronunciation 2, 7, and 8, which is of great significance compared with previous studies. Considering the above experimental considerations, our research can be deployed well in the mobile terminal, and the efficiency is high.

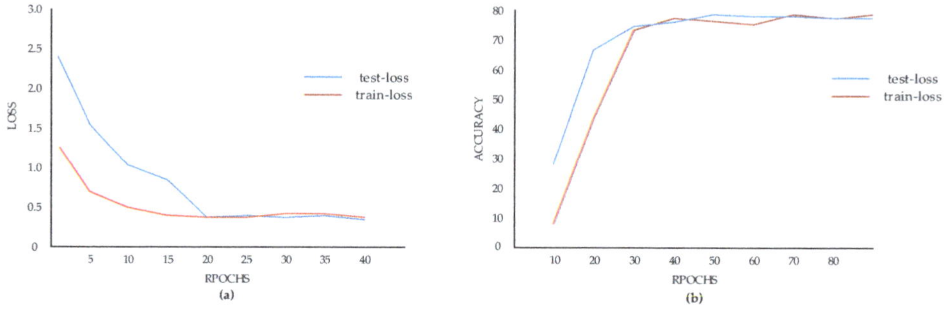

Figure 9. Comparison of our proposed model (**a**) Losses of each period in two networks. (**b**) Accuracy of each period in two networks.

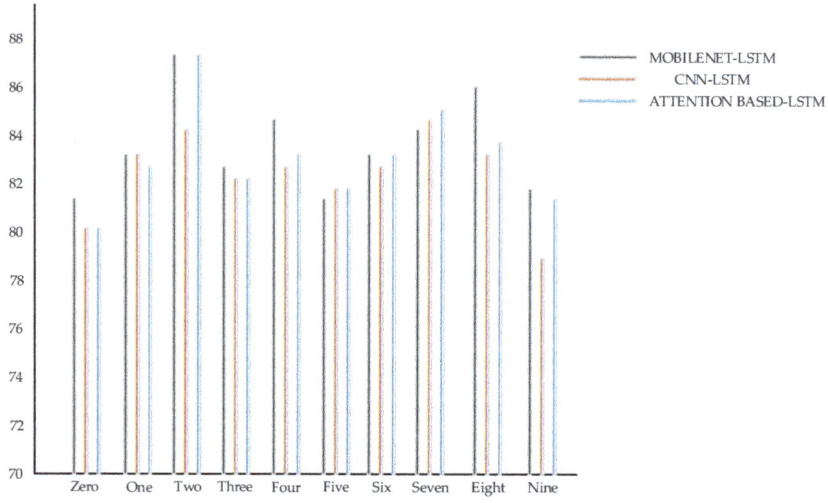

Figure 10. Recall of model and comparison model.

4. Conclusions

This paper concerns a lip language video obtained from Raspberry Pi. In order to optimize the recognition results and reduce redundant information, first, we extract the fixed-length frame sequence with our efficient and concise method, use MTCNN to correct the lip, and then use the lightweight MobileNets structure to train the model. Then, LSTM network is used to learn the sequence weights and sequence information between frame-level features. Finally, two full connection layers and one softmax layer are used to implement the classification. We independently established a dataset consisting of three men and three women. We recorded the pronunciation of English from 0 to 9. Each digital pronunciation was divided into independent video clips. We expanded the original dataset. Experimental results show that the mobile lip reading recognition system can effectively recognize words from the video, the complexity of the model is low, the amount of parameters has been reduced by 20 times, and the speed increased by 50%. This is the first mobile lip reading recognition system that uses a lightweight network in lip language research. It has reached the highest level of our research. We have also expanded the data to make it more versatile. However, our research of lip reading recognition is chronological, not aiming at a particular type of lip movement at a certain time. Therefore, the real-time performance is not good. In future research, we will focus on how to improve the speed of the recognition system based on time series and a train lip reading model on news video datasets, including news video samples from different environments to test our designed recognition system. According to the extended research of VR in the future, we will be more proficient in deploying and naming algorithms of mobile devices, so as to add multi-dimensional input to the VR scenes. For saving space of mobile devices and speeding up the operation and data-sharing, we will try to transfer data from raspberry pi by 5G (5th generation mobile networks) and utilize a server for algorithm identification and then return to the mobile devices, adding interactive virtual sensing technology to enable a wide range of facial recognition applications.

Author Contributions: Data curation, Y.L. and J.W.; Formal analysis, Y.L. and J.W.; Methodology, Y.L. and J.W.; Project administration, Y.L.; Resources, Y.L.; Supervision, Y.L.; Validation, J.W.; Visualization, J.W.; Writing—original draft, J.W.; Writing—review and editing, Y.L. and J.W.

Funding: This research was supported by the National Natural Science Foundation of China (61571013), by the Beijing Natural Science Foundation of China (4143061), by the Science and Technology Development Program of Beijing Municipal Education Commission (KM201710009003) and by the Great Wall Scholar Reserved Talent Program of North China University of Technology (NCUT2017XN018013).

Conflicts of Interest: The authors declare no conflict of interest.

References

1. Jaimes, A.; Sebe, N. Multimodal human–computer interaction: A survey. *Comput. Vis. Image Underst.* **2007**, *108*, 116–134. [CrossRef]
2. Loomis, J.M.; Blascovich, J.J.; Beall, A.C. Immersive virtual environment technology as a basic research tool in psychology. *Behav. Res. Methods Instrum. Comput.* **1999**, *31*, 557–564. [CrossRef] [PubMed]
3. Hassanat, A.B. Visual passwords using automatic lip reading. *arXiv* **2014**, arXiv:1409.0924.
4. Thanda, A.; Venkatesan, S.M. Multi-task learning of deep neural networks for audio visual automatic speech recognition. *arXiv* **2017**, arXiv:1701.02477.
5. Biswas, A.; Sahu, P.K.; Chandra, M. Multiple cameras audio visual speech recognition using active appearance model visual features in car environment. *Int. J. Speech Technol.* **2016**, *19*, 159–171. [CrossRef]
6. Scanlon, P.; Reilly, R. Feature analysis for automatic speech reading. In Proceedings of the IEEE Fourth Workshop on Multimedia Signal Processing, Cannes, France, 3–5 October 2001; pp. 625–630.
7. Matthews, I.; Potamianos, G.; Neti, C.; Luettin, J. A comparison of model and transform-based visual features for audio-visual LVCSR. In Proceedings of the IEEE International Conference on Multimedia and Expo, Tokyo, Japan, 22–25 August 2001; pp. 825–828.
8. Aleksic, P.S.; Katsaggelos, A.K. Comparison of low- and high-level visual features for audio-visual continuous automatic speech recognition. In Proceedings of the IEEE International Conference on Acoustics, Speech and Signal Processing, Montreal, QC, Canada, 17–21 May 2004; Volume 5, pp. 917–920.
9. Stork, D.G.; Wolff, G.; Levine, E. Neural network lip reading system for improved speech recognition. In Proceedings of the International Joint Conference on Neural Networks, Baltimore, MD, USA, 7–11 June 1992; pp. 285–295.
10. Shaikh, A.A.; Kumar, D.K.; Yau, W.C.; Azemin, M.C.; Gubbi, J. Lip reading using optical flow and support vector machines. In Proceedings of the 2010 3rd International Congress on Image and Signal Processing, Yantai, China, 16–18 October 2010; pp. 327–330.
11. Puviarasan, N.; Palanivel, S. Lip reading of hearing impaired persons using HMM. *Expert Syst. Appl.* **2011**, *38*, 4477–4481. [CrossRef]
12. Lu, Y.; Li, H. Automatic Lip-Reading System Based on Deep Convolutional Neural Network and Attention-Based Long Short-Term Memory. *Appl. Sci.* **2019**, *9*, 1599. [CrossRef]
13. Howard, A.G.; Zhu, M.; Chen, B.; Kalenichenko, D. MobileNets: Efficient Convolutional Neural Networks for Mobile Vision Applications. *arXiv* **2017**, arXiv:1704.0486.
14. Kamal, K.C.; Yin, Z.D.; Wu, M.Y.; Wu, Z.L. Depthwise separable convolution architectures for plant disease classification. *Comput. Electron. Agric.* **2019**, 8. [CrossRef]
15. Simonyan, K.; Zisserman, A. Very deep convolutional networks for large-scale image recognition. *arXiv* **2014**, arXiv:1409.1556v6.
16. Ma, M.; Wang, J. Multi-view Face Detection and Landmark Localization Based on MTCNN. In Proceedings of the 2018 Chinese Automation Congress (CAC), Xi'an, China, 30 November–2 December 2018.

17. Martins, A.; Astudillo, R. From softmax to sparsemax: A sparse model of attention and multi-label classification. In Proceedings of the International Conference on Machine Learning, New York, NY, USA, 19–24 June 2016; pp. 1614–1623.
18. Lu, Y.; Yan, J. Automatic Lip Reading Using Convolution Neural Network and Bi-directional Long Short-term Memory. *Int. J. Pattern Recognit. Artif. Intell.* **2019**. [CrossRef]
19. Edwin, J.; Greeshma, M.; Mithun Haridas, T.P.; Supriya, M.H. Face Recognition based Surveillance System Using FaceNet and MTCNN on Jetson TX2. In Proceedings of the International Conference on Advanced Computing & Communication Systems, Coimbatore, India, 15–16 March 2019.
20. Jia, X.; Gengming, Z. Joint Face detection and Facial Expression Recognition with MTCNN. In Proceedings of the 2017 4th International Conference on Information Science and Control Engineering, Changsha, China, 21–23 July 2017.
21. Howard, A.G.; Zhu, M.; Chen, B.; Kalenichenko, D.; Wang, W.; Weyand, T.; Andreetto, M.; Adam, H. MobileNets: Efficient Convolutional Neural Networks for Mobile Vision Applications. *arXiv* **2017**, arXiv:1704.04861.
22. Chollet, F. Xception: Deep Learning with Depthwise Separable Convolutions. In Proceedings of the 2017 IEEE Conference on Computer Vision and Pattern Recognition, Honolulu, HI, USA, 21–26 July 2017.
23. Sandler, M.; Howard, A.; Zhu, M.; Zhmoginov, A.; Chen, L.C. MobileNetV2: Inverted Residuals and Linear Bottlenecks. In Proceedings of the IEEE/CVF Conference on Computer Vision and Pattern Recognition, Salt Lake City, UT, USA, 18–23 June 2018.
24. Chen, H.Y.; Su, C.Y. An Enhanced Hybrid MobileNet. In Proceedings of the IEEE International Conference on Awareness Science and Technology, Fukuoka, Japan, 19–21 September 2018.
25. Michele, A.; Colin, V.; Santika, D.D. Santika MobileNet Convolutional Neural Networks and Support Vector Machines for Palmprint Recognition. *Procedia Comput. Sci.* **2019**, *157*, 110–117. [CrossRef]
26. Hochreiter, S.; Schmidhuber, J. Long short-term memory. *Neural Comput.* **1997**, *9*, 1735–1780. [CrossRef] [PubMed]
27. Gers, F.A.; Schmidhuber, J.; Cummins, F. Learning to forget: Continual prediction with LSTM. In Proceedings of the 9th International Conference on Artificial Neural Networks: ICANN'99, Edinburgh, UK, 7–10 September 1999.
28. Weilin, Z.; Huilin, X.; Zhen, Y.; Tao, Z. Bi-directionallong short-term memory architecture for person re-identification with modified triplet embedding. In Proceedings of the IEEE International Conference on Image Processing, Beijing, China, 17–20 September 2017.
29. Cho, K.; Courville, A.; Bengio, Y. Describing multimedia content using attention-based encoder-decoder networks. *IEEE Trans. Multimed.* **2015**, *17*, 1875–1886. [CrossRef]

applied sciences

MDPI

Article

Design, Application and Effectiveness of an Innovative Augmented Reality Teaching Proposal through 3P Model

Alejandro López-García [1], Pedro Miralles-Martínez [2,*] and Javier Maquilón [3]

1 DICSO Group, Department of Mathematics and Social Sciences Teaching, Faculty of Education, University of Murcia, 30100 Murcia, Spain; aloga@um.es
2 DICSO Group (P.I.), Department of Mathematics and Social Sciences Teaching, Faculty of Education, University of Murcia, 30100 Murcia, Spain
3 Institutional Evaluation, Accreditation, Competences and Teaching-Learning in EHEA, Department of Methods of Research and Diagnosis in Education, Faculty of Education, University of Murcia, 30100 Murcia, Spain; jjmaqui@um.es
* Correspondence: pedromir@um.es; Tel.: +34-868-88-7076

Received: 12 November 2019; Accepted: 4 December 2019; Published: 11 December 2019

Abstract: Augmented reality (AR) has evolved hand in hand with advances in technology, and today is considered as an emerging technique in its own right. The aim of our study was to analyze students' perceptions of how useful AR is in the school environment. A non-experimental quantitative design was used in the form of a questionnaire in which 106 primary sixth-grade students from six schools in the Region of Murcia (Spain) participated. During the study, a teaching proposal using AR related to the content of some curricular areas was put forward in the framework of the 3P learning model. The participants' perceptions of this technique were analyzed according to each variable, both overall and by gender, via a questionnaire of our own making, which had previously been validated by AR experts, analyzing its psychometric qualities. The initial results indicate that this technique is, according to the students, useful for teaching the curriculum. The conclusion is that AR can increase students' motivation and enthusiasm while enhancing teaching and learning at the same time.

Keywords: augmented reality; applications in subject areas; interactive learning environments; 3P model; primary education; educational technology

1. Introduction

Education is at a stage where the ways of accessing information and knowledge are changing and evolving at a dizzying pace due to a series of developing technologies and pedagogies (gamification, cloud computing, learning in networks, flipped classroom, Massive Open Online Course-MOOC, etc.). The idea of using latest generation mobile devices in the classroom is gaining new ground. It has been demonstrated that mobile learning can enhance digital literacy skills and serve, among other benefits, as a strong contextual and institutional support which monitors students' research as well as some aspects of their learning [1,2].

One of the main emerging concepts in this sector is augmented reality (AR). An argument supporting the growth of AR is the systematic review of Akçayır and Akçayır [3], which analyzed 68 research papers from the Social Sciences Citation Index (SSCI). Their study found that the number of articles published on AR applied to education had increased exponentially in the last four years. An important question is how it is perceived as a technique that can enhance learning in the framework of various theoretical models and how an instructional design can be put into practice via a coherent methodology. This study attempts to answer these questions in depth in order to intensify awareness of AR and boost its development, showing how satisfaction, motivation and other positive variables are manifested in the participants in a noticeable way after its implementation.

1.1. Conceptualization and Terminology

Caudell and Mizell [4], who coined the concept, define AR as a technology that enhances the user's field of vision with the information necessary to perform a task, thanks to computational processes that can transform and chart simple graphics in real time. Milgram and Kishino [5] add to this definition when they describe AR as any case in which a real environment is enhanced with virtual objects (computer graphics). Likewise, they present the term within a taxonomy (a virtuality continuum), in which all the possible types of viewing appear. Within this continuum, the real and virtual elements coexist in a single mixed reality space where AR is closer to the entirely real environment than to the entirely virtual environment. Another pioneer, Azuma [6], sees AR as a variation of virtual environments which enables the user to see reality through superimposed objects.

It is noteworthy that some authors [7,8] have contradicted the idea of presenting AR conceptually as a technology in the strict sense, since it can be based on technology or understood as a resource that can accompany technology or draw from it, which means that it is necessary to interpret it beyond this exclusively classificatory treatment. Hence, one approach to this issue could be that AR is an emerging technique which is mediated by technology and which enables the superimposition of virtual information on a real environment, thus facilitating access to the borders of mixed reality, which can be two- or three-dimensional.

1.2. Educational Experiences and Evidence

The usability of AR environments has experienced a tremendous upsurge in education in recent years. In this vein, some authors [3,9,10] have carried out systematic reviews in order to discover more about this technique, and the results point to AR being potentially able to support or enhance teaching and learning processes, concluding that its didactic use should grow in the coming years as there is an increase in the research, the expected technological developments and users' knowledge.

More specifically, there are studies on games [11–13], applications [14–16] and illustrated books [17,18], which use AR to facilitate functionalities that allow teachers to establish new ways of showing relationships and connections for learning, incorporating image and video animations to the illustration of their texts.

In relation to the above, and starting from the construction of new learning processes, AR may be advantageous for formal learning since it allows students to interact with the real world and the digital world at the same time, and in that way creating new exciting and refreshing classroom situations in which to acquire knowledge. Indeed, it seems that this technique is widely accepted and improves academic outcomes [19,20].

Other studies have shown that AR can have a positive impact on motivation, attention and attitude [9,21], on conflict resolution and comprehension [15,22] and on learning efficiency and performance [23,24]. Positive results have also been obtained with regards to the use of this technique in the design and use of 3D video and image markers [25] demonstrating that AR through markers is the most widely used version. Likewise, two areas that should be studied are the accessibility and usability of the learning experience [9]. There is no doubt that to definitively implement AR in educational establishments this technique's value for teaching and learning and its coexistence with other time-tested theoretical models and curricula must be demonstrated to and accepted by the educational community.

1.3. Learning Theories and Augmented Reality

In order to carry out experiences of this type in classrooms, a solid and consistent methodological approach is necessary with instructional design supported by accepted learning theories, which facilitate and justify the educational process itself.

During the review of the scientific literature, examples were identified among authors using AR to support implementing learning models and theories, such as the Situated Learning Theory [12,14,15], Kolb's Experiential Learning Theory [23], or the Constructivist Theory [12], among others.

Robust models that allow for the collection of relevant information on educational perception or satisfaction with regards to AR techniques also exist. To this effect, it is essential to adapt these theoretical principles to the educational process and to the product that will give us the information we desire. Some examples are, the Integrated Cognitive Affective Model of Learning with Multimedia (ICALM) by Plass and Kaplan [26], the Attention, Relevance, Confidence, and Satisfaction (ARCS) model of motivational design by Keller [27], which was developed in order to identify actions and approaches that would enable understanding the main motivational influences or solve problems about motivation for learning, following a systematic design process, and the Technology Acceptance Model (TAM) by Davis et al. [28]. Indeed, the effect of AR on motivation and learning for all these models has been studied [21,23,25].

Another possible area for studies about AR is the 3P learning model. Starting from the original model proposed by Dunkin and Bidle [29], Biggs [30] which adopts this approach, known as the 3P model, to describe and analyze the student body's perspective and learning in a system composed of three basic components: Presage, Process and Product (3P). According to this model, these three factors interact with each other in a tendency toward equilibrium, which represents the proper functioning and success of the teaching and learning process [30].

With all due caution, these theoretical models for learning provide methodological rigor and coherency to justify instructional design based on AR, taking all variables into account. As such, we believe it is necessary to increase the volume of investigation focused on these fields of study, raising diverse questions: What activities can be implemented in pre-adolescent stages to ensure students achieve a higher quality learning experience using AR? What are students' perceptions about the real applications of using this technique? Is it possible to implement AR from a theoretical model which improves learning effectiveness and motivation? To answer these questions, the idea of designing and implementing a teaching environment based on AR and, later, analyzing the students' degree of perception for this technique in primary education was conceived.

2. Materials and Methods

2.1. Aims of the Research

The overall aim is to evaluate how useful augmented reality is in improving the teaching and learning processes for sixth-grade students in primary education. This aim can be split into two specific goals:

1. To analyze the psychometric qualities of the questionnaire "Sixth-grade primary education students' perception of the usefulness of augmented reality" (PEURA-E).
2. To evaluate the usefulness of augmented reality as a teaching and learning technique in the framework of the 3P model, according to each variable, both overall and by gender.

2.2. Design and Participants

To meet these objectives, a non-experimental, quantitative design study based on a survey was chosen. There were 106 participants from the sixth-grade of six primary schools in the Region of Murcia, an autonomous community of south-east Spain; 58 were boys and 48 were girls. Non-probabilistic convenience sampling was used to select the centers [31]. In this regard, the following inclusion conditions or criteria were considered when selecting the participants:

- Both students and their parents had to be aware of the study's objectives and give prior informed consent.
- All participants had to be aged between 11 and 12 years old.

- More than 50% of the students selected needed to have a last generation mobile device.
- Participants had to be registered on the AR platform and download the application on their mobile device at the start of the sessions.

2.3. Integration of the Attention, Relevance, Confidence, and Satisfaction (ARCS) Model into the Augmented Reality (AR)-Based 3P (Presage, Process and Product) Model

A basic assumption of this study is the idea of an instructional design based on the Presage-Process-Product (3P) learning model, adopted by Biggs [30], with the addition of AR as a mediating element for learning. In this defining framework, everything related to the students' characteristics and the teaching context are included in the presage stage. The student learning process and the connection between the strategies used by the students, their motivation to learn, and the use of AR are included in the process stage. The component which deals with satisfaction and expected academic performance, if the remainder of the conditions are met, is included in the product stage. Figure 1 shows an adaptation of this model used to carry out our study.

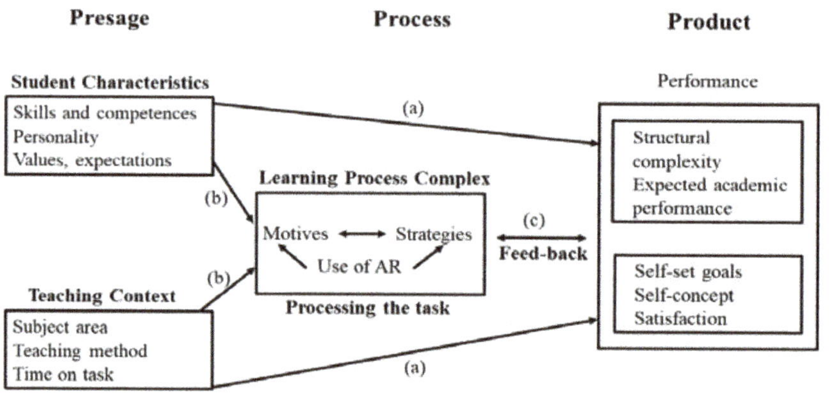

Figure 1. General model of student learning. Adapted from a previous study [32].

Regarding the teaching and learning processes, Keller [27] proposed that in every educational process it is essential to design an action plan that promotes student interest and attention while pointing out the relevancy of what has been learned, bestowing confidence in the achievement of learning objectives, and measuring satisfaction after overcoming the proposed challenges. This idea culminates in the ARCS model based on the categories of Attention, Relevance, Confidence and Satisfaction, as precursors to learning.

Consequently, this study intends to integrate the Keller ARCS model [27], in the Process stage of the Biggs 3P model [30]. To this end, a teaching proposal with activities based on AR was created following the ARCS model in order to understand the primary motivational influences in learning, in accordance with the four established categories.

This model is shown in Table 1, with a description of the relationship among the different strategies employed throughout the teaching experience and the implicit motivational categories.

Table 1. Motivational categories and strategies. Adapted from Keller [33].

Category	Definition	Motivational Strategies
Attention	Spark the students' interest and stimulate the curiosity to learn.	View images and videos Interact with virtual layers Manipulate the models Follow the web links Listen to the audio tracks
Relevance	Value the usefulness, applicability and impact of the content	Pose questions orally Learn search strategies Analyze the content's usefulness Show the relationship to the curriculum
Confidence	Foster positive expectations for success	Complete the interactive activities and surveys Repeat the action models Correct the activities Foster positive reinforcement
Satisfaction	Give their perspective after overcoming the academic challenges	Fill out the final questionnaire Discuss the overall experience Raise doubts and problems Propose improvements

2.4. Teaching Proposal

A teaching proposal was designed and implemented in the framework of the described models that took into consideration Royal Decree 198/2014 of 5 September 2014, which establishes the primary education curriculum in the Autonomous Community of Murcia, under the provisions of the LOMCE, the Education Act in force in Spain since 2013. Five subject areas from the block of core subjects were selected and a 55-min learning session was prepared for each of them addressing the AR content; 15 min were allocated to the teaching process, 25 min to the activities and 15 min to allow the students to experiment and clear up any doubts. Some photographs of the research can be seen in the supplementary document 1. As such, several interactive images were designed for each subject area. They included web links, images, audio tracks, videos and surveys about the described curricular contents so the information could be viewed on a mobile device or tablet through markerless or level image recognition.

We used the tool Layar Creator (version 7) [34] to prepare these AR environments. This software allows users to design images and incorporate virtual layers in a variety of formats. Figure 2 shows the process of creating an interactive image using the platform.

Figure 2. Tool for making an interactive image.

The teaching method for each session was based on showing the virtual information superimposed on the interactive images, as is shown in Figure 3. The content was explained, and the motivational strategies used in the different categories of the ARCS model were emphasized with the goal of guiding the students throughout the learning process. Next, the students were tasked with interacting with the layers, and completing the activities, either individually or in groups.

Figure 3. Example of augmented reality (AR) viewing.

Students were then given enough time to practice and create new images, autonomously, using the resources provided by the teaching staff and with the freedom to make their own models. Moreover, they were able to take the interactive images home or to the library and experiment with AR outside the school environment, researching and creating new proposals, on their own, using the resources from class and the contents taught, which are shown in Table 2.

Table 2. Classification of the resources used by subject area and curricular content.

Areas	Blocks	Contents	Resources
Spanish Language and Literature	Block 2: Reading	Text comprehension according to typology	Reading on screen with access to different texts. Layer with a web link to reading comprehension activities.
		Reading of different genres of text: descriptive, argumentative, expository, instructive, literary	Display of a strip of images of different types of text and practical examples. Web link to a content manager with activities for identifying texts.
		Strategies for reading comprehension: Dictionary	Layer with a web link to an academic dictionary. Layer with a blog of interactive activities to evaluate reading comprehension
		Journalistic and advertising texts. Information, opinion and advertising	Layer with a web link to the newspaper *La Verdad* from Murcia. Viewing of a video about journalistic and advertising texts. Interactive survey activity on screen about the resources explained in the video.
First Foreign Language	Block 3: Understanding written texts	Understanding simple written narratives using the present and future	Viewing of a video to study present and future tense grammar resources. Interactive survey activity on screen about the resources explained in the video, regarding peace and solidarity.
		Understanding written texts about ownership related to people and objects	Layer with a blog about the song "Imagine" by John Lennon dealing with the elements of ownership of people and objects and an assessment activity. Web link to an online translator as a quick strategy for reading comprehension.
		Nouns, pronouns, articles and demonstratives	Display of a strip of images about identity expressions (nouns, pronouns, articles and demonstratives) and vocabulary. Layer with interactive audio tracks to focus on the action of 'listening and repeating' the words being taught.
		Lexis related to daily routines and natural environment	Viewing of interactive images on screen of vocabulary related to daily routines and nature. Layer with an activity about the resources taught in the images that were shown.

Table 2. *Cont.*

Areas	Blocks	Contents	Resources
Social Sciences	Block 3: Living in society	Cultural manifestations in Spain	Display of a strip of images about the main artistic and historical monuments in Spain. Layer of an image that contains a web link to Wikipedia to learn about the history of the Kingdom of Spain.
		The European Union	Viewing of an interactive map of Europe, with its countries and capitals superimposed in AR. Web link to the official website of the European Union to study its history, countries and symbols. Display of a strip of explanatory images with questions about the European Union's institutions, government bodies, and symbols.
		Employability and entrepreneurship	Video about Spain, its entry into the European Union, its economic systems, and its main institutions and bodies. Interactive survey activity on screen about the resources explained in the video.
Natural Sciences	Block 2: Human beings and health	Harmful effects of alcohol and drug consumption	Layers with web links to different resources about alcohol and drugs and their harmful effects for health. Interactive activity on screen about the effects of these substances.
		Knowledge of oneself and others. Identity and personal autonomy Relating to others	Interactive activity about the construction of a pyramid that addresses the issues of personal autonomy and its relationship with healthy actions and routines. Layer with a web link that includes healthy behaviors, eating routines, and ways of caring for the human body with a self-evaluation activity.
		Decision making: criteria and consequences	Viewing of a video on screen that addresses making decisions and the consequences of good or bad behavior. Viewing of a strip of images about awareness and prevention of the consumption of alcohol, drugs and tobacco.
Mathematics	Block 2: Numbers	Reading and representing fractions.	Display of an introductory video on the topic of fractions. Display of a strip of images about the process of reading, ordering and representing fractions with explanations supported by examples.
		Order of simple fractions	Interactive questionnaire with short questions or exercises on ordering fractions.
		Fraction of a number and equivalent fractions	Layer with a blog that includes activities, self-evaluation exercises and games about fractions of a number and equivalent fractions.

From this table, it becomes apparent that curricular contents are taught through a series of learning actions or activities which favor experimenting with AR. Most of the tasks are supported by academic web portals, web resources, content managers, blogs or activities whose origin is the manipulation of AR, whether using images, videos, and audio tracks or the use of touch-sensitive interactive surveys.

2.5. Data Collection Tool

Regarding information collection, the PEURA-E questionnaire (see the tool in the supplementary document 2) whose language was adapted to the age and maturity of the participants was designed and validated to collect the data. The instrument contained 40 items classified in seven constructs: Teaching, Learning, Spanish Language and Literature, First Foreign Language, Social Science, Natural Sciences, and Mathematics. According to Krosnick and Presser [35], this is a potentially large number of questions for sixth-grade students. For this reason, two control questions (items 33 and 39) were included in order to stand out when answering and prevent students from responding mechanically. However, the data from these two questions were not included in the final analysis so as not to alter the results of the research regarding perceived usefulness.

Regarding content, it is composed of closed-ended questions presented in the form of a Likert scale with five options, ranging from 1 (strongly disagree) to 5 (strongly agree), in addition to a nominal question at the beginning, to indicate gender. More specifically, the questionnaire contains several questions soliciting information for more than one dimension or aspect, called double-barreled questions [36]. It should be noted that participants were advised that all the conditions of each item must be met for an answer to be given a positive rating, such that if the student felt any specific criterion was not met for the item, he or she was free to give the entire item a negative rating.

3. Results

3.1. Analysis and Description of Data

3.1.1. Aim 1

To address the first aim, the analysis of the psychometric properties of the PEURA-E questionnaire, the tool underwent a validation process. Three experts in AR validated the tool using an official rating scale designed by Serrano [37]. This table appears in the supplementary document 3. The content was validated by defining the items in the questionnaire according to relevant criteria related to the participants in the research and to their area of expertise. The scale was structured by organizing the items into 5 blocks (see Table 3).

Table 3. Classification of the rating scale used by blocks and items.

Blocks	Dimensions	Items
1	Presentation of the questionnaire	1 to 7
2	Instructions for completing the questionnaire	8 to 10
3	Structure and overall design of the questionnaire	11 to 20
4.1	Suitability of the lexis/language	Analysis of each item
4.2	Suitability of the response options	Analysis of each item
5	Overall rating of the questionnaire's suitability	21 to 23

Table 4 shows the mean scores for the items in the scale and for each of the dimensions. Note that the scores range from 1 (strongly disagree) to 4 (strongly agree).

Table 4. Content validity of the "Sixth-grade primary education students' perception of the usefulness of augmented reality" (PEURA-E) questionnaire according to the experts.

Blocks	Dimensions	Mean
1	Presentation of the questionnaire	3.71
2	Instructions for completing the questionnaire	3.22
3	Structure and overall design of the questionnaire	3.57
4.1	Suitability of the lexis/language	3.96
4.2	Suitability of the response options	4.00
5	Overall rating of the questionnaire's suitability	3.89

The experts' evaluation was highly positive (see supplementary document 4). Nevertheless, some modifications in the wording of certain items and in some aspects of the form were needed to obtain the final version. Regarding reliability, the questionnaire was analyzed to verify the internal consistency of the PEURA-E tool. The covariance of the items was measured with Cronbach's alpha, which is commonly used in questionnaires that have a range of answers for each item [31,38]. Internal consistence was acceptable ($\alpha = 0.927$). Finally, to evaluate the tool's viability and identify any possible errors, a pilot study with a group of students of the same age and characteristics as the present study was carried out using the teaching approach and the questionnaire validated by experts. After the pilot study, which served as a trial run, only minimal modifications to the questionnaire were necessary.

3.1.2. Aim 2

To respond to objective 2, which referred to assessing the usefulness of AR as a teaching and learning technique in the framework of the 3P model, according to each variable, both overall and by gender, the students filled out the questionnaire individually and the data was organized by blocks or theoretical constructs with the goal of clarifying and organizing the results.

Table 5 shows the descriptive statistics for the overall perception students have of the usefulness of AR for the teachers' teaching and for their own learning.

Table 5. Students' overall scoring of the teaching and learning constructs.

Overall	Total	Strongly Disagree	Disagree	Neither Disagree Nor Agree	Agree	Strongly Agree	Md.	M	Sd.
Teaching									
1. AR allows teaching to happen via discovery.									
Frq.	106	3	3	8	29	63	5.00	4.38	0.951
%	100	2.8	2.8	7.5	27.4	59.4			
2. AR can be another way for the teacher to teach knowledge as well as using books and note taking									
Frq.	106	1	0	9	31	65	5.00	4.50	0.734
%	100	0.9	0	8.5	29.2	61.3			
3. Teachers can teach better if they also use AR in the classroom									
Frq.	106	3	5	11	48	39	4.00	4.08	0.957
%	100	2.8	4.7	10.4	45.3	36.8			
4. AR can be used by teachers to build knowledge adapted to each area									
Frq.	106	2	1	13	38	52	4.00	4.29	0.862
%	100	1.9	0.9	12.3	35.8	49.1			
Learning									
5. AR can increase students' attention									
Frq.	106	5	2	12	39	48	4.00	4.16	1.025
%	100	4.7	1.9	11.3	36.8	45.3			
6. AR can increase students' motivation									
Frq.	106	4	6	8	30	58	5.00	4.25	1.067
%	100	3.8	5.7	7.5	28.3	54.7			
7. AR helps students to understand the contents better									
Frq.	106	4	8	13	38	43	4.00	4.02	1.087
%	100	3.8	7.5	12.3	35.8	40.6			
8. Activities using AR encourage students to participate more in class									
Frq.	106	6	6	12	43	39	4.00	3.97	1.108
%	100	5.7	5.7	11.3	40.6	36.8			
9. AR can help students to work in collaboration									
Frq.	106	0	4	20	42	40	4.00	4.11	.843
%	100	0	3.8	18.9	39.6	37.7			
10. AR can improve the quality of students' learning and studying									
Frq.	106	4	3	18	32	49	4.00	4.12	1.039
%	100	3.8	2.8	17.0	30.2	46.2			

Frq.: Frequency; %: Percentage.

The perspective for the teaching construct reveals high values, with a mean score between 4.08 (Sd. = 0.957) and 4.50 (Sd. = 0.734) out of 5.00, and medians of 4.00 and 5.00 points for the four items. The most notable rating was for item 2, for which 61.3% of students selected the option strongly agree, and therefore consider that AR can complement teaching performed with books and notes.

For the learning construct the scores were slightly lower but still high, with the lowest rating being 3.97 (Sd. = 1.108) and the highest 4.25 (Sd. = 1.067). Item 6 stands out, with 83% of students responding agree or strongly agree, showing a firm majority in support of the idea that AR can improve students' motivation. This was also the only item whose median was 5 points, while the others returned a value of 4.

As for the overall score for the usefulness of AR with respect to the curriculum, Table 6 shows that there are two items in each area whose values were respectively the highest and the lowest, according to the proposed contents that were taught. We have had to select results due to spatial limitations, but all the items can be seen in the supplementary document 5.

It can be seen that most of the items achieved mean scores of over 4.00 points, except for items 14 and 20, which were rated 3.99 (Sd. = 0.951) and 3.93 (Sd. = 1.106), respectively. The median was between 4.00 and 5.00 points for all the items. Also of note is the fact that in two of the five areas the items referring to the usefulness of learning these types of content with (items 16 and 34) received the maximum rating for their constructs (4.45; Sd. = 0.794 and 4.64; Sd. = 0.679, respectively). The highest percentages were in Social Sciences (item 23) and Natural Sciences (item 34), where strongly agree was over 70%. In contrast, strongly disagree was very low in all the areas with no item receiving over 5% and many at 0%.

Table 6. Students' overall ratings for the curriculum construct.

Overall	Total	Strongly Disagree	Disagree	Neither Disagree Nor Agree	Agree	Strongly Agree	Md.	M	Sd.
Spanish Language and Literature									
16. Using AR to learn these types of curricular contents seems useful to me									
Frq.	106	0	3	11	27	65	5.00	4.45	0.794
%	100	0	2.8	10.4	25.5	61.3			
14. AR activity can be used to deal with concepts of information, opinion and advertising in journalistic texts and advertising									
Frq.	106	2	5	21	42	36	4.00	3.99	0.951
%	100	1.9	4.7	19.8	39.6	34.0			
First Foreign Language									
19. The activity can facilitate rapid strategies for text understanding, e.g., the AR translator									
Frq.	106	0	2	11	37	56	5.00	4.39	0.751
%	100	0	1.9	10.4	34.9	52.8			
20. The strip of images in AR can be used to study nouns, pronouns, articles and demonstratives as well as area contents									
Frq.	106	5	5	23	32	41	4.00	3.93	1.106
%	100	4.7	4.7	21.7	30.2	38.7			
Social Sciences									
23. AR activity can help in viewing monuments in Spain as cultural heritage									
Frq.	106	0	2	3	26	75	5.00	4.64	0.635
%	100	0	1.9	2.8	24.5	70.8			
24. From what I saw in the video, Europe seems to be a continent with many resources									
Frq.	106	2	6	12	43	43	4.00	4.12	0.953
%	100	1.9	5.7	11.3	40.6	40.6			
Natural Sciences									
34. Using AR to learn these types of curricular contents seems useful to me									
Frq.	106	1	1	3	25	76	5.00	4.64	0.679
%	100	0.9	0.9	2.8	23.6	71.7			
31. The strip of images can help in learning about the importance of raising awareness about and preventing the consumption of alcohol, tobacco and drugs									
Frq.	106	0	3	8	33	62	5.00	4.45	0.758
%	100	0	2.8	7.5	31.1	58.5			
Mathematics									
36. The video presented in AR helps to show simple, easily readable and understandable graphic representations of fractions									
Frq.	106	0	2	5	37	62	5.00	4.50	0.680
%	100	0	1.9	4.7	34.9	58.5			
37. This proposal can help to work in AR to order simple fractions, calculate the fraction of a number or tackle equivalent fractions									
Frq.	106	4	2	8	39	53	4.50	4.27	0.961
%	100	3.8	1.9	7.5	36.8	50.0			

Figure 4 offers the mean scores for the items in each area. Figure 5 shows the mean values for the item that specifically measures the usefulness of AR in each area construct.

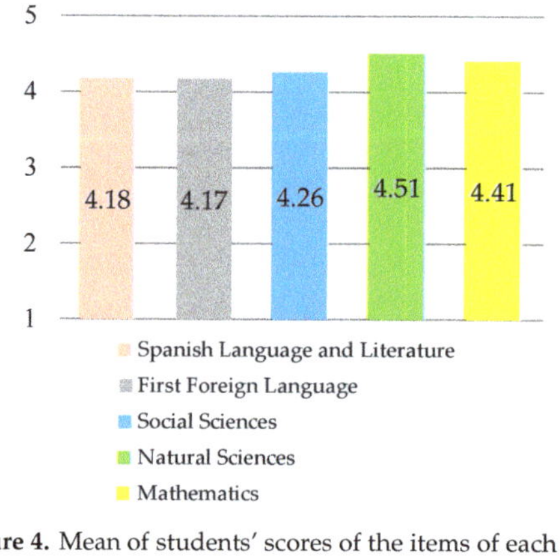

Figure 4. Mean of students' scores of the items of each area.

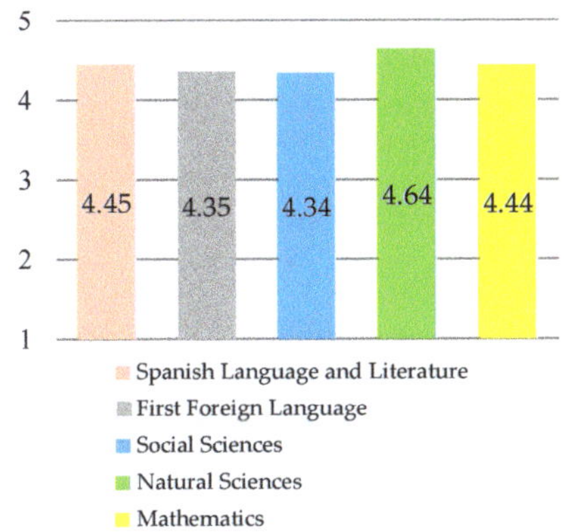

Figure 5. Mean of students' scores for the item "Using AR to learn these types of curricular contents seems useful to me", for each area.

As Figure 4 shows, Natural Sciences was the area with the highest construct mean (4.51), although it is notable that all the blocks were scored above 4.00 points. In the items in Figure 5 we see Natural Sciences (item 34) has the highest mean score (4.64; Sd. = 0.679), which coincides with the highest construct mean, again for Natural Sciences.

Finally, the overall analysis of the results on the students' perception of the useful AR according to the three constructs was repeated according to gender. We used the Kolmogorov–Smirnov test to make an exploratory analysis, which revealed that the distribution was not normal ($p < 0.05$). Backed by this normality test, we applied the non-parametric Mann–Whitney U test for two groups to compare whether opinions according to occupied equivalent positions.

For Teaching and Learning, females rated AR usefulness slightly higher than males in almost all the items. However, the curriculum constructs revealed differences in preferences depending on the item, although the overall opinion was highly positive for all the questions.

The analysis showed that in items 23 ($p = 0.035$) and 32 ($p = 0.040$), the distribution between genders varied, which allows us to state that there were significant differences in the responses ($p < 0.05$). Item 23 referred to whether AR can help students in studying the monuments of Spain as cultural heritage (Social Sciences), while item 32 asked whether AR helps in developing responsible behaviors in terms of healthy lifestyle, diet and proper functioning of the body (Natural Sciences). For the rest of the items there were no significant differences ($p > 0.05$).

4. Discussion

At the beginning of the article, we posed three questions: what activities could be implemented in pre-adolescent stages, so our students achieve higher quality learning using AR? What was the students' perception of the real applications of using this technique? Moreover, is it possible to implement AR within the framework of a theoretical model that would improve the effectiveness and motivation to learn? We believe, according to objectives, all these questions have received a relevant response. This was due to the design of the study itself, which took all possible details into account in its articulation process, in planning activities designed for the subject areas, in its theoretical and practical justification by means of an adaptation of the 3P and ARCS models, through the use of motivational strategies, and with the greatest possible reliability and validity in data collection to attempt to assess the student body's perceptions.

First, both the reliability and the validity study of the questionnaire content returned satisfactory results. Along these lines, the expert assessment and the pilot study provided ample assurance that the questionnaire complied with the proposed intervention model and the variables analyzed. Hence, the data collected in the questionnaire is suitable to respond to the research questions.

Secondly, after analyzing the students' assessment of AR's usefulness as a technique for teaching and learning within the framework of the 3P model in each construct overall and according to gender, the results were very positive for all constructs. Notwithstanding this, within each construct the assessment of some items were more notable than others.

The results from the teaching construct showed that AR can enable teaching to be undertaken as a discovery of contents where the teacher acts as a guide and allows knowledge acquisition to be adapted to the areas being taught. Specifically, the most promising results were found in item 2, which indicates that AR might provide another way, alongside books and notes, for the teacher to teach. One of the questions that now arises is whether AR will be able to provide short-term changes to teaching methodology, given that participants stated that one idea would be to alternate between traditional textbook teaching and notes and this new form of teaching.

From the analysis of the results for the learning construct, it appears that AR can heighten students' attention and motivation, helps them to understand content better, increases classroom participation, promotes student collaboration and enhances the quality of learning and studying. However, even though all scores were high, the highest were obtained in the assessment referring to motivation and attention, coinciding with other studies mentioned [9,21]. Specifically, the study carried out by Di-Serio et al. [21] also examined the ARCS model [27], analyzing four motivational factors and concluded that the attention and satisfaction categories are rated higher using AR as a learning environment in comparison to other settings. In our study, the attention factor was considered to capture students' interest in AR and stimulate their curiosity to learn, while the satisfaction category was used in completing the final questionnaire, resolving uncertainties, and proposing improvements.

Regarding the different areas of the core curriculum, the results were again interesting, especially for Natural Sciences, where the scores were highest for the use of AR for the content dealing with the prevention and consumption of drugs. It was also shown that this technique can be successful when teaching fractions in Mathematics as it facilitates their reading, interaction, comprehension and solution. Noteworthy results were also found in Social Sciences in the area of the country's cultural and historical heritage and when studying the main European capitals. Lastly, in Spanish Language and Literature, AR was reported to produce improvements in digital interaction and manipulation by working with different types of text. There is no doubt that working this way fosters interactivity and immediacy to teaching and learning of standard curricular contents.

In terms of gender, as mentioned earlier, the perception of the usefulness of AR only varied significantly for two items (23 and 32). Elsewhere, the male score was higher in some items, while in others it was the female score, although differences were not significant. The results for these two items can be explained reasoning that the boys were more interested in the AR proposal addressing Social Science content, while the girls were more motivated by the study of behavior, healthy living practices, good eating routines and guidelines for body care, due to greater interest in this topic than the boys. Indeed, during the study the boys were observed showing greater interest than the girls in cultural heritage. They constantly interacted with their AR layers and asked about the origin and historical meaning of specific monuments, while the girls asked more questions about food, viewed AR layers about healthy living practices and searched the internet for more content on these topics.

In short, the objectives proposed received very positive responses, which encourages the idea that these educational practices are viable. There are many benefits that AR can bring to education, as it allows unknown areas to be explored, as well as fostering new dynamic and engaging ways of teaching and learning. In fact, the main justification of our study to show that the use of AR also enables teaching which is completely contextualized within the curricular content, drawing from a theoretical model that supports and improves the learning process. As Biggs et al., [32] pointed out, student factors, the teaching context, the learning approaches used for the task, and expected outcomes or performance interact with each other forming a dynamic system that, if well defined, can favor new motivational strategies that enhance learning.

5. Limitations, Prospective and Practical Socio-Educational and Research Implications

Despite the optimistic findings, this research has been a subjective analysis of AR, so its real value in enhancing learning needs to be verified. We are aware that this requires more than just observations and opinions and should consider aspects regarding its real possibilities for implementation based on a controlled assessment of the learning that is going on. Moreover, this study does not collect information from all the variables involved within the theoretical models considered. Our intention is to supplement this investigation by studying the interaction of the presage, process, and product variables within the 3P model to see how they affect perceived acceptability and academic performance. At the same time, we must verify to what extent the motivational strategies adopted by a given student influence the student's perception of these environments. To this end, categories are being designed and will soon be implemented within a qualitative approach. We also want to acknowledge limitations such as the lack of a comparison group or a focus group, which would have provided additional insights into the students' experiences.

At practical, socio-educational and research levels, the findings provide a valid and reliable response to the objectives sought, and although future research should provide a more solid base, these results allow us to state that AR enjoys wide acceptance among primary school students and has a very relevant role to play in teaching and learning in the coming years. The development of this technique (higher levels of processing, sensors, etc.) will also affect its level of acceptance and use as a school resource, which will serve to demand that high-level politicians invest larger amounts of money and draw up new plans and projects to be implemented around this technique.

6. Conclusions

The Biggs 3P model and the Keller ARCS model have served to support our study and provide rigor, demonstrating that using AR as a mediating element can foster new learning strategies and that these affect motivational processes. However, for this to happen, baseline aspects such as the student's characteristics and skills or other contextual factors such as the subject area taught, the method used, or the time devoted to the task must be considered. All of these factors impact the degree of perceived usefulness and expected performance, and is directly related to the level of participant interest. These contributions are shown in Figure 6, through a flow chart that synthesizes the main conclusions of the paper.

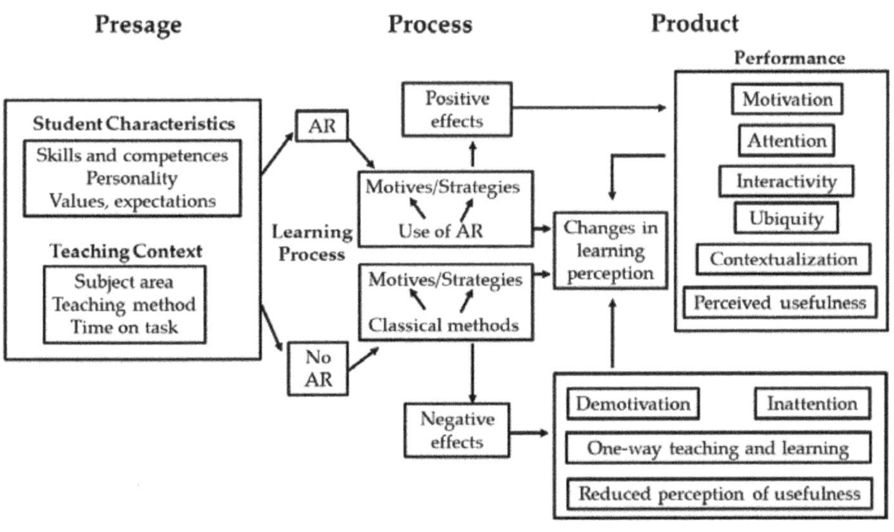

Figure 6. Flow chart to illustrate the concept of the contributions of this paper.

Regarding teaching, this group of schoolchildren finds AR to be a more satisfying and enjoyable complementary approach to more traditional teaching materials. Most students are used to working with books, photocopies or notes, and any new learning format which differs from these materials is perceived as appealing, useful, and interesting to them. As such, they would like the teacher to teach them ways to discover new content on their own and to use alternative methods based on the use of touch screens, online resources, interactive surveys or experimentation with virtual resources to make such processes more dynamic [39,40].

As regards their own learning, the students highlighted an increase in motivation thanks to their being able to tackle contents using the AR technique. Clearly, AR motivates the students and allows them to acquire knowledge in a more dynamic, fun and interactive way. Consequently, the justification for using AR must be supported by the promotion of actions based on hands on learning, interactivity and research, which are backed by powerful theoretical models capable of developing competent students who know how to deal with future real-life situations.

Concerning curriculum, we contend that the use of AR for the core subjects is very efficient and practical, since it provides for a more contextualized and organized way of learning whereby, starting from a theoretical model, AR is the mediating element of knowledge. The fact that Natural Sciences was so highly rated may be due to the content, as the subject matter tackled aroused a lot of interest among the pre-adolescent students when they are beginning to acquire some notions of alcohol, drugs, relationships with others and decision making. In fact, the attention and interest shown by students made it possible to address the motivational strategies at greater depth with respect to the categories being taught (ARCS model) and the proposed activities.

There is no doubt that AR should be implemented as a standard approach in formal teaching. The main reason for its inclusion in teaching and learning processes stems from its innovative character, as well as the benefits derived from its interactive nature, ease of use, immediacy, and the motivation it induces.

AR implies direct involvement in students' learning since students started to ask questions that highlighted their interests in knowing more about AR (What is it for? How do we use it? What materials should we use when working with it? How do I incorporate virtual layers?). Students found that learning with AR was something new and viable, and this would seem to be a clear invitation to professionals to think about other ways to teach and not just the traditional one.

Other advantages that AR point to in the short term include fostering better relations between teachers and students, more active and collaborative participation, better understanding and advances in the cohesion between teaching and methodological styles.

It is also important to highlight the key role that mobile AR plays in the environment, since it is an approach for both the classroom, with the guidance of teachers, and outside the school setting, as was witnessed when the students continued using AR outside of the classroom. The pervasiveness and flexibility that AR systems offer represent an improvement that should be taken advantage of in the educational sector to provide competence, commitment and stability to academic work.

Supplementary Materials: The following are available online at http://www.mdpi.com/2076-3417/9/24/5426/s1, Supplementary document 1: Some photographs of the research, Supplementary document 2: Questionnaire PEURA-E, Supplementary document 3: Scale for experts to evaluate content, Supplementary document 4: Evaluation of the contents by experts, Supplementary document 5: Evaluation of the usefulness of AR by students.

Author Contributions: Conceptualization, A.L.-G.; methodology, J.M.; software, A.L.-G.; validation, A.L.-G; formal analysis, A.L.-G., J.M. and P.M.-M.; investigation, A.L.-G.; resources, P.M.-M., J.M. and A.L.-G.; data curation, A.L.-G.; writing—original draft preparation, A.L.-G.; writing—review and editing, P.M.-M., A.L.-G. and J.M.; visualization, P.M.-M. and A.L.-G.; supervision, P.M.-M.; project administration, P.M.-M.; funding acquisition, P.M.-M. All authors have read and approved the final version of the paper.

Funding: This research was funded by the "Spanish Ministry for Science, Innovation and Universities. Secretary of State for Universities, Research, Development and Innovation", grant number PGC2018-094491-B-C33, "Spanish Ministry for Economy, Industry and Competitiveness and European Social Fund", grant number EDU2015-65621-C3-2-R (BES-2016-078837), and "Seneca Foundation. Regional Agency for Science and Technology", grant number 20874/PI/18. We would like to thank these bodies.

Acknowledgments: Our thanks to Stephen Hasler, John Meagher and Katy Mikes for their help in translating the paper.

Conflicts of Interest: The authors declare no conflict of interest.

References

1. Christensen, R.; Knezek, G. Readiness for integrating mobile learning in the classroom: Challenges, preferences and possibilities. *Comput. Hum. Behav.* **2017**, *76*, 112–121. [CrossRef]
2. Suárez, Á.; Specht, M.; Prinsen, F.; Kalz, M.; Ternier, S. A review of the types of mobile activities in mobile inquiry-based learning. *Comput. Educ.* **2018**, *118*, 38–55. [CrossRef]
3. Akçayır, M.; Akçayır, G. Advantages and challenges associated with augmented reality for education: A systematic review of the literature. *Educ. Res. Rev.* **2017**, *20*, 1–11. [CrossRef]
4. Caudell, T.P.; Mizell, D.W. Augmented Reality: An application of heads-up display technology to manual manufacturing processes. In Proceedings of the Twenty-Fifth Hawaii International Conference on System Sciences, Kauai, HI, USA, 7–10 January 1992; IEEE Computer Society: Washington, DC, USA, 1992; pp. 659–669. [CrossRef]
5. Milgram, P.; Kishino, F. A taxonomy of mixed reality visual displays. *IEICE Trans. Inf. Syst.* **1994**, *77*, 1321–1329.
6. Azuma, R. A survey of Augmented Reality. *Presence Teleoperators Virtual Environ.* **1997**, *6*, 355–385. [CrossRef]
7. Squire, K.; Klopfer, E. Augmented reality simulations on handheld computers. *J. Learn. Sci.* **2007**, *16*, 371–413. [CrossRef]
8. Wu, H.K.; Lee, S.W.-Y.; Chang, H.-Y.; Liang, J.-C. Current status, opportunities and challenges of augmented reality in education. *Comput. Educ.* **2013**, *62*, 41–49. [CrossRef]
9. Bacca, J.; Baldiris, S.; Fabregat, R.; Graf, S. Augmented Reality Trends in Education: A Systematic Review of Research and Applications. *Educ. Technol. Soc.* **2014**, *17*, 133–149.
10. Cheng, K.-H.; Tsai, C.-C. Affordances of Augmented Reality in Science Learning: Suggestions for Future Research. *J. Sci. Educ. Technol.* **2013**, *22*, 449–462. [CrossRef]
11. Gottlieb, O. Time travel, labour history, and the null curriculum: New design knowledge for mobile augmented reality history games. *Int. J. Herit. Stud.* **2018**, *24*, 287–299. [CrossRef]
12. Koutromanos, G.; Styliaras, G. The buildings speak about our city: A location based augmented reality game. In Proceedings of the 6th International Conference on Information, Intelligence, Systems and Applications (IISA), Corfu, Greece, 6–8 July 2015; IEEE Computer Society: Washington, DC, USA; pp. 1–6. [CrossRef]
13. Tobar-Muñoz, H.; Baldiris, S.; Fabregat, R. Augmented Reality Game-Based Learning: Enriching Students' Experience during Reading Comprehension Activities. *J. Educ. Comput. Res.* **2017**, *55*, 901–936. [CrossRef]
14. Chen, Y.; Zhou, D.; Wang, Y.; Yu, J. Application of Augmented Reality for Early Childhood English Teaching. In Proceedings of the International Symposium on Educational Technology (ISET), Hong Kong, China, 27–29 June 2017; IEEE Computer Society: Washington, DC, USA; pp. 111–115. [CrossRef]
15. Kamarainen, A.M.; Metcalf, S.; Grotzer, T.; Browne, A.; Mazzuca, D.; Tutweiler, M.S.; Dede, C. EcoMOBILE: Integrating augmented reality and probeware with environmental education field trips. *Comput. Educ.* **2013**, *68*, 545–556. [CrossRef]
16. Sáez-López, J.-M.; Sevillano-García, M.L.; Pascual-Sevillano, M.A. Application of the ubiquitous game with augmented reality in Primary Education. *Comunicar* **2019**, *61*. [CrossRef]
17. Weng, C.; Rathinasabapathi, A.; Weng, A.; Zagita, C. Mixed Reality in Science Education as a Learning Support: A Revitalized Science Book. *J. Educ. Comput. Res.* **2019**, *57*, 777–807. [CrossRef]
18. Yilmaz, R.M.; Kucuk, S.; Goktas, Y. Are augmented reality picture books magic or real for preschool children aged five to six? *Br. J. Educ. Technol.* **2017**, *48*, 824–841. [CrossRef]
19. Chang, Y.-L.; Hou, H.-T.; Pan, C.-Y.; Sung, Y.-T.; Chang, K.-E. Apply an augmented reality in a mobile guidance to increase sense of place for heritage places. *Educ. Technol. Soc.* **2015**, *18*, 166–178.
20. Joo-Nagata, J.; Martínez Abad, F.; García-Bermejo Giner, J.; García-Peñalvo, F.J. Augmented reality and pedestrian navigation through its implementation in m-learning and e-learning: Evaluation of an educational program in Chile. *Comput. Educ.* **2017**, *111*, 1–17. [CrossRef]
21. Di-Serio, A.; Ibañez, M.; Kloos, C. Impact of an augmented reality system on students' motivation for a visual art course. *Comput. Educ.* **2013**, *68*, 586–596. [CrossRef]

22. Dunleavy, M.; Dede, C.; Mitchell, R. Affordances and Limitations of Immersive Participatory Augmented Reality Simulations for Teaching and Learning. *J. Sci. Educ. Technol.* **2009**, *18*, 7–22. [CrossRef]
23. Huang, T.-C.; Chen, C.-C.; Chou, Y.-W. Animating eco-education: To see, feel, and discover in an augmented reality-based experiential learning environment. *Comput. Educ.* **2016**, *96*, 72–82. [CrossRef]
24. Radu, I. Augmented reality in education: A meta-review and cross-media analysis. *Pers. Ubiquitous Comput.* **2014**, *18*, 1533–1543. [CrossRef]
25. Wojciechowski, R.; Cellary, W. Evaluation of learners' attitude toward learning in ARIES augmented reality environments. *Comput. Educ.* **2013**, *68*, 570–585. [CrossRef]
26. Plass, J.L.; Kaplan, U. Emotional Design in Digital Media for Learning. In *Emotions, Technology, Design & Learning*; Tettegah, S.Y., Gartmeier, M., Eds.; Elsevier: New York, NY, USA, 2016; pp. 131–161. [CrossRef]
27. Keller, J.M. Development and Use of the ARCS Model of Motivational Design. *J. Instr. Dev.* **1987**, *10*, 2. [CrossRef]
28. Davis, F.D.; Bagozzi, R.P.; Warshaw, P.R. User acceptance of computer technology: A comparison of two theoretical models. *Manag. Sci.* **1989**, *35*, 982–1003. [CrossRef]
29. Dunkin, M.J.; Biddle, B.J. *The Study of Teaching*; Holt, Rinehart and Winston: New York, NY, USA, 1974.
30. Biggs, J. What do inventories of students' learning processes really measure? A theoretical review and clarification. *Br. J. Educ. Psychol.* **1993**, *63*, 3–19. [CrossRef]
31. McMillan, J.H.; Schumacher, S. *Research in Education: A Conceptual Introduction*, 5th ed.; Longman: New York, NY, USA, 2001.
32. Biggs, J.B.; Kember, D.; Leung, D.Y.P. The revised two-factor Study Process Questionnaire: R-SPQ-2F. *Br. J. Educ. Psychol.* **2001**, *71*, 133–149. [CrossRef]
33. Keller, J.M. *Motivational Design for Learning and Performance: The ARCS Model Approach*; Springer: New York, NY, USA, 2010. [CrossRef]
34. Layar (Version 7) [Computer Software]. Layar and Blippar Group: London, UK. Available online: http://www.layar.com/ (accessed on 4 November 2019).
35. Krosnick, J.A.; Presser, S. Question and Questionnaire Design. In *Handbook of Survey Research*; Marsden, P.V., Wright, J.D., Eds.; Emerald Group: Bingley, UK, 2010; pp. 263–313.
36. Schwarz, N. Self-reports: How the questions shape the answers. *Am. Psychol.* **1999**, *54*, 93–105. [CrossRef]
37. Serrano, F.J. El cuestionario como instrumento de obtención de datos en la Investigación sobre Educación Matemática. In *Seminario Sobre Investigación en Didáctica Matemática*; Sociedad Educación Matemática: Badajoz, Spain, 2008.
38. O'Dwyer, L.; Bernauer, J. *Quantitative Research for the Qualitative Researcher*; Sage: Riverside, CA, USA, 2014.
39. Miralles, P.; Gómez, C.J.; Monteagudo, J. Percepciones sobre el uso de recursos TIC y «mass-media» para la enseñanza de la historia. Un estudio comparativo en futuros docentes de España-Inglaterra. *Educ. XXI* **2019**, *22*, 187–211. [CrossRef]
40. Miralles-Martínez, P.; Gómez-Carrasco, C.J.; Arias-González, V.B.; Fontal-Merillas, O. Digital resources and didactic methodology in the initial training of History teachers. *Comunicar* **2019**, *XVII*, 45–56. [CrossRef]

applied sciences

MDPI

Article

Aroma Release of Olfactory Displays Based on Audio-Visual Content

Safaa Alraddadi [1,*], Fahad Alqurashi [1], Georgios Tsaramirsis [2,*], Amany Al Luhaybi [3] and Seyed M. Buhari [2]

1 Computer Science Department, King Abdulaziz University, Jeddah 21589, Saudi Arabia; fahad@kau.edu.sa
2 Information Technology Department, King Abdulaziz University, Jeddah 21589, Saudi Arabia; mesbukary@kau.edu.sa
3 College of Computing in Al-Qunfudhah, Umm Al Qura University, Makkah 24382, Saudi Arabia; amluhaybi@uqu.edu.sa
* Correspondence: salraddadi0009@stu.kau.edu.sa (S.A.); gtsaramirsis@kau.edu.sa (G.T.)

Received: 23 October 2019; Accepted: 12 November 2019; Published: 14 November 2019

Abstract: Variant approaches used to release scents in most recent olfactory displays rely on time for decision making. The applicability of such an approach is questionable in scenarios like video games or virtual reality applications, where the specific content is dynamic in nature and thus not known in advance. All of these are required to enhance the experience and involvement of the user while watching or participating virtually in 4D cinemas or fun parks, associated with short films. Recently, associating the release of scents to the visual content of the scenario has been studied. This research enhances one such work by considering the auditory content along with the visual content. Minecraft, a computer game, was used to collect the necessary dataset with 1200 audio segments. The Inception v3 model was used to classified the sound and image dataset. Further ground truth classification on this dataset resulted in four classes: grass, fire, thunder, and zombie. Higher accuracies of 91% and 94% were achieved using the transfer learning approach for the sound and image models, respectively.

Keywords: audio classification; olfactory display; deep learning; transfer learning; inception model

---

1. Introduction

The auditory and visual information of computer games is easy to obtain through software means by capturing screenshots and recording audio. It is as easy to recognize the events and the characters in the game as it is to hear the character and the soundtrack in the game. However, the olfactory information related to the game cannot be obtained through the media, either television or any other device, due to the challenges of comparing it digitally with visual and auditory information [1,2].

Olfactory displays have recently been used with virtual reality applications where it imitates reality and allows user interaction with an imaginative world by specific interaction devices [3]. However, the association between virtual content and scents is application specific and cannot be used in other applications. Studies have shown that the information obtained through the sense of smell is lesser than that obtained through the senses of hearing and sight [4]. At the same time, the olfactory information enhances the senses and immersion in reality more than the other senses. Nonetheless, the sense of smell is still the least used to enrich user experience in the virtual world. The literature review covers many studies that have developed olfactory displays that release scents based on a specific time.

Most of the current approaches either have no direct association with the virtual content (releasing scents based on preset timers) or are specific to an application. This makes them inappropriate for gaming and virtual reality applications as it is not possible to predict the user's actions and release the appropriate scents. Recent research [5] associated virtual artifacts with scents, thus allowing olfactory displays to be used in highly dynamic applications. The work presented in this article builds on [5] by enabling the release of scents based on visual and audio information.

The proposed system uses image recognition classified from [5] and pairs it (Logical OR operator) with a new audio classifier. Transfer learning with Inception v3, which takes the log-Mel spectrogram of a short audio sample as input, is used to recognize the sound. While it is easy for humans to associate sounds with a specific scenario [6], it is challenging for machines, as it requires a significant amount of audio data and can easily be disturbed by undesired noise. In this research, noise was considered as unclassified sounds played at the same time as classified (labeled) sounds, and has a direct negative impact on the accuracy of the recognition.

This study contributes to the areas of gaming and virtual reality as it adds the option of scents to be released based on audio as well as recognized images. This is an important addition as sometimes, some virtual elements are auditory, but with limited or no visual information. For example, it might be raining in the game, but the user cannot see it as it is outside their field of view or due to low lighting conditions. As long as the user can hear the rain, the scent will still be released.

The rest of this paper is organized as follows. Section 2 reviews the related work of olfactory displays and sound recognition techniques. Section 3 describes the methodology of the proposed system. Section 4 presents the data analysis and discusses the experimental results. Finally, the study concludes in Section 5.

2. Literature Review

The literature review is divided into two sections. The first section discusses how recent studies have used convolutional neural networks (CNNs) for sound recognition and justifies the use of CNN in the current research. The second section presents the latest developments in olfactory displays.

2.1. Sound Recognition

In recent years, studies have shown that the CNN model outperforms traditional methods in different taxonomic tasks including sound recognition. For sound recognition, the most common auditory features such as raw waveform, log-Mel spectrogram, or Mel frequency cepstral coefficient (MFCC) are used to train the deep CNN.

A novel end-to-end system to classify raw sound with two conventional layers was proposed in [7]. The experimental results showed that the combination of the proposed model and log-Mel-CNN exceeded the state-of-the-art log-Mel-CNN model with 6.5% improvement in the classification accuracy. However, the model is inappropriate to learn the complex structure of audio due to the presence of only two conventional layers.

Transfer approach called SoundNet used to transfer knowledge from visual recognition network was presented in [8]. The aim was to train a CNN that classified raw audio waveforms from unlabeled videos. The experimental result showed that SoundNet achieved an acoustic classification accuracy of 97%. However, if the CNN is trained on a large scale dataset (around two million samples), it can achieve a similar accuracy.

A very deep conventional network with 34 weight layers that processes the raw audio waveform directly was proposed in [9]. The model applied batch normalization on each output layer while residual learning skipped some fully connected layers and down sampling accurately in the initial layer. All of these contributed to avoiding difficulty in the trained model as well as providing low computational cost. The result showed that the CNN deep architecture outperformed CNN with the log-Mel spectrogram with a 71.8% accuracy.

Another study proposed CNN architecture with three conventional layers to classify sound signals using the log-Mel spectrogram as features to learn the model [10]. Furthermore, different types of audio data augmentation techniques such as time stretching (fast or slow audio), pitch shifting (higher or lower pitch of audio), dynamic range compression (compresses audio sample), and background noise (mix sample sounds with another sound that contains background from different acoustics) were used to overcome the problem of a lack of data. However, the performance improved in terms of the classification accuracy only in some types of augmentation, while it remained non-progressive in others. This CNN architecture classified short audio by using a log-Mel spectrogram with the same features as that used in [6]. Moreover, the training procedure with two phases non-fully trained and fully trained, improved the accuracy by reaching 86.2% as well as outperformed the accuracy of the Gaussian mixture modeling-Mel frequency cepstral coefficient (GMM-MFCC) by 6.4%.

A fully connected CNN model for partly labeled audio based on a trained large scale dataset (audio set) using the log-Mel spectrogram as input was introduced in [11]. Moreover, a CNN model was used as the framework to transfer and learn audio representation (spectrogram) using different methods where the accuracy of the proposed model reached up to 85%.

In order to overcome the difficulty of distinguishing sounds that come from various sources as well as the missing labels of these sounds, the authors in [12] proposed a deep CNN called AENet. The model processes large temporal input with the data augmentation technique called equalized mixture data augmentation (EMDA), which mixes sounds that belong to the same class and modified frequency of the audio sample by boosting and attenuating in a particular band. Moreover, it applied transfer of learning to extract audio features from AENet and combine them with visual features. The authors claimed that combining AENet features with visual features significantly improved its performance than that by combining MFCC with visual features.

A small number of systems have used spatial features extracted from binaural recordings. In order to obtain the advantages from feature engineering approaches (i-vector) and feature learning methods (CNN), the authors in [13] proposed a multichannel i-vector by computing MFCC for both channels in the audio sample. In addition, they built a CNN model similar to VGG-net (invented by the Visual Geometry Group) architecture that takes spectrogram features as the input. Moreover, combining two models was performed using the score vision technique, which creates the probability scores of each method and then fuses these scores. The performance of this hybrid approach achieved state-of-the-art and obtained first rank in the DCASE-2016 (Detection and Classification of Acoustic Scenes and Events 2016) challenge. However, this approach requires a large set of trainable parameters, which is not possible with our small dataset.

The authors in [14] proposed a CNN that consisted of eight convolutional layers and two fully connected layers using two spectrogram representations, the log-Mel spectrogram and gammatone spectrogram, as input. Traditional data augmentation methods were used to generate a new audio sample such as time stretch and pitch shift, in addition to applying the Mixup method on the training data by mixing two samples randomly selected within or without the same class. It was claimed that Mixup improved performance by 1.5% on the ESC-10 [15] dataset, 2.4% on the ESC-50 [15] dataset, and 2.6% on the UrbanSound8k dataset [16].

Most CNN models need a huge dataset in order to recognize the sound correctly. This makes them difficult to apply on limited datasets. Therefore, we will apply the transfer learning method to recognize sound samples in this research.

2.2. Olfactory Displays

Olfactory displays are devices designed to release scents into the environment. They are classified into two types: "wearable", which are placed either on-body or on-head, and "environmental", which are placed in the physical environment [17].

A wearable and fashionable olfactory necklace called Essence was designed in [18]. The Essence is able to release scents automatically based on data from the virtual context such as the location and current time of the users as well as on physiological data such as brain activity and heart rate. Moreover, the necklace can be activated manually, and the intensity of scents can be controlled through the stretch necklace thread. The results of the user experience show that the device is small enough and comfortable to be worn in most daily life activities. However, the device was unable to release multilabel scents at a time, and released one scent for one case based on the chosen user.

A smelling screen is an olfactory display embedded in a Liquid-Crystal Display (LCD) screen to generate and distribute odor along the screen based on the image shown [19]. The proposed device consists of four fans located at the corners of the screen to generate airflow that collides multiple times. Then, the airflow blows toward the user through tubing from the airflow collision point, which is considered as the odor source. However, the time between releasing the scents and its recognition by the user is not synchronized due to the delay of the scent reaching the olfactory organ.

An olfactory display named inScent [20] can be worn as a necklace that enables the user to receive scented notifications. The device was built to hold eight aromas where the scent is exchangeable through a small cartridge. Scents are triggered either manually by the users, or remotely by the instructor via an Android application based on the correlated scenario like scent that reflects the emotional link of the message sender and generates scents by using heating and a small fan to blow airflow toward the user. However, heating a wearable device may cause discomfort to the wearer.

Another study [21] proposed a thermal/heating approach to distribute scents from generation of the olfactory aromas. The device unit was built to hold eight aroma dispensers; each one containing a capillary tube, speed control fan, gas sensor to measure the release rate, and temperature to control the heating elements. The user controls the intensity of the aroma and fan speed as well as selects the aroma to be released by a software application. However, the heating approach might destroy the chemical components, which may limit the range of odors.

The authors in [5] presented a placed-in environment olfactory display that released six scents based on the visible content displayed by using an Inception v3 model for image recognition. However, visual elements were only associated with scents.

Overall, wearable devices can cause discomfort to users, thus hindering immersion into the virtual world. In contrast, environmental olfactory displays do not share this issue, but tend to have synchronization issues.

3. Methodology

3.1. System Overview

The proposed system consists of a Windows application that records the sounds and transforms raw sound into a log-Mel spectrogram while simultaneously taking screenshots from a game called Minecraft [22]. Conceptually, the proposed approach can be applied to any application as long as the classifiers have been trained to associate scents with its visual and auditory virtual phenomenon. The approach was applied on Minecraft as a proof of concept. The image classifier [5] and the sound classifier operate separately and identify which scents are to be released. Their results are then merged (union) and passed to the application that will inform the olfactory display. Only classes with an accuracy of 90% or more will be released. The olfactory display used in [5] was also used in this research. It is worth noting that this research focused on adding the capability of releasing scents based on an audio-visual virtual phenomenon and not on the development of an olfactory display. Figure 1 illustrates the system overview.

3.2. Dataset

The dataset consisted of 1200 audio segments that were distributed equally between four classes: grass, fire, thunder, and zombie. We selected these sounds based on the sound popularity in the Minecraft game as well as the availability of scents. The duration of all audio samples was four seconds with a 44.1 kHz sampling frequency and single audio channel (mono). Due to the similarity of the sound and to avoid overfitting, two deformation methods were used directly on the segments to generate a new sample. First, time stretching (TS) was applied to fast and slow audio samples using the Librosa function [23] (librosa.effects.time_stretch). In order to change the stretch, we used two speed factors of 0.5 and 2. Second, pitch shifting (PS) was applied to the high and low pitch of samples through the use of the Librosa function [23] (librosa.effects.pitch_shift) to change the pitch randomly.

All audio segments were then converted into a log-Mel spectrogram and used as input representation to train the network. We extracted log-Mel features from raw wave sounds by applying a short-time Fourier transform (STFT) over 40 ms windows with 50% overlap and Hamming windowing. We took the absolute value of each bin to square it and applied a 60-band Mel-scale filter bank. Finally, we computed the logarithmic conversion of the Mel energies using the Librosa library [23]. The log-Mel spectrogram was used to train the network without the need to combine features. Figure 2 illustrates a sample of the log-Mel spectrogram.

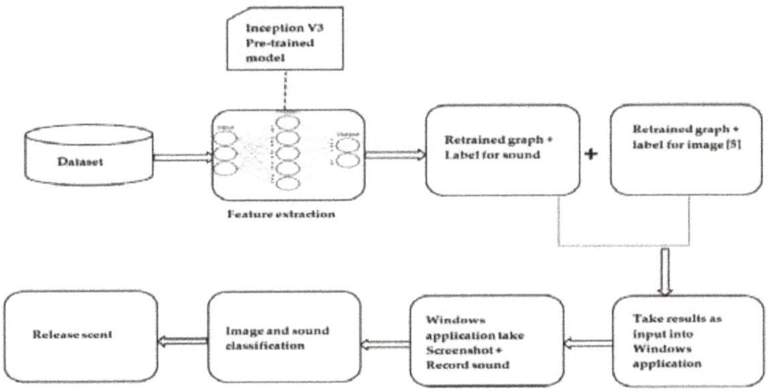

Figure 1. System overview.

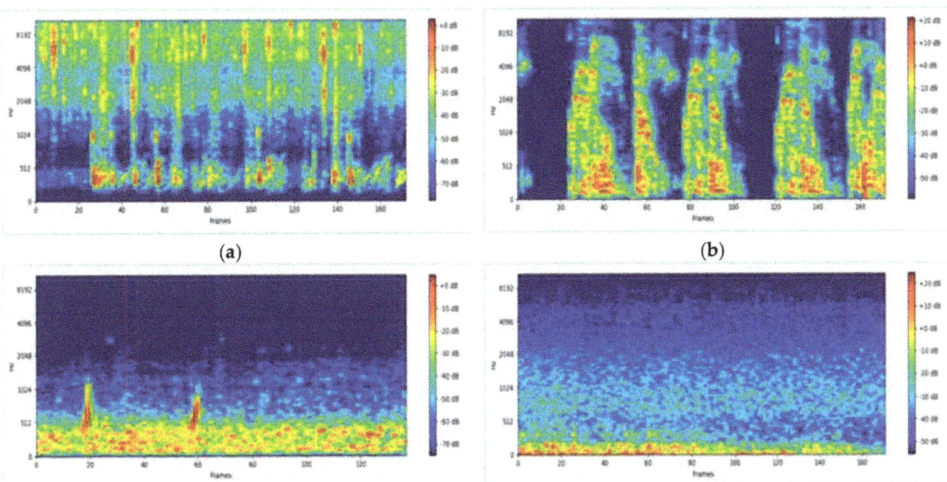

(a) (b) (c) (d)

Figure 2. Sample of log-Mel spectrogram to train the model. (**a**) log-Mel of fire; (**b**) log-Mel of zombie; (**c**) log-Mel of ocean; (**d**) log-Mel of thunder.

3.3. *Transfer Learning*

Training the deep convolutional network from initialization requires a huge dataset to learn discernable features. The limited availability of data makes automatic image recognition impossible. In such cases, transfer learning makes CNN able to recognize images successfully by transferring knowledge from a model trained on a huge dataset into the target model, which is used for the new task.

Recently, many CNN architectures with a deep layer have been developed. In this research, Inception v3 [24] was adopted as a pre-trained model because of its ability to reduce computational complexity by using different sizes of convolutional filters (e.g., 1 × 1, 3 × 3, and 5 × 5) in the same layer and then sending them to the next layer to detect new features. In [25,26], one of these filters had to be chosen to be used first, followed by the max pooling layer, then this operation was repeated with the hope of detecting new features. However, this operation is computationally intensive due to the many operations that occur in each neuron. Despite the complexity of the architecture in Inception v3, it achieved extraordinary performance in terms of accuracy. Inception v3 was trained on an ImageNet dataset [27] that contained 1.2 million images with more than 1000 labels. Inception v3 extracted the features of ImageNet by using a CNN with fully connected layers and a SoftMax layer to classify images based on the ImageNet labels. The transfer learning used all convolutional layers and pooling layers in Inception v3 to extract the input features of the log-Mel spectrogram. Then, it removed the top layer (SoftMax) that classified the original dataset and trained the new layer with our task. Finally, the new model classified the images based on the labels of the new dataset. The process of transfer learning is illustrated in Figure 3.

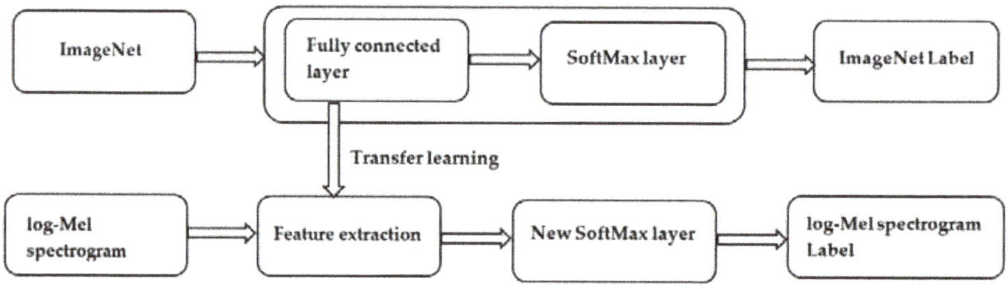

Figure 3. Transfer learning.

4. System Evaluation and Results

The work was evaluated in two stages. First, the capability of the model to identify the various audio classes was tested using a separate dataset of log-Mel spectrograms. Second, the integrated approach, consisting of both the image (re-used from [5]) and the proposed audio classifiers were tested for their consistency. The model was retrained using TensorFlow [28] on an Intel core i7-4720HQ processor with 16.0 GB memory. The dataset was trained with the slandered learning rate of 0.01, the iteration was set as 20,000, and the batch size of each iteration was equal to 100.

The retrained model was evaluated with 30 log-Mel spectrograms for each class. The classification result was obtained from the confusion matrix, as shown in Figure 4. As we can infer, the ocean was the least accurately recognized class by the system. The reason behind this is that the ocean in the game contains other creatures and their sounds overwhelm the sound of the ocean. On the other hand, the model predicts thunder sounds successfully because the sound of thunder is very loud and clear.

The model was evaluated before integrating the application by computing the accuracy, precision, recall, and f1 score for each category from a confusion matrix with 0.5 as the threshold value by using Equations (1)–(4). The prediction of the lowest value was not accepted because the application cannot release the aroma if the prediction is lower than 90%. The performance measurements in Table 1 were computed by using the following equations:

$$Accuracy = \frac{True\ Positives + False\ Negatives}{Total\ Number\ of\ Samples} \quad (1)$$

$$Precision = \frac{True\ Positives}{True\ Positives + False\ Positives} \quad (2)$$

$$Recall = \frac{True\ Positives}{True\ Positives + False\ Negatives} \tag{3}$$

$$F1\ Score = 2 \times \frac{Precision \times Recall}{Precision\ + Recall} \tag{4}$$

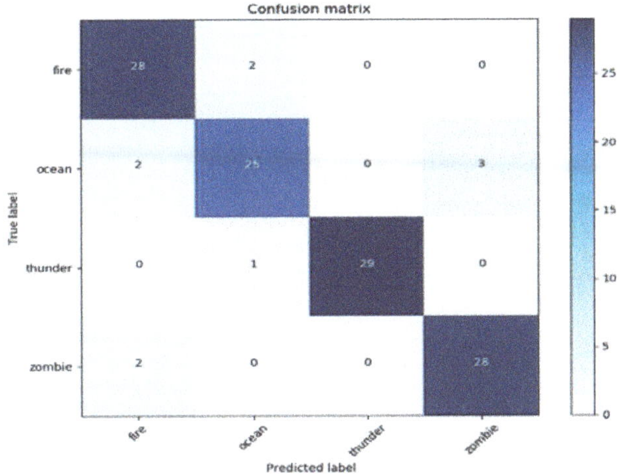

Figure 4. Confusion matrix for the retrained model evaluated based on the log-Mel spectrograms for fire, ocean, thunder, and zombie.

Table 1. Performance Measure for retrained Sound Model before Integrated into Application.

Category	Accuracy	Precision	Recall	F1 score
Fire	0.95	0.8	0.9	0.8
Ocean	0.93	0.8	0.8	0.8
Thunder	0.99	1	0.9	0.9
Zombie	0.95	0.9	0.9	0.9
Average	0.95	0.8	0.8	0.8

As can be seen from Table 1, the accuracy of the thunder outperformed other categories at 99% because it has a high sound that overshadows any other sounds around it, while ocean had less accuracy among the other categories with 93%. Overall, the average accuracy of the model was 95%; this was satisfied in this application due to the limited sounds in the game. The other statistical measures of precision, recall, and F1 score reached 0.9 in most cases, which was satisfied.

4.1. Performance Sound and Image Classifier in the Application

Evaluation performance of the integrated sound classifier with the Windows application was conducted, with audio samples recorded every ten seconds and converted into log-Mel spectrograms. At the same time (10 s), the application took screenshots and passed them to the image classifier. We compared the two classifiers to measure accuracy, precision, recall, and F1 score for the fire, zombie, and ocean categories using a confusion matrix with 0.5 as the threshold value. The predications that scored less than the threshold value were rejected. The following Figure 5 shows a sample of the accuracy of both classifiers within the application at the same time.

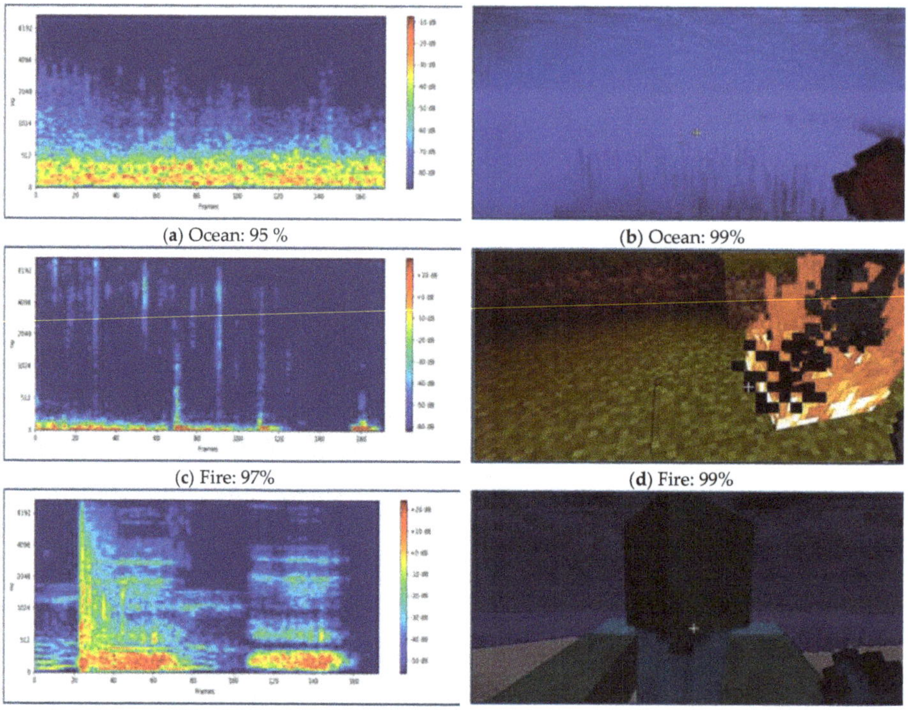

(**a**) Ocean: 95 % (**b**) Ocean: 99%

(**c**) Fire: 97% (**d**) Fire: 99%

(**e**) Zombie: 96% (**f**) Zombie: 96%

Figure 5. The accuracy of the sound and image classifier inside the application at the same time. (**a**) The accuracy of the ocean sound; (**b**) The accuracy of the ocean image; (**c**) The accuracy of the fire sound; (**d**) The accuracy of the fire image; (**e**) The accuracy of the zombie sound; (**f**) The accuracy of the zombie image.

The confusion matrixes for sound and image classifier performance within the application are shown in Figures 6 and 7, respectively. As can be seen, the ocean was the most misclassified category because the ocean in Minecraft contains other audio sources such as zombies and other creatures, which overlap the sound of the ocean. Additionally, two sounds from fire were classified as the ocean because the lava sound (a type of fire in the game) is similar to the sound of the ocean. In contrast, the ocean in the image classifier were classified correctly. Zombie was the most misclassified class in the image classifier with five images in the fire class because in the game, the zombie burns if exposed to the sun. However, in the sound classifier, zombie was classified with all classes correctly except fire, which was misclassified with two images due to the overlap of fire sounds with zombie sounds when the zombie was burning.

The accuracy performance for both classifiers are illustrated in Tables 2 and 3. It can be seen that the accuracy of the sound classifier decreased after being integrated into the application. It is believed this is occurs in circumstances when players move very fast from one scene to another, which makes the sounds overlap and become difficult to recognize. Overall, the average accuracy of the audio classifier (91%) was less than the accuracy of the image classifier (94%). This is because, unlike images, the game produces multiple sounds at the same time (e.g., the sound of the ocean and a pack of zombies), which cannot be predicted. Nevertheless, in some cases such as fire and zombie, the accuracy outperformed fire and zombie in the image classifier. Thus, the smells are released based on the classifier that represents the highest accuracy. Furthermore, the recall result of the image classifier was 0.9, which outperformed the result of the sound classifier. Finally, the average results of precision and F1 score were 0.8 for both classifiers, which were satisfactory in this application. It is worth noting that the two classifiers are complementary and do not complete each other. Additionally, the audio classifier could identify additional virtual phenomenon (e.g., thunder), even if it is not in the field of view of the player.

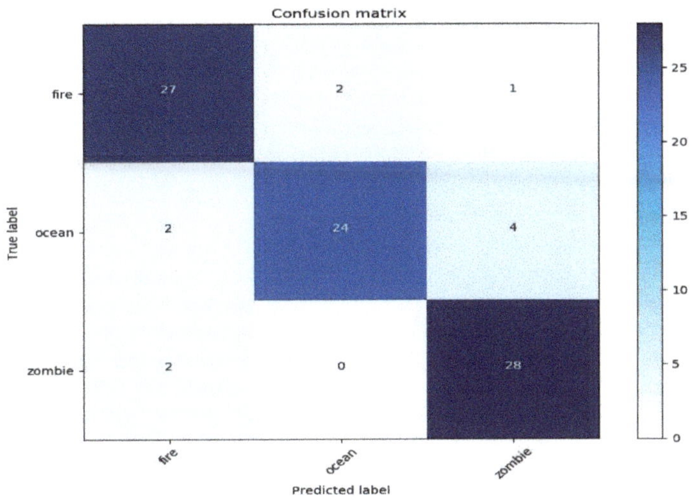

Figure 6. Confusion matrix for the evaluated sound within the application based on the log-Mel spectrogram for fire, ocean, and zombie.

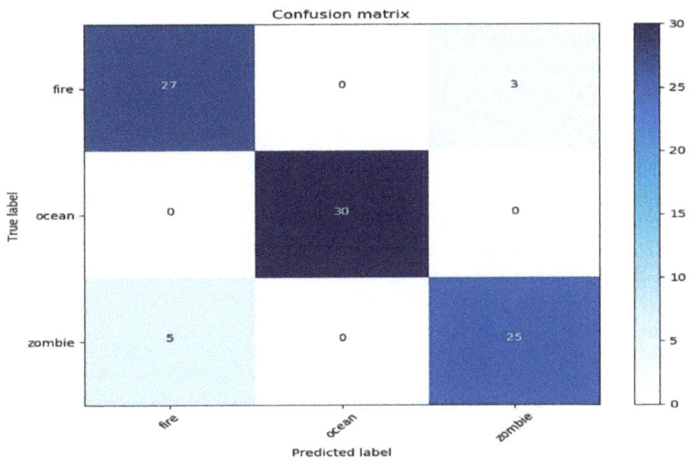

Figure 7. Confusion matrix for the evaluated image within the application based on the log-Mel spectrogram for fire, ocean, and zombie.

Table 2. Performance Measure for Sound Model within Application.

Category	Accuracy	Precision	Recall	F1 score
Fire	0.92	0.8	0.9	0.8
Ocean	0.91	0.9	0.8	0.8
Zombie	0.92	0.8	0.9	0.8
Average	0.91	0.8	0.8	0.8

Table 3. Performance Measure for Image Model within Application.

Category	Accuracy	Precision	Recall	F1 score
Fire	0.91	0.8	0.9	0.8
Ocean	1	1	1	1
Zombie	0.91	0.8	0.8	0.8
Average	0.94	0.8	0.9	0.8

4.2. User Experience

In order to test the impact on the user's experience, we conducted an experiment with five participants. The device was placed under the monitor, at the front of the users. Initially, the device was set to release new scents every three seconds. The synchronization between the game event and the release was acceptable, however, the scent persisted in the air for far longer. However, even after six minutes of game play (average), users could still differentiate the released aromas. Thus, they reported that the atmosphere was uncomfortable. In order to improve the user experience, we modified the release code to prevent an aroma being released more than once per minute. This improved the user experience, but still lacked a way to clean the previously released scent, which was proven to be a major drawback as the new aromas mixed with the old ones. Preventing the release of a new scent for 10 s (used in this research) resulted in a better overall user experience, but at the cost of a lot of missed releases, revealing a trade-off. While out of scope of this research, it is the belief of the authors that a new algorithm to decide when to release a new scent, based on the last release as well as the different persistence rates of various aromas, will have a positive impact on the user experience.

5. Conclusions

This study proposed an approach that combined audio and visual contents to automatically trigger scents through an olfactory device using deep learning techniques. The log-Mel spectrogram sound identification model was built based on a pre-trained Inception v3 model. Moreover, a Windows application was designed to record audio and convert it to a log-Mel spectrogram as well as take a screenshot of the same scene at the same time. In addition, the application controls the release of scents that are identified based on the highest accuracy. The accuracy of the integrated sound model with the application reached 91%, however, the accuracy was lower due to various sound recording situations. For example, sounds may overlap and become difficult to recognize. While the accuracy of the image outperformed that of the sound, sometimes it was misclassified. The sound and image models complement each other: in case one misrecognizes the scene, the higher accuracy will prevail, or the absence of either of them from a scene. The proposed approach can be applied to different virtual environments as long as scents can be associated with visual and auditory content. Further work is required to associate scents automatically with more sounds and images. Additionally, the approach can be tested with other games or virtual reality applications.

Author Contributions: Methodology, S.A., G.T., and A.A.L.; software S.A., A.A.L.; validation, S.A., and G.T.; formal analysis, S.A.; investigation, S.A., and G.T.; data curation, S.A., and A.A.L.; writing—original draft preparation, S.A.; writing—review and editing, S.A., G.T., and S.M.B.; supervision, F.A.

Funding: This research received no external funding.

Conflicts of Interest: The authors declare no conflicts of interest.

References

1. Hashimoto, K.; Nakamoto, T. Tiny Olfactory Display Using Surface Acoustic Wave Device and Micropumps for Wearable Applications. *IEEE Sens. J.* **2016**, *16*, 4974–4980. [CrossRef]
2. Hashimoto, K.; Nakamoto, T. Stabilization of SAW atomizer for a wearable olfactory display. In Proceedings of the 2015 IEEE International Ultrasonics Symposium (IUS), Taiwan, China, 21–24 October 2015; pp. 1–4.
3. Steuer, J. Defining virtual reality: Dimensions determining telepresence. *J. Commun.* **1992**, *42*, 73. [CrossRef]

4. Kadowaki, A.; Noguchi, D.; Sugimoto, S.; Bannai, Y.; Okada, K. Development of a High-Performance Olfactory Display and Measurement of Olfactory Characteristics for Pulse Ejections. In Proceedings of the 2010 10th IEEE/IPSJ International Symposium on Applications and the Internet, Seoul, Korea, 19–23 July 2010; pp. 1–6.
5. Al Luhaybi, A.; Alqurashi, F.; Tsaramirsis, G.; Buhari, S.M. Automatic Association of Scents Based on Visual Content. *Appl. Sci.* **2019**, *9*, 1697. [CrossRef]
6. Valenti, M.; Squartini, S.; Diment, A.; Parascandolo, G.; Virtanen, T. A convolutional neural network approach for acoustic scene classification. In Proceedings of the 2017 International Joint Conference on Neural Networks (IJCNN), Anchorage, AK, USA, 14–19 May 2017; pp. 1547–1554.
7. Tokozume, Y.; Harada, T. Learning environmental sounds with end-to-end convolutional neural network. In Proceedings of the 2017 IEEE International Conference on Acoustics, Speech and Signal Processing (ICASSP), New Orleans, LA, USA, 5–9 March 2017; pp. 2721–2725.
8. Aytar, Y.; Vondrick, C.; Torralba, A. Soundnet: Learning sound representations from unlabeled video. In Proceedings of the Advances in Neural Information Processing Systems, Barcelona, Spain, 5–10 December 2016; pp. 892–900.
9. Dai, W.; Dai, C.; Qu, S.; Li, J.; Das, S. Very deep convolutional neural networks for raw waveforms. In Proceedings of the 2017 IEEE International Conference on Acoustics, Speech and Signal Processing (ICASSP), New Orleans, LA, USA, 5–9 March 2017; pp. 421–425.
10. Salamon, J.; Bello, J.P. Deep Convolutional Neural Networks and Data Augmentation for Environmental Sound Classification. *IEEE Signal Process. Lett.* **2017**, *24*, 279–283. [CrossRef]
11. Kumar, A.; Khadkevich, M.; Fügen, C. Knowledge Transfer from Weakly Labeled Audio Using Convolutional Neural Network for Sound Events and Scenes. In Proceedings of the 2018 IEEE International Conference on Acoustics, Speech and Signal Processing (ICASSP), Calgary, AB, Canada, 15–20 April 2018; pp. 326–330.
12. Takahashi, N.; Gygli, M.; Gool, L.V. AENet: Learning Deep Audio Features for Video Analysis. *IEEE Trans. Multimed.* **2018**, *20*, 513–524. [CrossRef]
13. Eghbal-zadeh, H.; Lehner, B.; Dorfer, M.; Widmer, G. A hybrid approach with multi-channel i-vectors and convolutional neural networks for acoustic scene classification. In Proceedings of the 2017 25th European Signal Processing Conference (EUSIPCO), Nairobi, Kenya, 28 August–2 September 2017; pp. 2749–2753.
14. Zhang, Z.; Xu, S.; Cao, S.; Zhang, S. Deep Convolutional Neural Network with Mixup for Environmental Sound Classification. In Proceedings of the Pattern Recognition and Computer Vision, Guangzhou, China, 23–26 November 2018; pp. 356–367.
15. ESC: Dataset for Environmental Sound Classification. Available online: https://dataverse.harvard.edu/dataset.xhtml?persistentId=doi:10.7910/DVN/YDEPUT (accessed on 9 November 2019).
16. UrbanSound8k. Available online: https://urbansounddataset.weebly.com/urbansound8k.html (accessed on 9 November 2019).
17. Murray, N.; Lee, B.; Qiao, Y.; Muntean, G.-M. Olfaction-Enhanced Multimedia: A Survey of Application Domains, Displays, and Research Challenges. *ACM Comput. Surv.* **2016**, *48*, 1–34. [CrossRef]
18. Amores, J.; Maes, P. Essence: Olfactory Interfaces for Unconscious Influence of Mood and Cognitive Performance. In Proceedings of the 2017 CHI Conference on Human Factors in Computing Systems, Denver, CO, USA, 6–11 May 2017; pp. 28–34.
19. Matsukura, H.; Yoneda, T.; Ishida, H. Smelling Screen: Development and Evaluation of an Olfactory Display System for Presenting a Virtual Odor Source. *IEEE Trans. Vis. Comput. Graph.* **2013**, *19*, 606–615. [CrossRef] [PubMed]
20. Dobbelstein, D.; Herrdum, S.; Rukzio, E. inScent: A wearable olfactory display as an amplification for mobile notifications. In Proceedings of the 2017 ACM International Symposium on Wearable Computers, Maui, Hawaii, 11–15 September 2017; pp. 130–137.
21. Covington, J.A.; Agbroko, S.O.; Tiele, A. Development of a Portable, Multichannel Olfactory Display Transducer. *IEEE Sens. J.* **2018**, *18*, 4969–4974. [CrossRef]
22. Minecraft. Available online: https://minecraft.net/en-us/?ref=m (accessed on 6 April 2019).
23. Librosa. Available online: https://librosa.github.io/librosa/ (accessed on 28 March 2019).
24. Szegedy, C.; Vanhoucke, V.; Ioffe, S.; Shlens, J.; Wojna, Z. Rethinking the Inception Architecture for Computer Vision. In Proceedings of the 2016 IEEE Conference on Computer Vision and Pattern Recognition (CVPR), Las Vegas, NV, USA, 27–30 June 2016; pp. 2818–2826.

25. Krizhevsky, A.; Sutskever, I.; Hinton, G.E. ImageNet Classification with Deep Convolutional Neural Networks. *Commun. ACM* **2017**, *60*, 84–90. [CrossRef]
26. Simonyan, K.; Zisserman, A. Very Deep Convolutional Networks for Large-Scale Image Recognition. Available online: https://arxiv.org/abs/1409.1556 (accessed on 25 June 2019).
27. ImageNet. Available online: http://www.image-net.org (accessed on 20 March 2019).
28. TensorFlow. Available online: https://www.tensorflow.org/ (accessed on 9 March 2019).

applied sciences

MDPI

Article

Construction Hazard Investigation Leveraging Object Anatomization on an Augmented Photoreality Platform

Hai Chien Pham [1], Nhu-Ngoc Dao [2], Sungrae Cho [2], Phong Thanh Nguyen [3] and Anh-Tuan Pham-Hang [4,*]

1 Applied Computational Civil and Structural Engineering Research Group, Faculty of Civil Engineering, Ton Duc Thang University, Ho Chi Minh City 7000000, Vietnam; phamhaichien@tdtu.edu.vn
2 School of Computer Science and Engineering, Chung-Ang University, Seoul 06974, Korea; dnngoc@uclab.re.kr (N.-N.D.); srcho@cau.ac.kr (S.C.)
3 Department of Project Management, Ho Chi Minh City Open University, Ho Chi Minh City 7000000, Vietnam; phong.nt@ou.edu.vn
4 School of Computer Science and Engineering, International University—Vietnam National University HCMC, Block 6, Ward Linh Trung, Thu Duc District, Ho Chi Minh City 7000000, Vietnam
* Correspondence: anhtuanphamhang@gmail.com; Tel.: +84-902437625

Received: 26 September 2019; Accepted: 18 October 2019; Published: 23 October 2019

Abstract: Hazard investigation education plays a crucial role in equipping students with adequate knowledge and skills to avoid or eliminate construction hazards at workplaces. With the emergence of various visualization technologies, virtual photoreality as well as 3D virtual reality have been adopted and proved advantageous to various educational disciplines. Despite the significant benefits of providing an engaging and immersive learning environment to promote construction education, recent research has also pointed out that virtual photoreality lacks a 3D object anatomization tools to support learning, while 3D-virtual reality cannot provide a real-world environment. In recent years, research efforts have studied virtual reality applications separately, and there is a lack of research integrating these technologies to overcome limitations and maximize advantages for enhancing learning outcomes. In this regard, the paper develops a construction hazard investigation system leveraging object anatomization on an Interactive Augmented Photoreality platform (iAPR). The proposed iAPR system integrates virtual photoreality with 3D-virtual reality. The iAPR consists of three key learning modules, namely Hazard Understanding Module (HUM), Hazard Recognition Module (HRM), and Safety Performance Module (SPM), which adopt the revised Bloom's taxonomy theory. A prototype is developed and evaluated objectively through interactive system trials with educators, construction professionals, and learners. The findings demonstrate that the iAPR platform has significant pedagogic methods to improve learner's construction hazard investigation knowledge and skills, which improve safety performance.

Keywords: construction hazard; safety education; photoreality; virtual reality; anatomization

1. Introduction

The construction industry has been recognized as a dangerous and hazardous industry throughout the world [1,2]. Despite providing a significant contribution to a country's development, construction accidents have accounted for a very high rate among various industries [3]. For instance, safety statistics reveal that construction accounts for 20–40 percent of the occupational fatal accidents in spite of employing just around 6–10 percent of the workforce. Throughout the world, approximately 60,000 construction fatalities occur per year, which corresponds to one injury every nine minutes [4]. Consequently, fatalities and injuries cause many cost overruns and delays, which negatively impact the project safety performance [5]. To reduce construction accidents at the workplace, it is very important to provide graduates with hazard education at the tertiary level, so that they have professional hazard knowledge and skills [6,7].

In the past decades, virtual reality has been acknowledged as a state-of-the-art technology to improve hazard education [8]. The 3D CAD models are developed using computer to provide a 3D virtual reality (3D-VR) platform [9]. The advantage of 3D-VR is that it provides virtual construction sites, where students can interact with the educator and other learners to obtain hazard knowledge and skills [10]. With an interactive learning environment, learners will be motivated and engage sufficiently with their construction hazard investigation and recognition [11]. Moreover, recent efforts [12] have developed 3D-VR based object anatomization models for anatomizing and analyzing complicated hazard case studies. This object anatomization approach has benefits not only in construction hazard education [12] but also in other disciplines [13–15]. Despite its advantages, 3D-VR lacks a real-word environment [16], in which graduates can experience practical construction sites to investigate hazards.

To improve the real-world experience, 360 degree panoramic Virtual Photoreality (360VP) has emerged as a potential pedagogic tool for learners to experience real-word construction workplace environments [17]. The advantage of 360VP is that it provides an immersive learning platform, where the learner can move flexibly and observe the scenes to investigate hazards as they would experience in real construction sites [18]. Due to the greater immersion and higher degree of realism, 360VP assists learners in experiencing an emotional and cognitive presence at the scene. Moreover, the 360VP platform has been proved to be energy-efficient [19] and cost-efficient compared to the 3D-VR [20]. Despite the prominent advantages, the current applications of 360VP still lack 3D object anatomization tool to enhance the learning outcomes. Moreover, in recent years, researchers have focused on adapting virtual reality technologies separately, and lack of research integrates 360VP with 3D-VR to eliminate limitations and maximize the advantages of these technologies for promoting educational purposes.

In response to this status-quo, this research proposes a construction hazard investigation system, which leverages 3D object anatomization on an Interactive Augmented Photoreality (iAPR) platform. The proposed iAPR system augments a 360VP platform by integrating 3D-VR object anatomization technologies in order to create a learning environment, which reflects a real-word construction workplace. The iAPR consists of three key learning modules including Hazard Understanding Module (HUM), Hazard Recognition Module (HRM), and Safety Performance Module (SPM), which adapt Bloom's taxonomy learning theory for hazard investigation knowledge and skill development of learners. A prototype is developed with hazard cases that often occur in real construction workplaces. In addition, the effectiveness of iAPR in improving the hazard investigation is validated using before-after experimental studies. Due to the prominent visibility characteristic of 360VP and 3D-VR technologies using in the iAPR system, learners can only investigate the construction hazards through "sight" sense. Thus, recognizing and evaluating hazards, which need to use other human senses such as smell, touch, hearing, are out of research scope.

2. Related Work

2.1. Bloom's Taxonomy for Construction Hazard Education

Bloom's cognitive taxonomy was an attempt to improve educational objectives regarding assessment and testing of teaching materials [21]. It provided an organizational structure including six categories, namely knowledge, comprehension, application, analysis, synthesis and evaluation, which are arranged from the simplest to the most complex. Subsequently, many theorists have developed improvements in the domains of human learning such as effective domain [22] and psychomotor domain [23–25]. In 2001, Anderson and Krathwohl [26] proposed a revised Bloom's taxonomy which improved from uni-dimension to two-dimension. In the knowledge dimension, the authors classified four types: factual, conceptual, procedural, and metacognitive knowledge. In the cognitive process dimension, six levels were changed into verb format: 1-remembering, 2-understanding, 3-applying,

4-analyzing, 5-evaluating, and 6-creating [21]. With the combination of two dimensions, educators could change from passive views of learning towards active engagement in meaningful learning [26]. Bloom's taxonomy and its revised version became the standard for designing educational curricula [27]. For example, Thambyah et al. [28] recommended twenty four learning outcomes based on the revised Bloom's taxonomy for the final year project of undergraduate engineering. In addition, it provided for teachers a common language to compare and discuss between two different subject areas, understand how these subjects overlap, and deliver the conceptual and practical knowledge concurrently [29]. Thus, Bloom's taxonomy could widely influence the educational systems as a whole, through teacher preparing programs, assessment programs, and educational research [30]. A survey of the education literature also revealed that several studies attempted to apply and implement the taxonomy in many domains of education including computing, medical and nursing, music, and engineering [31].

In the construction hazard domain, there is a potential to apply Bloom's taxonomy for educational enhancement. The recent assessment methods reveal a gap in evaluating the implementation level of safety knowledge and skills in practice [32]. To overcome these gaps, appropriate approaches and solutions based on cognitive and awareness processes should be proposed to achieve the learning outcomes and maximize the collective amount of knowledge for each objective. Using Bloom's taxonomy, Kaskutas et al. [33] identified the gap between fall prevention training and the favorite learning methods of apprentices, and then designed new curricula. Endroyoa et al. [34] built an occupational safety and health model to achieve learning outcomes according to Bloom's taxonomy. Pedro et al. [35] proposed a context-based assessment system to improve visualization in teaching safety knowledge. Moreover, there is a growing tendency to apply modern technologies to construction hazard education. Building anatomy modelling [36] and social virtual reality integrated into construction safety system [5] are some examples. Although most studies have their own evaluation system, the application of Bloom's taxonomy could be beneficial in helping educational programs better meet their learning objectives.

2.2. *Virtual Reality in Construction Hazard Investigation*

Virtual Reality (VR) has been applied widely in the context of hazard education and training in recent years [37]. Several VR applications have been developed for hazard identification for targeted users such as designers, site workers, construction students, safety managers, among others. For example, these research efforts include Design-for-Safety-Process (DFSP) systems [38], Cave Automatic Virtual Environments (CAVEs) [39], System for Augmented Virtuality Environments (SAVEs) [40], Visualized Safety Management System (VSMS) [1], Multiuser Virtual Safety Training System (MVSTS) [41], among others. Le et al. developed a learning framework based on an online social VR system, which includes role-playing, dialogic learning, and social interaction for construction safety and health education [5]. Following this, researches have proved the advantages of VR in providing interactive and experiential learning environments [42,43], which are very important to motivate and engage learners in obtaining hazard knowledge and skills [44]. In an effort to develop VR systems that can provide close-to-reality visibility, recent studies have adopted the 360VP technology, which captures 360 panorama images of real construction sites [17]. The 360VP enhances the real-world learning environment by presenting the dynamic nature of construction sites in reality to improve the hazard investigation knowledge skills of construction students and professionals. Furthermore, 360VP has demonstrated improvements in not only hazard education and training [45], but also energy-efficiency [19]. Since VR or 360VP have their own limitations, researches have tried to integrate some VR technologies in order to enhance the advantages of these VR technologies so that hazard education and training can be improved.

3. Framework

The iAPR learning framework consists of three modules including Hazard Understanding Module (HUM), Hazard Recognition Module (HRM), and Safety Performance Module (SPM) in order to improve the construction hazard investigation knowledge and skills of learners (see Figure 1). These modules adopt six Bloom's levels of critical thinking development through the iAPR platform.

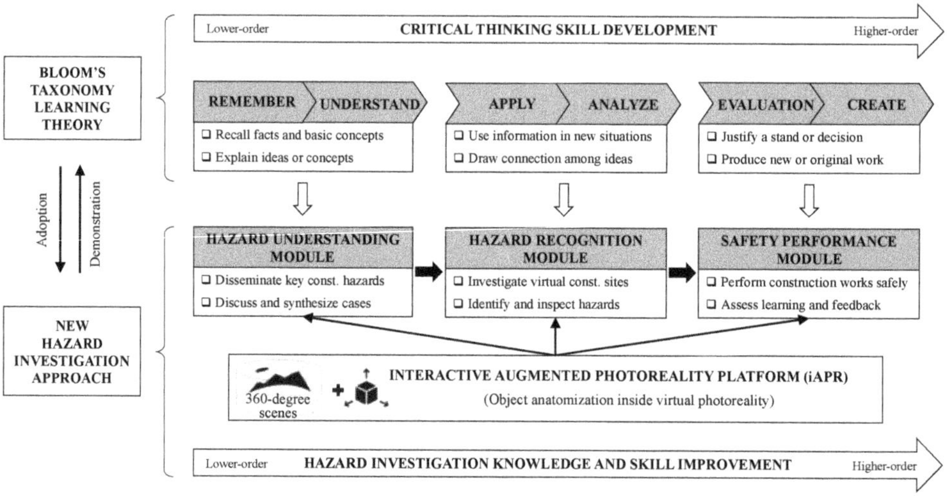

Figure 1. Learning framework.

In particular, HUM aims to help learners remember and understand hazards that often happen in real construction sites. To do this, the educator disseminates key construction hazards in HUM. Following educator instruction, the learners participate in an online discussion to discuss and analyze the hazard cases provided by the educators. During discussion, the learners can easily ask questions to the educator as well as other learners through the chatroom (see Figure 2a) in the iAPR platform. Moreover, hazard-related e-materials (object anatomization, links to images, videos, animation, and e-documents) are supported and can be uploaded through the chatroom in order to provide comprehensive information of the hazard case being discussed. The educator explains the contents in detail, and then synthesizes each hazard case to ensure that all learners thoroughly understand the lesson and hazard-related issues before moving to the next step. As depicted in Figure 1, HUM applies the first two levels of Bloom's taxonomy learning theory for construction hazard education.

Next, HRM focuses on assisting learners to apply the hazard knowledge they learned in the HUM in order to recognize new hazards. To do this, HRM provides an iAPR platform where learners experience virtual construction sites to inspect hazards. In particular, the learners play the role of a safety manager to investigate potential hazards, which are embedded by hotspots (see Figure 2b) in the virtual jobsite. Each hotspot includes hazard information and e-materials related to hazardous scenarios in order to help the learners recognize potential hazards. During hazard investigation in the iAPR environment, the online chatroom assists learners establish online meetings to discuss and share e-materials with other learners when analyzing difficult hazardous situations. This illustrates a common approach that a safety manager needs to adopt in reality when facing difficult problems during construction. While inspecting the recognized hazard, the learners are required to address the root causes, and then propose prevention methods for eliminating the hazard. Furthermore, by clicking on an anatomy function (the second function from the right in the hotspot), a 3D anatomy popup window (see on the right of Figure 2b) appears, assisting learners to perform prevention methods by anatomizing 3D objects (scaffolding, working platform, temporary safety handrails, mobile ladders, safety barriers, etc.). For example, if a construction hazard arises due to improper erection or dismantling of scaffolding, the learners are required to perform proper scaffolding erection and dismantling for ensuring safety. After finishing a hazard inspection, the learners are required to investigate other hazardous scenarios in the virtual panoramic construction site by themselves in order to consolidate their hazard knowledge. HRM adopts two application and analysis levels of Bloom theory for improving the hazard investigation and recognition skills of learners.

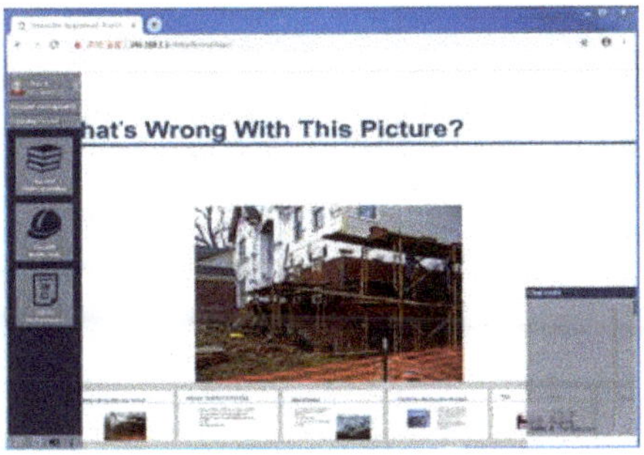

(**a**) Hazard Understanding Module (HUM)

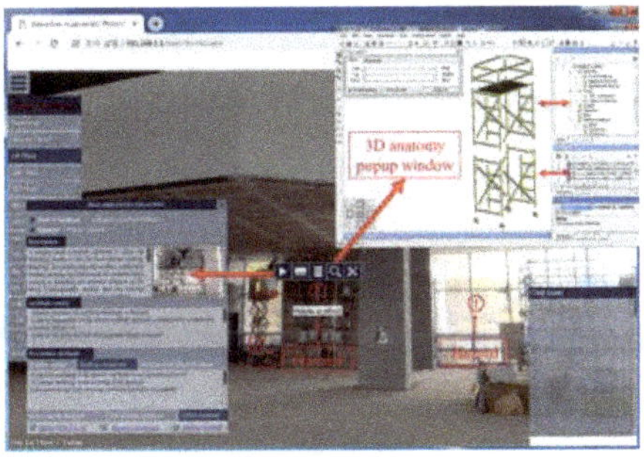

(**b**) Hazard Recognition Module (HRM)

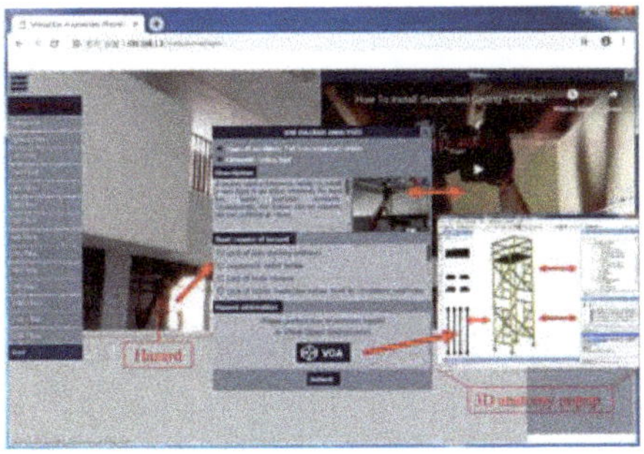

(**c**) Safety Performance Module (SPM)

Figure 2. The Interactive Augmented Photoreality platform (iAPR) prototype interfaces.

Finally, the purpose of SPM is to evaluate the hazard investigation knowledge and skills that learners obtained from the previous modules. To do this, the learners are required to finish a game-based testing on the iAPR platform (see Figure 2c). Through an individual account, each learner acts as a safety manager to navigate a virtual construction site for analyzing dangerous situations. After accurately identifying a potential hazard that is embedded by a alarm sign, a popup window appear, showing a Job Hazard Analysis (JHA) report. The learner has to answer all questions in the JHA, including accident type, hazard description, root causes, as well as prevention methods to eliminate the hazard. Especially, a 3D anatomy popup window (see on the right of Figure 2c) would appear when the user clicks on the Virtual Object Anatomization (VOA) function in the JHA. This function assists the learner to easily propose prevention methods by anatomizing 3D objects such as mobile scaffold, guardrails, etc. For example, as illustrated in Figure 2c, learners are required to install a mobile scaffold step by step in order to prevent a fall from temporary ladder in this recognized dangerous case. Through this, learners can improve their hazard elimination skills. After that, they continue to investigate other potential hazards. Due to the significant importance of hazard investigation skills for preventing construction accidents in practice, the learners cannot move to the next step if they have not accurately investigated and inspected all the hazards in the current step. The iAPR system automatically records the game-based testing performance and feedback of learners for assessing their hazard investigation knowledge and skills. As shown in Figure 1, SPM demonstrates the adoption of the last two Bloom's levels for educating construction hazards.

4. System Architecture

From a systematical perspective, the iAPR architecture is designed by adopting a standard web-app model including three major components: iAPR browser-based client, web service server, and central database; see Figure 3. In the iAPR system architecture, the server and database are implemented in data center on the cloud, while the client is locally installed on user devices. Accordingly, the connection between the server and database is internal. Meanwhile, the client and server communicate with each other via an external networking infrastructure (e.g., Internet).

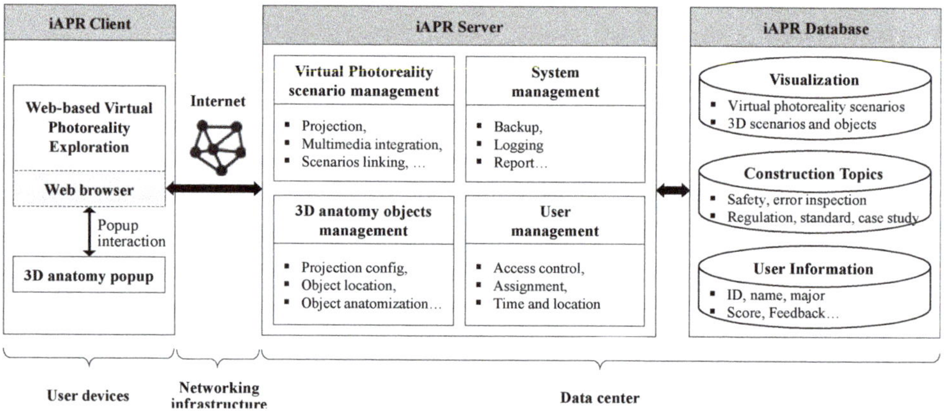

Figure 3. System architecture.

In the iAPR system, the web service server plays a key role and performs four core functions, namely virtual photoreality scenario, 3D anatomy of objects, system management, and user management. In particular, the VP scenario management function provides digital image rendering and projection to reconstruct the captured real environment on the screen. By using multiple deterministic anchors on the surface of the projected sphere, multimedia contents and inter-scene linking objects can be integrated to generate an interactive environment for human sightseeing behavior. On the other hand, the 3D anatomy object management function enables users to perform object projection configuration, multi-level anatomization, and integration on the VP scenario. By default, these functions are activated by user requests through interaction with the corresponding objects inside the VP scenes. In addition, typical system and user management functions are developed to monitor the system operations (e.g., backup, logging, and reporting) and user access (e.g., access control, role assignment, and time&date and location), respectively.

On the iAPR database server side, three databases are designed to manage visualization, construction topics, and user information data. Among these databases, the visualization volume contains VP scenarios, multimedia contents, and 3D object components. The construction topics volume stores the learning data such as safety, error inspection, construction regulation, standard, and case studies. Lastly, the user information volume manages user ID, name, education major, subject score and feedback, etc.

These iAPR servers are located on the cloud to ensure high service availability, and they are accessed from user devices through a networking infrastructure [46]. It is worth noting that the web-based application is designed to work on the Internet protocol (IP) stack; therefore, the iAPR system is able to adapt to almost all popular networking technologies such as cellular and WiFi access networks [47,48]. In the user devices, the iAPR client mainly operates using the built-in web browser to access the iAPR system. When a VP scene is loaded, its integrated 3D objects are temporally cached on the dedicated memory. The 3D objects are called by using popup interaction in the web browser to initiate a 3D anatomy window.

The iAPR system architecture design is advantageous from several perspectives as follows:

- The adoption of Web-app model allows the iAPR application to provide services to a variety of user device classes regardless of their current operation systems (iOS, Android, Windows, and Linux) and processing capability.
- The Central server deployment and module-based function development enables easy maintenance and upgrade of system functions and contents.
- The IP-based service implementation facilitates either local or remote user access technologies via heterogeneous networking infrastructures.

5. System Evaluation

5.1. Prototype Development

In order to evaluate the advantages of the proposed iAPR system, a prototype of the iAPR was implemented by adopting the above-mentioned system evaluation scheme. In particular, the iAPR server was developed by utilizing the Krpano 1.19 framework [49], which provides stable and open interfaces for VP applications. Krpano 1.19 is certified to adapt to the latest version of webvr engine in various web browsers. Moreover, the Krpano enables advanced web-app programming languages such as Hypertext Markup Language version 5 (HTML5), Cascading Style Sheets version 3 (CSS3), eXtensible Markup Language (XML), and JavaScript. On the other hand, mySQL server [50] is used for iAPR database management. The iAPR system was installed on a Raspberry Pi Model B [51] equipped with a 64GB SD card. The Raspberry Pi server connects to a WiFi access point, which provides both local and Internet access for user devices; see Figure 4. In the user devices, the iAPR client consists of a built-in web browser for permanent 360VP scenario access and an NGRAIN interface [52] for 3D anatomy popup window per user demands. The iAPR client is installed on Windows OS. Technical details of the iAPR system setup are summarized in Table 1. For VP data preparation, Samsung Gear 360 camera [53] is used to capture real scenes from construction sites while multimedia videos are uploaded to either an authorized YouTube channel or a local server to get their links.

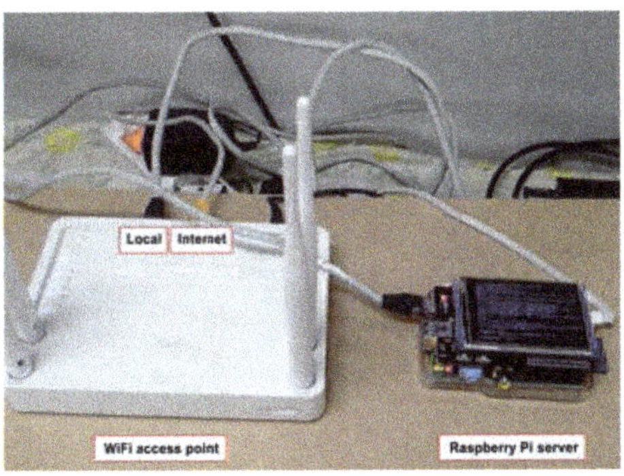

(a) iAPR server

(b) iAPR client

Figure 4. The iAPR system prototype implementation.

Table 1. System prototype configuration.

	iAPR Server	iAPR Client
Hardware	Raspberry Pi Model B	Laptops and PCs
OS	Linux	Windows
Software	Krpano (application) & mySQL (database)	Bult-in web browser & NGRAIN software
Network	Ethernet link	Ethernet/WiFi access

5.2. Case Study

After the system setup stage, case studies related to construction hazard topics are developed for the iAPR prototype in order to evaluate the learning framework and the proposed system. A virtual high-rise building jobsite, which includes fourteen floors and a basement, was chosen for the construction hazard investigation. According to Occupational Health and Safety Administration (OSHA) statistic, "Fatal Four" [54] (falls, struck by object, electrocutions, and caught-in/between accidents) constitute the highest rate of construction accidents in a year, therefore hazard case studies related to "Fatal Four" are chosen for the iAPR trial system, which are summarized in Table 2.

Table 2. Case studies for construction hazard investigation.

No.	Fatal Four	Potential Hazards	Virtual Scenario
1	Falls	Fall from mobile scaffold	1st floor
2	Falls	Fall from 2nd floor to ground floor due to lack of guardrails	2nd floor
3	Falls	Falling from temporary ladder during installation of ceiling panels	3rd floor
4	Falls	Fall from stair due to lack of temporary handrails at the edge of floor	5th floor
5	Falls	Fall from Boatswain's chair due to lack of an independent lifeline	11th floor
6	Falls	Fall into opening of stair at the 8th floor due to lack of barriers	8th floor
7	Struck by object	Struck-by falling object due to lack of safety nets	Roof
8	Struck by object	Bricks falling from height on worker's head without safety helmet	7th floor
9	Struck by object	Metal pipes falling on worker's head during lifting operation	9th floor
10	Electrocution	Electrocution when installing an air-conditioner	10th floor
11	Electrocution	Electrocution when using hand tool	Basement
12	Caught-in/between	Worker is trapped during lift maintenance	12th floor
13	Caught-in/between	Worker is caught between a truck and concrete due to toppling over of precast concrete building unit	Ground floor

Firstly, the educators and learners log into the iAPR prototype through their own ID account, and then click on the HUM function (see Figure 2a) to start the hazard lesson. After that, the educator delivers online slides providing key hazards that often occur during high-rise building construction. With the educator's explanation and synthesis, the learners take part in "question and answer" activities with the educator and other learners until they thoroughly remember and understand all construction hazards in the lesson. Next, the learners log into the HRM function (see Figure 2b) to navigate a virtual high-rise building jobsite and investigate potential hazards. Following the educator's guidance, the learners inspect all hazard cases studies in a virtual construction site to acquire knowledge and skills. Finally, the learners play a testing simulation game using the SPM function (see Figure 2c) in order to assess their performance.

5.3. Evaluation Methodology

In order to address the advantages and limitations of the proposed iAPR system, the evaluation scheme comprises of usability and effectiveness stages, as depicted in Figure 5.

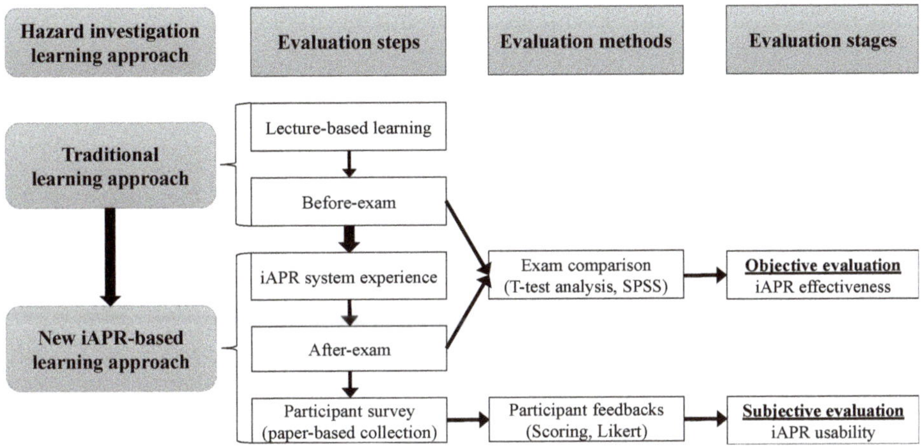

Figure 5. System evaluation scheme.

Following the evaluation steps, the educators disseminated the hazard investigation lessons to 40 learners through lecture-based learning. This is a traditional learning approach where educators provide lectures using a whiteboard and slides. After that, a before-exam was carried out to evaluate the learning outcome of learners before they used the iAPR platform in the subsequent steps. Thereafter, the learners were required to investigate and obtain hazard knowledge and skill using the iAPR platform. After new iAPR-based learning approach, these learners were examined by conducting an after-exam to evaluate the effectiveness of the proposed iAPR system.

Moreover, to validate the system usability, ten educators, ten construction professionals, and 40 students were asked to experience the iAPR platform. After that, all the participants took part in a survey through questionnaires and interviews, which evaluated the proposed system according to the following criteria [11,19,55,56]: (1) Sense of being in the construction jobsite; (2) Ease of hazard investigation and inspection; (3) Real-world visibility of construction jobsite; (4) Support of 3D object anatomization; (5) Interactiveness with virtual environment; and (6) Learning motivation and engagement of users. Finally, all their feedbacks were scored and analyzed using a five-point Likert scale, which range from 1 point for strongly disagree to 5 points for strongly agree.

5.4. Evaluation Results

Table 3 provides the average results of two exams, which were attempted by 40 four-year construction students before and after using the iAPR system for learning. To objectively compare the learning outcome of the learners, a paired sample T-test (called the dependent sample *t*-test) was developed to determine whether the mean difference between the two exams was statistically significant. The null hypothesis was that the mean difference between the two exam scores is equal, while the alternative hypothesis was that the mean difference between the two exam scores is not equal. The SPSS.20 statistics software was utilized to statistically analyze the before-exam and after-exam scores at the 5% significance level. According to Table 3, the mean value and standard deviation are 76.250 and 4.770, respectively for the before-exam, while they are 80.125 and 4.001, respectively, for the after-exam. Since the Sig. (2-tailed) value of 0.001 (see Table 4) is less than the significance level of 0.05, the null hypothesis was rejected. Because of this, it is concluded that there is a statistically significant difference between the mean scores of the two exams. Meanwhile, the effectiveness evaluation results in Table 3 reveal that learners using the iAPR system for construction hazard investigation would have higher scores (80.125) than those who do not utilize the proposed platform for learning

(76.250). Therefore, it proves that the proposed iAPR system can assist learners in improving hazard investigation knowledge and skill.

Table 3. Effectiveness evaluation result.

	N	Mean	Standard Deviation
Before-exam	40	76.250	4.770
After-exam	40	80.125	4.001

Table 4. Paired Samples Test.

	Paired Differences					t	df	Sig. (2-Tailed)
	Mean	Std. Deviation	Std. Error Mean	95% Confidence Interval of the Difference				
				Lower	Upper			
Pair: Before-After exams	−3.87500	5.48746	0.86764	−5.62997	−2.12003	−4.466	39	0.000

Figure 6 depicts the results of iAPR usability evaluation, focusing on the six aforementioned criteria. The interviews and questionnaires related to the criteria for usability evaluation were conducted with all participants including the ten educators, ten construction managers, and 40 learners right after they experienced the iAPR prototype. For the first question "your sense of being in the construction site", which was adopted from [56], all the participants totally agreed that they had a good sense of being in a real construction workplace. Moreover, the users stated that the iAPR design is intuitive so that they could easily investigate and inspect the potential hazards by using mouses (for PCs, laptops) or their own fingers (for smart devices). The functions and tools designed in the iAPR prototype are user-friendly and similar to popular applications in various operating environments such as PCs and laptops as well as smart devices (e.g., ipad, tablets); therefore, it is easy for users to interact with the virtual environment. Regarding real-world visibility, the learners emphasized the advantage of the 360VP technology, which could present the virtual jobsite more realistically than other 3D-VR technologies. They also pointed out that the support of 3D object anatomization function in the iAPR is very necessary to assist learners in proposing prevention methods, which are very important for hazard elimination. Finally, the participants agreed that the iAPR motivates and engages them in learning construction hazard investigation and inspection.

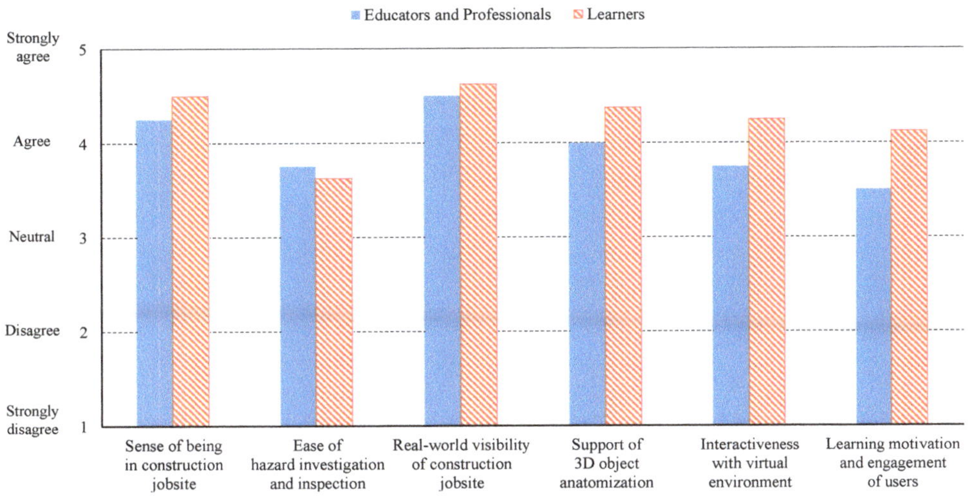

Figure 6. Usability evaluation.

6. Conclusions

Hazard investigation education is very important for equipping learners with adequate knowledge and skills to reduce potential hazards at construction jobsites. Therefore, research efforts have studied and adapted the state-of-the-art technologies such as 360VP and 3D-VR to promote construction hazard education in recent years. Despite the significant benefits of providing an engaging and immersive learning environment, 360VP does not have 3D object anatomization tools to support learning, while the 3D-VR limitation is a lack of providing a real-world environment. Meanwhile, it is a fact that a lack of research integrates these technologies in order to eliminate limitations and maximize advantages for enhancing learning outcomes. Thus, this research objective focuses on developing a construction hazard investigation system entitled iAPR, which integrates 360VP and 3D-VR technologies by leveraging 3D object anatomization on a 360VP platform. The proposed iAPR system consists of three key learning modules, namely HUM, HRM, and SPM, which adopt the revised Bloom's taxonomy theory to improve hazard education. The iAPR prototype were developed, and the usability of the system was evaluated by users including educators, construction professionals, and learners. Moreover, a comparison between before-exam and after-exam results is carried out objectively in order to evaluate the effectiveness of the iAPR platform. Preliminary findings prove that the iAPR has significant pedagogic methods to improve the learner's construction hazard investigation knowledge and skills, which improves safety performance.

Despite the proposed advantages of the iAPR system, future works need to further investigate the following perspectives:

- Regarding an adaptability and reality, an extended study of deploying the iAPR system on various wearable devices should to be conducted, e.g., head-mounted-displays, Microsoft Hololens, Google glass, etc.
- From usability and popularity perspectives, cost efficiency should be considered in terms of initial investigation, and maintenance as well as human labor.
- In application and utilization points of view, it is necessary to conduct an in-depth investigation of how much improvement in learning outcome the learners would obtain in different types of construction such as bridges, tunnels, damps, etc.
- In terms of systematical implementation and elasticity, since this research proposes and validates the iAPR system in a prototype, a full-scale system of iAPR should be deployed in order to comprehensively validate the system performance for a large number of learners.

Author Contributions: Conceptualization, H.C.P. and A.-T.P.-H.; Data curation, H.C.P. and A.-T.P.-H.; Methodology, H.C.P.; Software, N.-N.D., S.C. and A.-T.P.-H.; Supervision, H.C.P. and A.-T.P.-H.; Validation, H.C.P., P.T.N. and A.-T.P.-H.; Writing—original draft, H.C.P., N.-N.D. and A.-T.P.-H.; Writing—review & editing, S.C.

Conflicts of Interest: The authors declare no conflict of interest.

References

1. Park, C.S.; Kim, H.J. A framework for construction safety management and visualization system. *Autom. Constr* **2013**, *33*, 95–103. [CrossRef]
2. Zaira, M.M.; Hadikusumo, B.H. Structural equation model of integrated safety intervention practices affecting the safety behaviour of workers in the construction industry. *Saf. Sci.* **2017**, *98*, 124–135. [CrossRef]
3. Le, Q.T.; Lee, D.Y.; Park, C.S. A social network system for sharing construction safety and health knowledge. *Autom. Constr.* **2014**, *46*, 30–37. [CrossRef]
4. Raheem, A.A.; Hinze, J.W. Disparity between construction safety standards: A global analysis. *Saf. Sci.* **2014**, *70*, 276–287. [CrossRef]
5. Le, Q.T.; Pedro, A.; Park, C.S. A social virtual reality based construction safety education system for experiential learning. *J. Intell. Robot. Syst.* **2015**, *79*, 487–506. [CrossRef]
6. Wybo, J.L.; Van Wassenhove, W. Preparing graduate students to be HSE professionals. *Saf. Sci.* **2016**, *81*, 25–34. [CrossRef]
7. Arezes, P.M.; Swuste, P. Occupational health and safety post-graduation courses in Europe: A general overview. *Saf. Sci.* **2012**, *50*, 433–442. [CrossRef]
8. Wang, P.; Wu, P.; Wang, J.; Chi, H.L.; Wang, X. A critical review of the use of virtual reality in construction engineering education and training. *Int. J. Environ. Res. Public Health* **2018**, *15*, 1204. [CrossRef]
9. Le, Q.T.; Pedro, A.; Lim, C.; Park, H.; Park, C.; Kim, H. A framework for using mobile based virtual reality and augmented reality for experiential construction safety education. *Int. J. Eng. Educ.* **2015**, *31*, 713–725.
10. Gheisari, M.; Foroughi Sabzevar, M.; Chen, P.; Irizzary, J. Integrating bim and panorama to create a semi-augmented-reality experience of a construction site. *Int. J. Constr. Educ. Res.* **2016**, *12*, 303–316. [CrossRef]
11. Le, Q.T.; Pedro, A.; Pham, H.C.; Park, C.S. A Virtual World Based Construction Defect Game for Interactive and Experiential Learning. *Int. J. Eng. Educ.* **2016**, *32*, 457–467.
12. Pham, H.C.; Pedro, A.; Le, Q.T.; Lee, D.Y.; Park, C.S. Interactive safety education using building anatomy modelling. *Univers. Access Inf.* **2017**. [CrossRef]
13. McLachlan, J.C.; Bligh, J.; Bradley, P.; Searle, J. Teaching anatomy without cadavers. *Med. Educ.* **2004**, *38*, 418–424. [CrossRef] [PubMed]
14. Petersson, H.; Sinkvist, D.; Wang, C.; Smedby, Ö. Web-based interactive 3D visualization as a tool for improved anatomy learning. *Anat. Sci. Educ.* **2009**, *2*, 61–68. [CrossRef] [PubMed]
15. Sugand, K.; Abrahams, P.; Khurana, A. The anatomy of anatomy: A review for its modernization. *Anat. Sci. Educ.* **2010**, *3*, 83–93. [CrossRef] [PubMed]
16. Eiris Pereira, R.; Gheisari, M. Site Visit Application in Construction Education: A Descriptive Study of Faculty Members. *Int. J. Constr. Educ. Res.* **2017**. [CrossRef]
17. Eiris, R.; Gheisari, M.; Esmaeili, B. PARS: Using augmented 360-degree panoramas of reality for construction safety training. *Int. J. Environ. Res. Public Health* **2018**, *15*, 2452. [CrossRef]
18. Pham, H.C.; Dao, N.N.; Pedro, A.; Le, Q.T.; Hussain, R.; Cho, S.; Park, C.S. Virtual Field Trip for Mobile Construction Safety Education using 360-degree Panoramic Virtual Reality. *Int. J. Eng. Educ.* **2018**, *34*, 1174–1191.
19. Pham, H.C.; Dao, N.N.; Kim, J.U.; Cho, S.; Park, C.S. Energy-Efficient Learning System Using Web-Based Panoramic Virtual Photoreality for Interactive Construction Safety Education. *Sustainability* **2018**, *10*, 2262. [CrossRef]
20. Jeelani, I.; Albert, A.; Azevedo, R.; Jaselskis, E.J. Development and testing of a personalized hazard-recognition training intervention. *J. Constr. Eng. Manag.* **2016**, *143*, 04016120. [CrossRef]
21. Bloom, B.S. *Taxonomy of Educational Objectives: The Classification of Educational Goals: Cognitive Domain;* Longman: London, UK, 1956.

22. Kratwohl, D.R.; Bloom, B.S.; Masia, B.B. *Taxonomy of Educational Objectives, the Classification of Educational Goals–Handbook II: Affective Domain*; McKay: New York, NY, USA, 1964.
23. Simpson, E.J. *The Classification of Educational Objectives, Psychomotor Domain*; Department of Health, Education, and Welfare, Office of Edcn.: Boston, MA, USA, 1974.
24. Dave, R. *Developing and Writing Behavioural Objectives*; Educational Innovators Press: Tucson, AZ, USA, 1975.
25. Harrow, A.J. *A Taxonomy of the Psychomotor Domain: A Guide for Developing Behavioral Objectives*; Addison-Wesley Longman Ltd.: Boston, MA, USA, 1972.
26. Anderson, L.W.; Krathwohl, D.R.; Airasian, P.W.; Cruikshank, K.A.; Mayer, R.E.; Pintrich, P.R.; Raths, J.; Wittrock, M.C. *A Taxonomy for Learning, Teaching, and Assessing: A Revision of Bloom's Taxonomy of Educational Objectives, Abridged Edition*; Longman: White Plains, NY, USA, 2001.
27. Murtonen, M.; Gruber, H.; Lehtinen, E. The return of behaviourist epistemology: A review of learning outcomes studies. *Educ. Res. Rev.* **2017**, *22*, 114–128. [CrossRef]
28. Thambyah, A. On the design of learning outcomes for the undergraduate engineer's final year project. *Eng. J. Eng. Educ.* **2011**, *36*, 35–46. [CrossRef]
29. Sharunova, A.; Butt, M.; Kresta, S.; Carey, J.; Wyard-Scott, L.; Adeeb, S.; Blessing, L.; Qureshi, A. Cognition and transdisciplinary design: An educational framework for undergraduate engineering design curriculum development. *Eng. Educ.* **2017**, *16*, 27. [CrossRef]
30. Agarwal, P.K. Retrieval practice & Bloom's taxonomy: Do students need fact knowledge before higher order learning? *J. Educ. Psychol.* **2018**. [CrossRef]
31. Hew, K.F.; Cheung, W.S. Use of Web 2.0 technologies in K-12 and higher education: The search for evidence-based practice. *Educ. Res. Rev.* **2013**, *9*, 47–64. [CrossRef]
32. Hussain, R.; Pedro, A.; Lee, D.Y.; Pham, H.C.; Park, C.S. Impact of safety training and interventions on training-transfer: targeting migrant construction workers. *Int. J. Occup. Saf. Ergon.* **2018**. [CrossRef]
33. Kaskutas, V.; Dale, A.M.; Lipscomb, H.; Gaal, J.; Fuchs, M.; Evanoff, B. Changes in fall prevention training for apprentice carpenters based on a comprehensive needs assessment. *J. Saf. Res.* **2010**, *41*, 221–227. [CrossRef]
34. Endroyo, B.; Yuwono, B.E.; Mardapi, D. Model of learning/training of Occupational Safety & Health (OSH) based on industry in the construction industry. *Procedia Eng.* **2015**, *125*, 83–88. [CrossRef]
35. Pedro, A.; Pham, H.C.; Kim, J.U.; Park, C.S. Development and Evaluation of Context-based Assessment System for Visualization-Enhanced Construction Safety Education. *Int. J. Occup. Saf. Ergon.* **2018**. [CrossRef]
36. Pham, H.C.; Le, Q.T.; Pedro, A.; Park, C.S. Visualization Based Building Anatomy Model for Construction Safety Education. In Proceedings of the The 6th International Conference on Construction Engineering and Project Management (ICCEPM), Busan, Korea, 11–14 October 2015; pp. 1–5.
37. Pedro, A.; Chien, P.H.; Park, C.S. Towards a Competency-based Vision for Construction Safety Education. iN *IOP Conference Series: Earth and Environmental Science*; IOP Publishing: London, UK, 2018; Volume 143, p. 012051.
38. Hadikusumo, B.; Rowlinson, S. Integration of virtually real construction model and design-for-safety-process database. *Autom. Constr.* **2002**, *11*, 501–509. [CrossRef]
39. Perlman, A.; Sacks, R.; Barak, R. Hazard recognition and risk perception in construction. *Saf. Sci.* **2014**, *64*, 22–31. [CrossRef]
40. Albert, A.; Hallowell, M.R.; Kleiner, B.; Chen, A.; Golparvar-Fard, M. Enhancing construction hazard recognition with high-fidelity augmented virtuality. *J. Constr. Eng. Manag.* **2014**, *140*, 04014024. [CrossRef]
41. Li, H.; Chan, G.; Skitmore, M. Multiuser virtual safety training system for tower crane dismantlement. *J. Comput. Civ. Eng.* **2012**, *26*, 638–647. [CrossRef]
42. Goulding, J.; Nadim, W.; Petridis, P.; Alshawi, M. Construction industry offsite production: A virtual reality interactive training environment prototype. *Adv. Eng. Inf.* **2012**, *26*, 103–116. [CrossRef]
43. Sacks, R.; Perlman, A.; Barak, R. Construction safety training using immersive virtual reality. *Constr. Manag. Econ.* **2013**, *31*, 1005–1017. [CrossRef]
44. Lin, K.Y.; Son, J.W.; Rojas, E.M. A pilot study of a 3D game environment for construction safety education. *J. Inf. Technol. Constr.* **2011**, *16*, 69–84.
45. Moore, H.F.; Eiris, R.; Gheisari, M.; Esmaeili, B. Hazard Identification Training Using 360-Degree Panorama vs. Virtual Reality Techniques: A Pilot Study. In *Computing in Civil Engineering 2019: Visualization, Information Modeling, and Simulation*; American Society of Civil Engineers Reston: Reston, VA, USA, 2019; pp. 55–62.

46. Na, W.; Dao, N.N.; Kim, J.; Ryu, E.S.; Cho, S. Simulation and Measurement: Feasibility Study of Tactile Internet Applications for mmWave Virtual Reality (VR). *ETRI J.* **2019**, in press.
47. Dao, N.N.; Lee, J.; Vu, D.N.; Paek, J.; Kim, J.; Cho, S.; Chung, K.S.; Keum, C. Adaptive resource balancing for serviceability maximization in fog radio access networks. *IEEE Access* **2017**, *5*, 14548–14559. [CrossRef]
48. Dao, N.N.; Sa'ad, U.; Vu, V.C.; Tran, Q.D.; Ryu, E.S.; Cho, S. A Softwarized Paradigm for Mobile Virtual Networks: Overcoming a Lack of Access Infrastructure. *IEEE Veh. Technol. Mag.* **2018**, *13*, 106–115. [CrossRef]
49. Krpano 1.19. Available online: https://krpano.com/ (accessed on 24 March 2019).
50. mySQL. Available online: https://dev.mysql.com/ (accessed on 24 March 2019).
51. Raspberry Pi 3 Model B. Available online: https://www.raspberrypi.org/products/raspberry-pi-3-model-b/ (accessed on 24 March 2019).
52. NGRAIN 5.1.1. Available online: http://ngrain.com (accessed on 24 March 2019).
53. Samsung Gear 360. Available online: https://www.samsung.com/global/galaxy/gear-360/ (accessed on 24 March 2019).
54. Occupational Health and Safety Administration, United States Department of Labor. Available online: https://www.osha.gov/oshstats/commonstats.html (accessed on 24 March 2019).
55. Virvou, M.; Katsionis, G. On the usability and likeability of virtual reality games for education: The case of VR-ENGAGE. *Comput. Educ.* **2008**, *50*, 154–178. [CrossRef]
56. Usoh, M.; Catena, E.; Arman, S.; Slater, M. Using presence questionnaires in reality. *Presence Teleoperators Virtual Environ.* **2000**, *9*, 497–503. [CrossRef]

applied sciences

MDPI

Article

Superpixel-Based Feature Tracking for Structure from Motion

Mingwei Cao [1,2], Wei Jia [1,2], Zhihan Lv [3], Liping Zheng [1,2] and Xiaoping Liu [1,2,*]

1 School of Computer Science and Information Engineering, Hefei University of Technology, Hefei 230009, China
2 Anhui Province Key Laboratory of Industry Safety and Emergency Technology, Hefei 230009, China
3 School of Data Science and Software Engineering, Qingdao University, Qingdao 266071, China
* Correspondence: liu@hfut.edu.cn

Received: 4 July 2019; Accepted: 22 July 2019; Published: 24 July 2019

Abstract: Feature tracking in image collections significantly affects the efficiency and accuracy of Structure from Motion (SFM). Insufficient correspondences may result in disconnected structures and incomplete components, while the redundant correspondences containing incorrect ones may yield to folded and superimposed structures. In this paper, we present a Superpixel-based feature tracking method for structure from motion. In the proposed method, we first propose to use a joint approach to detect local keypoints and compute descriptors. Second, the superpixel-based approach is used to generate labels for the input image. Third, we combine the Speed Up Robust Feature and binary test in the generated label regions to produce a set of combined descriptors for the detected keypoints. Fourth, the locality-sensitive hash (LSH)-based k nearest neighboring matching (KNN) is utilized to produce feature correspondences, and then the ratio test approach is used to remove outliers from the previous matching collection. Finally, we conduct comprehensive experiments on several challenging benchmarking datasets including highly ambiguous and duplicated scenes. Experimental results show that the proposed method gets better performances with respect to the state of the art methods.

Keywords: feature tracking; superpixel; structure from motion; three-dimensional reconstruction; local feature; multi-view stereo

---

1. Introduction

In recent years, structure from motion (SFM) has received much attention from the computer vision and graphics communities. SFM is a collection of technologies, which is able to reconstruct 3D point-cloud model, and can estimate camera parameters (including intrinsic and extrinsic parameters) from image sequences [1]. A classic SFM framework usually consists of camera calibration, feature tracking, camera pose estimation, triangulation, and bundle adjustment [2]. It is well-known that SFM plays an important role in many research areas [3], such as augment reality, multi-view stereo [4], image-based localization [5], 3D reconstruction, image-based navigation [6], place recognition, autonomous driving, camera localization, and geographic information system (GIS) [7,8]. Based on different focuses, different types of SFM technologies have been proposed, such as incremental SFM, Global SFM, and Hybrid SFM [9].

Among existing Incremental SFMs, Bundler [10] is prestigious, which is a standard implementation of SFM, in which the scale invariant feature transform (SIFT) [11] is adopted to detect keypoints, then resulting in a highly computational cost. With the development of Graphics Process Unit (GPU), Wu et al. [12] implemented a GPU accelerated SIFT named SIFTGPU to reduce the computation time of feature tracking. Based on the SIFTGPU, Wu et al. developed a fast SFM system called Visual SFM (VSFM) [12], thus resulting in a significantly improvement in the aspect of time efficiency. In addition to the promising speed, the VSFM is user friendly due to its Graphic User Interface (GUI), and can not only work with multi-view stereo (MVS), such as the patch-based multi-view stereo (PMVS) [13], but also can be combined with Poisson surface reconstruction [14] to produce textured model of the scene. Dong et al. [15] developed a robust and real-time camera tracking system based on keyframes, called ACTS, for multi-view 3D reconstruction. The ACTS system consists of offline and online modules, both two modules work together can quickly recover the point-cloud model of the scene, and estimate camera's parameters containing intrinsic and extrinsic parameters. After a series of improvements on the ACTS, which is extended to work in large-scale surroundings [16]. Ni et al. [17] proposed a hierarchical SFM in a divide and conquer manner by using the bipartite graph structure of the scene. COLMAP [18] is an excellent incremental SFM implementation that contains many novel techniques such as scene augmentation, re-triangulation, and depth-fusion approach. All SFMs mentioned before use SIFT or SIFT's variants to locate keypoints and compute descriptors, other excellent local features may be ignored. Zach et al. [19] is the first time to use Speeded Up and Robust Features (SURF) [20] to detect keypoints and compute descriptors for feature for SFM, then leading a significantly boosting on speed.

Agarwal et al. [21] consider that feature tracking method may largely affect the quality of SFM. For example, if the captured image data contains few features, or many repeating features, the matching precision of feature tracking cloud be decreased significantly. To improve the problem of repeating features, some incomplete approaches has been proposed, such as loop constraint-based approach [22] where the observed redundancy in the hypothesized relations is used to reason the repetitive visual structures in the scene. Fan et al. [23] proposed to utilize the low distortion constraint approach to match pairs of interest points and then obtained feature correspondences from the matched pairs of interest points. Roberts et al. [24] found that the geometric ambiguities are usually caused by the presence of repeated structures and then proposed an expectation maximization (EM)-based algorithm that estimate camera poses and identifies the false match-pairs with an efficient sampling method to discover plausible data association hypotheses. Snavely et al. [25] presented a novel approach to solving the ambiguous problems by considering the local visibility structure of the repeated features and then presented a network theory-based method to score the repeated features. Recently, Ceylan et al. [26] designed an optimization framework for extracting repeated features in images of urban facades, while simultaneously calibrating the input images and estimating the 3D point-cloud model using a graph-based global analysis. Although some novel approaches have been proposed for the problem of ambiguous structures, they only work in the symmetric scenes.

To defend the ambiguous problem, we have paid much attention to investigate deeply the existing works [9,27,28], the following reasons may cause to produce ambiguous point-cloud model, that is repeated feature, untextured region where few keypoints can be found. As a result, we propose a superpixel segmentation-based feature tracking method for repeated and untextured scenes. Considering the simplicity, the superpixel-based feature tracking is abbreviated as "SPFT". The SPFT consists of feature detection, superpixel segmentation, and Markov Random Field (MRF)-based superpixel matching. Owing to the used superpixel segmentation, the SPFT can find sufficient keypoints in untextured scenes. Moreover, the SPFT can be considered as a general framework for feature tracking, which can be integrated with various local feature approaches such as SIFT, SURF, KAZE [29], and MSD [30]. Several challenging experiments made in Section 5 can efficiently prove the effectiveness and efficiency of the SPFT.

The main contributions of this work are summarized as follows:

- A Superpixel-based feature tracking method is proposed to locate keypoints and produce feature correspondences. The SPFT method has the fast speed and high matching confidence. Thus, SPFT can largely improve the quality of point-cloud model produced by SFM system.
- A combined descriptor extractor is proposed for producing robust descriptions for the detected keypoints. The proposed descriptor is robust to image rotation, lighting changes, and even can distinguish repeated features.

- We conduct a comprehensive experiment on several challenging datasets to assess the SPFT method, and comparison with the state-of-the-art methods. According to the evaluation, some valuable remarks are presented, which can be as a guide for developers and researchers.

The rest of this paper is organized as follows: related work is presented in Section 2. The proposed method is described in Section 3. In Section 4, a prototype 3D reconstruction system based on SFM is presented. Experimental results are given in Section 5. The conclusions and final remarks are given in Section 6.

2. Related Work

In this section, we will briefly review existing feature tracking methods and various SFM frameworks for better understanding the proposed feature tracking method.

2.1. Feature Tracking

Over the past years, many feature tracking methods has been proposed in the field of 3D reconstruction. The existing methods can be roughly divided into two categories, KLT-like approaches [31], and detection-matching framework (DMF)-based methods [32]. For the former, they compute displacement of keypoints between consecutive video frames when the image brightness constancy constraint is satisfied, and image motion is fairly small. However, KLT-like methods are only suitable to video data [33] in which each image frames have same resolution. To defend the drawbacks of KLTs, the DFM-based methods been proposed. In general, the DMF consists of keypoint detection, descriptor computing, and descriptor matching. For example, Snavely et al. [34] proposed a simple feature tracking method in which the SIFT and Brute-Force-Matching (BFM) were used to locate keypoints and to match descriptors respectively. Zhang et al. [35] developed a segment-based feature tracking method for camera tracking, the method can efficiently track non-consecutive video or image frames by the backend feature matching.

Moreover, researches proposed many novel local features to replace the SIFT in feature tracking procedure, such as speed up robust features (SURF) [20], Oriented Fast and Rotated Brief (ORB) [36], Binary Robust Invariant Scalable keypoints (BRISK) [37], maximally stable extremal regions (MSER) [38], and KAZE [29], features from accelerated segment test (FAST) [39], AGAST [40] and center surround detectors (CenSurE) [41]. Among these detectors, FAST and AGAST have fast speed, which are widely used in some real-time environments such as large scale simultaneous localization and mapping (SLAM) systems [42]. But they easily suffer from image rotation due to the local feature without main direction. To address this issue, Leutenegger et al. [37] proposed the BRISK detector, which is an invariant version of AGAST in multiple scale spaces. Unfortunately, BRISK has a low repeatability, which can further aggravate the drift problem in the process of feature tracking. Recently, binary descriptor has attracted much attention from the field of 3D reconstruction, such as local difference binary (LDB) [43,44], learned arrangements of three patch codes descriptors (LATCH) [45], boosting binary keypoint descriptors (BinBoost) [46], fast retina keypoint (FREAK) [47], and KAZE [29], etc. However, these binary descriptors can easily produce same descriptor in the scene with repeating structures according to [43]. Thus, the resulting ambiguous descriptors may further aggravate the ambiguity of feature matching especially in outdoors.

In addition to ambiguity, the existing local features have expensive computational cost. Even for binary local features, such as ORB, BRISK, the computational costs are also very high in large-scale scenarios. To accelerate the feature tracking method, Wu et al. [48] developed a SIFTGPU routine, which is the parallel implementation of the SIFT on GPU devices, then the SIFTGPU can achieve 10 times acceleration than that of original SIFT. Thus, the SIFTGPU is widely used in various computer tasks including SFM, simultaneous localization and mapping (SLAM), and robotic navigation. Inspired by SIFTGPU, Graves et al. [49] developed KLTGPU routines using OpenCL, then resulting in a 92% reduction in runtime compared to a CPU-based implementation. Cao et al. [50] proposed a GPU-accelerated feature tracking (GFT) method for SFM-based 3D reconstruction, which has a 20 times faster than that of SIFTGPU. Xu et al. [51] designed a GPU-accelerated image matching method with improved Cascade Hashing named CasHash-GPU, in which a disk-memory-GPU data exchange approach is proposed to optimize the load order of data, so the proposed method is able to deal with big data. According to their experiments, the CasHash-GPU can achieve hundreds of times faster than the CPU-based implementation.

2.2. Structure from Motion

Recent years, many 3D multi-view 3D reconstruction systems based on SFM technique have been proposed. For example, Snavely et al. [10] designed and implemented an excellent 3D reconstruction system, called Bundler, to reconstruct spare point-cloud model from unordered image collections. In the Bundler system, the authors employ scale invariant feature transform (SIFT) [11] to detect keypoints and compute descriptors, and use brute-force matching (BFM) strategy to match descriptors for image pair. However, owing to the usage of SIFT and BFM, the Bundler system has high computation cost. To save the computation time for 3D reconstruction based SFM, Wu et al. [12] developed a Visual SFM (VSFM) system based on Bundler, which use SIFTGPU to detect keypoints and compute descriptors for saving computation time. Micusik et al. [52] presented a novel SFM pipeline, which estimates motion and wiry 3D point clouds from imaged line segments across multiple views. The proposed SFM system tackle the problem of unstable endpoints by using relaxed constraints on their positions, both during feature tracking and in the bundle adjustment stage. Sweeney et al. [53] introduced the distributed camera model for 3D reconstruction based on SFM technique, in which, the proposed model describes image observations in terms of light rays with ray origins and directions rather than pixels. As a result, the camera model can describe a single camera or multiple cameras simultaneously as the collection of all light rays observed.

Based on the successes in solving for global camera rotations using averaging technique, Kyle et al. [54] proposed a simple, effective method for solving SFM problems by averaging epipolar geometries. The proposed unstructured SFM system (1DSFM) can overcome several disadvantages of existing sequential SFM. Moulon et al. [55] proposed a novel global calibration approach based on the global fusion of relative motions between image pairs for robust, accurate and scalable SFM. After an efficient contrario trifocal tensor estimation, the authors define an efficient translation registration method to recover accurate positions. Besides accurate camera position, Moulon et al. use KAZE [29] feature to detect keypoints in feature tracking, then resulting in a high-precision score. Based on optimized viewgraph, Chris et al. [56] designed and implemented an excellent SFM system, named Theia-SFM, to produce compact and accurate point-cloud model for both indoor and outdoor scenes. To recover the location of an object, Goldstein et al. [57] designed a scalable SFM system by utilizing ShapeFit and ShapeKick, even in the presence of adversarial outliers. Cohen et al. [58] proposed a novel solution for 3D reconstruction based on SFM to reconstruct the inside and the outside of a building into a single model by utilizing the semantic information, in which, novel cost function is proposed to determine the best alignment. To solve the degeneracies introduced by rolling shutter camera models, Albl et al. [59] show that many common camera configurations such as cameras with parallel readout directions, become critical and allow for a large class of ambiguities in 3D reconstruction based on SFM technique.

With the development of the depth camera, such as Kinect and RealSense, many RGBD datasets are publicly available for 3D reconstruction. Xiao et al. [60] developed RGBD-SFM system to produce dense point cloud model from RGBD images. Recently, Cui et al. [61] hold that SFM methods can be broadly categorized as incremental or global according to their ways to estimate initial camera poses. They proposed a unified framework to tackle the issues of efficiency, accuracy, and robustness, and developed a hybrid structure from motion (HSFM) system.

3. SLIC Method

Superpixel was first proposed by Ren et al. [62], and was used for image segmentation. In general, a superpixel in the image is a group of pixels that have continuous depths. The following properties for the superpixel are generally desirable: Superpixels should adhere well to image boundaries, and Superpixels should be fast to compute, memory efficient, and simple to use. Therefore, in the recent years, many superpixel algorithms, such as simple linear iterative clustering (SLIC) [63], superpixels extracted via energy-driven sampling (SEEDS) [64], Lattices [65], and GMMSP [66], have been proposed for various applications.

In this paper, the superpixel algorithm is selected as a preprocess step to segment tiny regions, as shown in Figure 1, the SLIC is the best choice due to its two important properties: (1) The number of distance calculations in the optimization is dramatically reduced by limiting the search space to a region proportional to the superpixel size. This reduces the complexity to be linear in the number of pixels N and independent of the number of superpixels k. (2) A weighted distance measure combines color and spatial proximity, while simultaneously providing control over the size and compactness of the superpixels. By default, the only parameter of the SLIC algorithm is k, which is the desired number of approximately equally-sized superpixels. For a given color image in the CIELAB color space, to get superpixel segmentations the following steps are required:

Step 1: Initialize cluster centers $C_i = \begin{bmatrix} l_i & a_i & b_i & x_i & y_i \end{bmatrix}^T$, which are sampled on the regular grid spaced S pixels apart.

Step 2: Move the cluster centers to the lowest gradient position in a 3×3 neighborhood.

Step 3: Compute the distance E between each cluster center C_k and pixel i in a $2S \times 2S$ region around C_k, if $D < d(i)$ then set $d(i) = D$, $l(i) = k$.

Step 4: Compute new cluster centers C_k' and residual error E.

Step 5: Repeat Step 3 and Step 4 until the residual E less than the threshold.

(**a**) Input

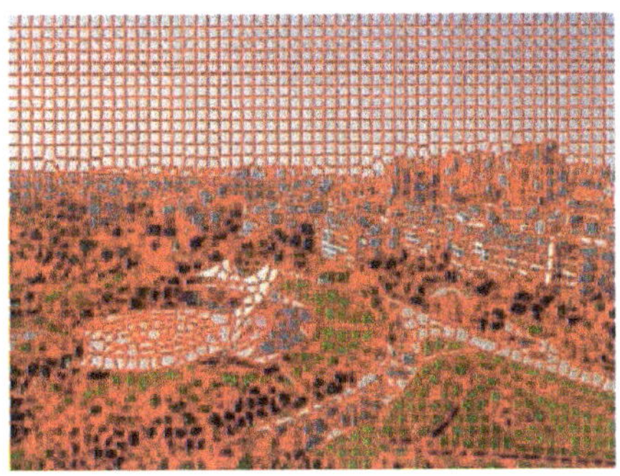

(**b**) Visual results of superpixels

Figure 1. Illustration for superpixels.

4. The Proposed Method

To improve the quality of SFM, we propose a superpixel-based feature tracking method (SPFT), which consists of feature detection, descriptor computing, feature matching, and outliers removing. The flowchart of SPFT is depicted in Figure 2. For given an image, we first use SLIC algorithm to segment it to obtain non-overlapping regions C_i, and then use SIFTGPU feature detector to locate keypoints K_j, thus the total keypoints $K_i' = \left\{C_i \cup K_j \middle| i = 1 \cdots N, j = 1 \cdots M\right\}$. Second, use ORB feature to describe the detected keypoints K_i', and use SLIC labels to compute a patch-based description, then resulting a combined descriptor. Third, use k nearest neighboring method (KNN) to match the combined descriptors between the reference image and the query image. Finally, we use cross-check to remove incorrect matches from the KNN matching, then resulting in a set of correct correspondences as shown in Figure 2.

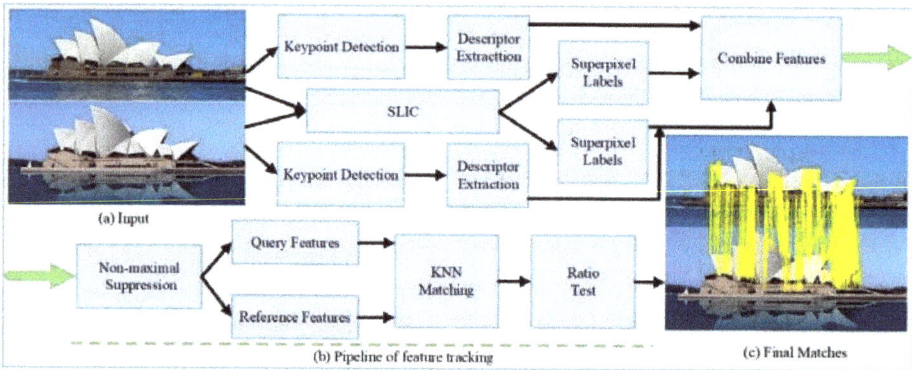

Figure 2. Flowchart of the superpixel-based feature tracking (SPFT) method.

4.1. *Joint Keypoint Detector*

To accelerate the speed of feature tracking, we propose a Joint Keypoint detector (JKD) that is based on FAST detector, as described in [39]. The JKD consists of two major stages: learning keypoint and superpixel-based keypoint location—each of which, in turn, takes several steps. In the stages of the learning keypoint, the input image is first convoluted. The output of convolution, known as the integral image, is then used as the basis of the scale-space analysis. The responses obtained from the scale-space analysis are utilized to detect the keypoints, $kp_i(x, y)$. In the stage of superpixel-based keypoint location, the SLIC is used to segment the input image to several labels, and then those labels have their center position, $cp_i(x, y)$. Finally, combine the $kp_i(x, y)$ and $cp_i(x, y)$, we can get the final keypoints, $k_i(x, y)$, via non-maximal suppression. The pipeline of the JKD keypoint detector is shown in Figure 3.

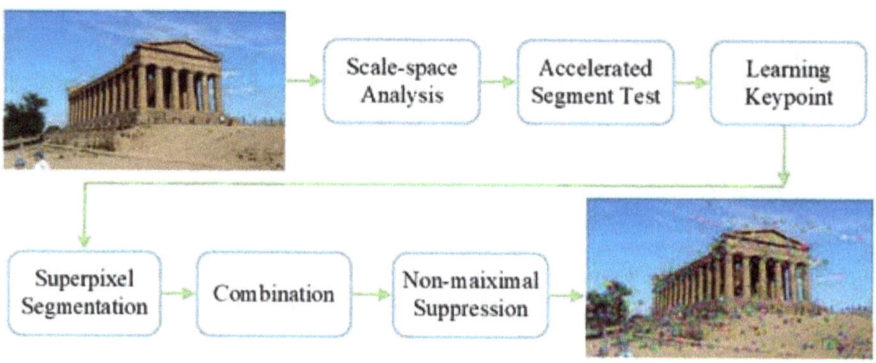

Figure 3. Joint keypoint detection.

Let $O(x, y)$ represents a candidate keypoint, and $N_O(x, y)$ represents the 7×7 neighbors of $O(x, y)$. Compute the DOG image of $R_O(x, y)$ to get $DOG_O(x, y)$ by Equation (1)

$$DOG_O(x, y) = G(x, y, k\sigma) - G(x, y, \sigma) \tag{1}$$

where k is a constant, $G(x,y,\sigma) = \frac{1}{2\pi\sigma}exp\left(-\frac{x^2+y^2}{2\sigma^2}\right)$ represents Gaussian density function with variance σ. Changing the value of σ, a set of DOG image is obtained as $DOG_{set}(x,y) = \{dog_1, \cdots, dog_5\}$ where 5 DOG images is constructed only for saving computation time.

For each location on the given DOG image I, the pixel at that position relative to O can have one of three states:

$$S_{O\to N} = \begin{cases} d, I_{o\to n} \leq I_o - t \\ s, I_o - t < I_{o\to n} < I_o + t \\ b, I_{o\to n} > I_o + t \end{cases} \tag{2}$$

where, $S_{O\to N}$ is a correlation between pixel o and n. d denotes darker, s denotes similar, and b denotes brighter. t is a threshold with a tiny value.

For all $O \in N_o$, the N_o can be divided into three subsets N_d, N_s, and N_b by computing $S_{O\to N}$. Use ID3 [67] algorithm to choose the first pixel n to compare with the candidate keypoint $O(x,y)$, and decide whether $O(x,y)$ is keypoint or not according to the entropy $H(O)$ of K_O.

$$\mathrm{H}(O) = (c+\bar{c})\log_2(c+\bar{c}) - c\log_2 c - \bar{c}\log_2 \bar{c} \tag{3}$$

where $\mathrm{c} = \left|\{p|K_p \text{ is true}\}\right|$ represents the number of keypoints and $\bar{c} = \left|\{p|K_p \text{ is true}\}\right|$ represents the number of non-keypoints.

If the selected n belongs to O_d and produce the max value of $H(O)$, then O_d can be further divided into the following five categories: O_{dd}, $O_{\overline{dd}}$, O_{ds}, O_{db}, $O_{\overline{db}}$. For O_s, divide it into O_{sd}, $O_{\overline{sd}}$, O_{ss}, O_{sb}, $O_{\overline{sb}}$. The process is applied recursively on all five subsets until $H(O)$ equals to zero. The candidate keypoint can be detected according to the value of K_O.

$$\mathrm{O}(x,\ y) = \begin{cases} true, K_O = 1 \\ false, K_O = 0 \end{cases} \tag{4}$$

where O is a keypoint if K_o is one. Repeat above process until all input images processed over, then a set of FAST keypoints can be obtained as follows:

$$K_{fast} = \{k_i|i = 1, \cdots, n\} \tag{5}$$

However, the keyoints detected by FAST are often distributed not average, then the resulting point-cloud models are discontinuous.

To avoid the in-averaging distributed of FAST keypoints, we use superpixel segmentation approach as a post-process step to find many small regions. Thus, the centers of the regions are selected as the candidate keypoints. For a given image in CIELAB color space, the candidate keypoints, K_{slic}, could be obtained by SLIC algorithm as described in Section 3.

$$K_{slic} = \left\{k^j_{slic}\middle|j = 1, \cdots, m\right\} \tag{6}$$

Once, the K_{fast} and K_{slic} are computed, the combined keypoints can be obtained as follows:

$$K_{find} = \left\{k^j_{slic} \cup k^i_{fast}\middle|j \in [1,m] \wedge i \in [1,n]\right\} \tag{7}$$

To choose high-quality keypoints that have maximal responses, we use non-maximal suppression (NMS) [39] to eliminate the unstable keypoints that have minimal responses. The NMS is defined as

$$\mathrm{V} = max\left(\sum_{x\epsilon S_s}|I_{o\to x} - I_O| - t,\quad \sum_{x\epsilon S_d}|I_O - I_{O\to x}| - t,\right) \tag{8}$$

As a result, by suppression the low-quality keypoints, the final keypoints that locate by the JKD is

$$K_{final} = \left\{k_j \middle| \, j \in [1,\ m+n]\right\} \tag{9}$$

It should be note that the number of keypoints by JKD is vary, which depends on the value of σ in Equation (1). Thus, we can change σ to obtain more keypoints for special applications such as dense simultaneous localization and mapping (SLAM) [68] and face recognition [69,70].

4.2. Joint Descriptor Computing

The robustness of descriptor is very important to achieve robust feature tracking, which has been analyzed deeply in [43]. According to the last recent evaluation work made by Zhu et al. [71], the SURF feature has desirable performance on aspect of matching speed and precision. However, the SURF feature easily suffers from affine transform, this may break the compactness of point-cloud model when it is used in 3D reconstruction system. To improve the quality of 3D reconstruction system, we propose a joint computing procedure that include SURF and binary test [36], the former is use to describe the keypoints located in the texture areas, then the latter is used in the textureless areas. For convenience, we called the proposed feature descriptor as joint feature descriptor (JFD), the pipeline for computing a JFD feature descriptor is depicted in Figure 4, in which it is run on GPU device for accelerating. In addition to the matching precision and fast speed, the proposed JFD feature is also robust to various perturbations such as noise, illumination or contrast change.

For the k_j located in the texture areas, we first use SURF feature to compute a vector of 64 dimensional which is an normalized gradient statistics extracted from a spatial grid R divided into 4×4 regions. These subregions are referred to as $\mathrm{R} = \left\{R_{i,j} \middle| 1 \le i, j \le 4\right\}$. According to [20], the weighted gradient at point (u, v) is defined as,

$$\begin{pmatrix} d_x(u,v) \\ d_y(u,v) \end{pmatrix} = R_{-\theta_k}\begin{pmatrix} D_x^{L_k} \\ D_y^{L_k} \end{pmatrix}\varphi(x,y) \times G_1(u,v) \tag{10}$$

where $D_x^{L_k}$ and $D_y^{L_k}$ denote first order box filters, which are used to compute the gradient components.

To this end, the SURF uses first order statistical results on vertical and horizontal gradient responses to produce the good description that achieves the best performance between accuracy and efficiency, then the resulting statistical vector with respect to $R_{i,j}$ can be calculated by the following formula,

$$\mu_k(i,j) = \begin{pmatrix} \sum_{u,v}^{R_{i,j}} d_x(u,v) \\ \sum_{u,v}^{R_{i,j}} d_y(u,v) \\ \sum_{u,v}^{R_{i,j}} \left|d_x(u,v)\right| \\ \sum_{u,v}^{R_{i,j}} \left\lceil d_y(u,v)\right\rceil \end{pmatrix}, i, j \in [1,4] \tag{11}$$

The SURF descriptor of k_i can be directly computed by concatenating the $\mu_k(i, j)$, which is defined as

$$\mu_k = vstack(\mu_k(i,j)) \tag{12}$$

where $vstack(\cdot)$ is function that represents stacking the matrix in vertical direction.

To improve the invariance to linear transform, the SURF descriptor should be normalized to a unit vector by L2 normal, the enhanced SURF descriptor can be calculated by the following formula

$$\mathrm{SURF}(k_i) = \mu_k / \|\mu_k\|_2 \tag{13}$$

However, for the keypoints distributed in textureless regions, we use binary test to produce robust descriptors in the neighbor regions that labeled by the superpixel-based segmentation. The binary test τ in [36] is defined as

$$\tau(L, x, y) = \begin{cases} 0, p(x) \geq p(y) \\ 1, p(x) < p(y) \end{cases} \tag{14}$$

where $p(x)$ represents the intensity of p at a point (x, y). Thus, the resulting feature vector is defined as

$$v_n(k_i) = v_n(p(x, y)) = \sum_{1 \leq i \leq n} 2^{i-1}(\tau(L_i, x_i, y_i)) \tag{15}$$

Note that n is set to 32 for saving computation time in the whole experiment, thus the resulting feature vector has 32 binary elements.

To this end, the JKD descriptor can be obtained by concatenating the SURF(k_i) and $v_n(k_i)$, then resulting a 96 dimensional of feature descriptor.

$$\text{JKD}(k_i) = concat(\text{SURF}(k_i), v_n(k_i)) \tag{16}$$

Owing to the JKD is hybrid type, namely it not only includes float type elements, but also contains binary type ones, thus, we need urgently a novel matching approach to match them.

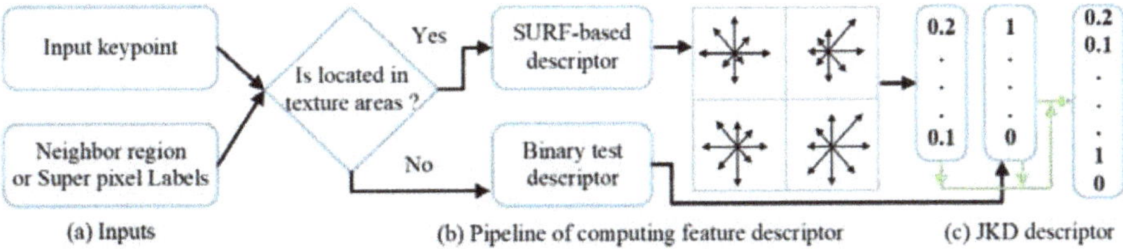

Figure 4. Flowchart of joint keypoint computing.

4.3. Fast Descriptor Matching

Feature matching aims to measure the similarity between the two feature descriptors. The float-type descriptors, such as SIFT, SURF et al. usually use Euclidean distance (L2 distance) to measure the similarity of two feature descriptors [11]. For binary descriptors such as BRISK [37] and LGHD [72], the Hamming distance is used [43]. Because our descriptor is hybrid type that not only includes float-type elements, but also contains binary-type ones. Thus, we use two metrics to measure the similarity of the proposed feature descriptors, namely Hamming distance and Euclidean distance as shown in Figure 5.

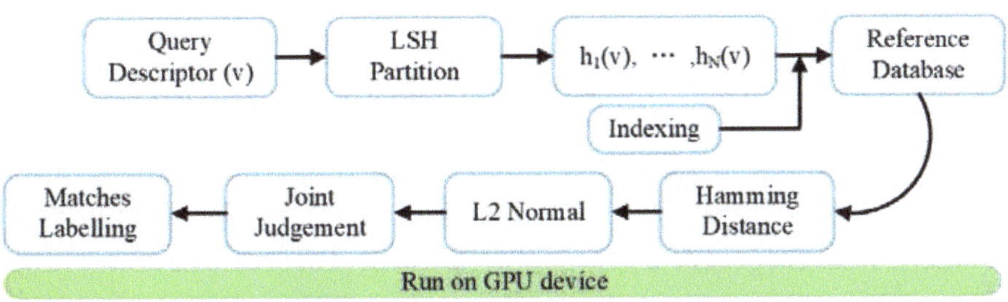

Figure 5. Flowchart of descriptor matching.

The former is utilized to measure similarity of superpixel-based feature descriptors, then the latter is exploited to handle float-type feature descriptors. For the given two binary-type descriptors,

$DB_r = \left\{d_r^1, \cdots, d_r^n\right\}$, $D_{Bq} = \left\{d_q^1, \cdots, d_q^n\right\}$, then the similarity between DB_r and DB_q can be calculated by the simple bitwise operation.

$$MS_{q,r} = DB_r \ xor \ DB_q \tag{17}$$

where *xor* denotes XOR operation which returns the number of different elements between DB_r and DB_q.

However, for float-type feature descriptor, $DF_q = \{q_1, \cdots, q_m\}$ and $DF_r = \{r_1, \cdots, r_m\}$, we use the Euclidean distance (L2 normal) to estimate the similarity of them, the matching confidence can be calculated as

$$C_{qr} = \|\mathrm{p}(q_i) - \mathrm{p}(r_i)\|, \ i \in [1, m] \tag{18}$$

where $\mathrm{p}(q_i)$ and $\mathrm{p}(r_i)$ denote the descriptor for keypoint q_i and keypoint r_i respectively.

Once, the metrics are defined, we can simply loop the above procedure until the feature descriptors in the feature database is processed over, then every feature descriptor in query feature database has two potentially corresponding candidates. Let $\mathrm{p}(r_i)$ and $\mathrm{p}\left(r_j\right)$ denote the candidates with respect to the query descriptor $\mathrm{p}(q_i)$, then we can judge whether the matching is successful by the following formula.

$$C_f = \frac{\|\mathrm{p}(q_i) - \mathrm{p}(r_i)\|}{\|\mathrm{p}(q_i) - \mathrm{p}\left(r_j\right)\|} \tag{19}$$

If $c < 0.7$, the $\left\langle q_i, r_j\right\rangle$ is a correct match. Base on the hybrid matching approach, we can use the Brute-Force-Match (BFM) [73] to find a candidate for each query keypoint.

However, BFM-based KNN approach is a greedy algorithm and has an expensive computational cost. If the matching method is utilized in large-scale 3D reconstruction, then the process of recovering 3D model is very slow. Thus, we must improve the computation efficiency of BFM-based KNN to accelerate the feature tracking method. After a deep investigation in descriptor matching methods [74,75], we found that local sensitive hash (LSH) [51,76] is an efficient approach to achieve descriptor matching. Thus, the LSH is utilized to match feature descriptors. The core of LSH algorithm is an approximate approach to compute k-nearest neighbors, which use N hash functions $h_1(\cdot), \cdots, h_N(\cdot)$ to transform the D-dimensional space R^D into a lattice space L^D, and the original each data is distributed into one lattice cell:

$$H(v) = \{h_1(v), \cdots, (v)\} \tag{20}$$

where v denotes a vector of query descriptor.

To this end, the LSH-based KNN can use the L2 distance to measure the similarity between the query descriptor and the reference descriptor.

Algorithm 1 Superpixel-based feature tracking scheme

Input: image sequences, $I = \{I_1, I_2, \cdots, I_N\}$.
Output: a set of matching pairs, $S = \left\{\left\langle k_{ij}, k_{hc}\right\rangle \middle| i, h \in [1, N]\right\}$.

Step1: Compute keypoints for each image in $\{I_1, I_2, \cdots, I_N\}$, then resulting in a set of keypoints, $\{k_1, k_2, \cdots, k_m\}$.

Step2: Compute feature descriptor for each located keypoint, if they are located in texture areas, then use Equations (11) and (12) to obtain robust description, otherwise, use binary test that defined in Equation (15) to describe the keypoints.

Step3: Construct hash tables via Equation (20), the large set of JKD descriptors is distributed into many lattice cells independently.

Step4: For $\mathrm{JKD}(k_i)$ and $\mathrm{JKD}\left(k_j\right)$ the similarity can be measured by Equations (17) and 19. If those formulas are true, $\mathrm{JKD}(k_i)$ and $\mathrm{JKD}\left(k_j\right)$ are considered matching.

Step5: Repeat Step4 for any two keypoints in $\{k_1, k_2, \cdots, k_m\}$ the resulting matching pairs is $S = \left\{\left\langle k_{ij}, k_{hc}\right\rangle \middle| i, h \in [1, N]\right\}$.

---

5. Experimental Results

The proposed SPFT is developed in C++, NVIDIA CUDA SDK 10.0 and OpenCV SDK 4.0, on a PC with Intel i7 CPU processor 3.40 GHz and 32.0GB memory. We have evaluated the SPFT method on several challenging dataset, and have compared it with the state-of-the-art methods, including HPM [77], ROML [78], MODS [79], ENFT [35] and SuperPoint [80]. It should be noted that SuperPoint is deep learning-based approach to feature detection and descriptor computing, and is published on the European Conference on Computer Vision in 2018.

5.1. Evaluation of Colosseum Dataset

We have evaluated the performance of the SPFT on the Colosseum dataset which is constructed by the authors of this paper. Samples of the Oxford benchmark are shown in Figure 6 where the lighting of every images is different to each other, and they also have many repeated features and structures. In the whole process of experiment, we use a standard evaluation metric to measure the performance for each method. The evaluation metric is defined as:

$$Precision = \frac{\#correct\ matches}{\#tentative\ matches} \tag{21}$$

where *#correct matches* stands for the number of correct matches, *#tentative matches* represents the number of raw matches, namely does not have any post-process steps such as RANSAC, cross-check and ratio-test.

Figure 6. Samples from Colosseum dataset.

5.1.1. Matching Precession

Figure 7 presents visualized results for each method, the green lines denotes correct matches. The HPM obtained the minimal number of feature correspondences. The number of feature correspondences from ROML is more than that of HPM. The number of feature correspondences of SuperPoint is the second place. The SPFT has the maximal number of feature correspondences. According to the common sense in the field of 3D reconstruction, the more the number of feature correspondences, the denser the point-cloud model from 3D reconstruction system. Thus, the SPFT can significantly increase the density of the reconstructed point-cloud model.

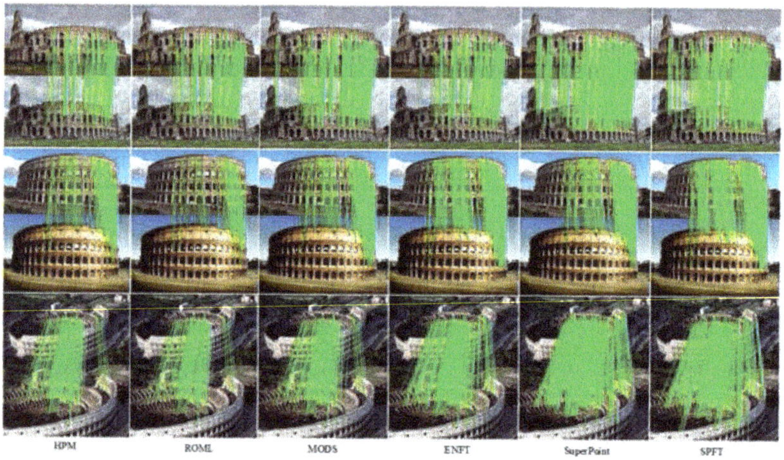

Figure 7. Matching results for the Colosseum dataset.

Moreover, we have tallied up the matching precision for each feature tracking method, the statistical results are depicted in Figure 8. Among those methods [77–80], the proposed SPFT has the highest matching precision, namely the matching precision is 85.6%; The SuperPoint is in the second place; The ENFT is in the third place; and the matching precision of HPM is the lowest. The matching performance of MODS is better that that of HMP and ROML. According to this experiment, we have the following valuable findings: (1) ENFT have robustness to rotation change due to the usage of SIFT feature; (2) The viewpoint change has a significantly impact on the matching precision of feature tracking method; (3) The scale-space has heavily impact on the matching precision because the number of keypoints in multiple scale spaces is more than that of the keypoint detector in single scale space; (4) Superpixel-based segmentation can be used to find potentially keypoints that in the textureless regions. As a result, the matching precision of the SPFT is largely attributed to the usage of multiple scale spaces and superpixel segmentation; (5) Deep learning-based method, such as SuperPoint, can improve the matching precession in the single scale space of the image.

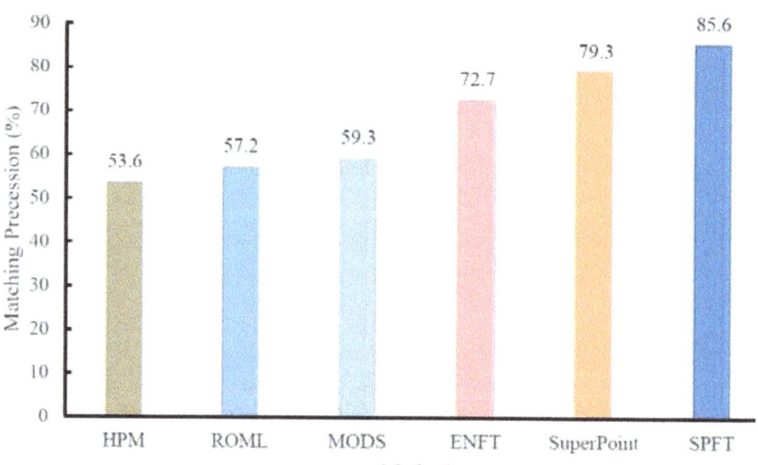

Figure 8. Averaging matching precessions for the evaluated methods, where the SPFT has the best performance, the SuperPoint is in the second place.

5.1.2. Computation Time

Computational cost is one of the most evaluation metrics for feature tracking methods, thus, we have collected the computation times for each compared feature tracking method according to the assessment that conducted on Colosseum dataset. The statistical results of computation times for each method are depicted in Figure 9 where the computation time is the sum of that spend on the whole pipeline including keypoint detection, descriptor computing, and feature matching. We can clearly see that the SPFT has the fastest speed, the averaging computation time is 6.7 s. The ENFT is in the second place, its averaging computation time is 9.2 s. Among those compared methods, the ROML has the lowest speed, which requires 21.3 s averagely for image pairs matching. After deeply investigation for ROML, we found that the main reason attributed to the highest computational cost of ROML is implementation in MATLAB routines. We hold that the ROML may be significantly accelerated when implementation in C++ programming language. As shown in Figure 9, the speed of the proposed SPFT is about 3 times faster than that of ROML, and is about 2–3 times faster than that of HPM and MODS. According to the statistical results of matching precision and computation time, we can conclude that the SPFT feature has the best performance in both accuracy and efficiency. In addition to ROML, the SuperPoint has the lowest speed, the averaging time is 18.2 s according to the experiment.

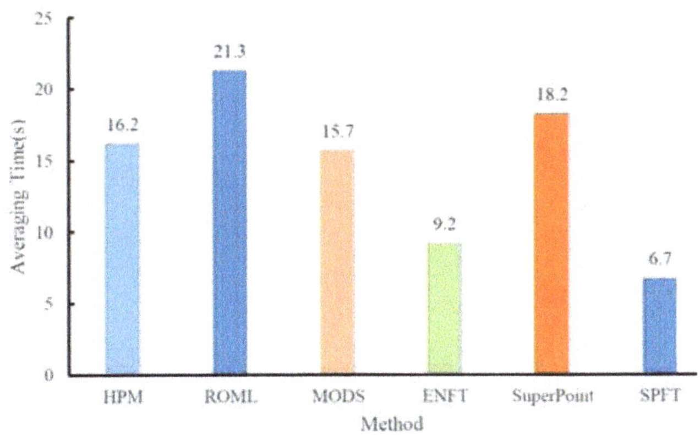

Figure 9. Averaging times for the evaluated methods.

5.2. Evaluation on HFUT Dataset

In the field of 3D reconstruction, if a feature tracking method is integrated into a 3D reconstruction system, which can produce high-quality point-cloud model, we consider the feature tracking method as a good approach to 3D reconstruction. Based on this judgement criteria, we create a new dataset captured by Canon EOS 550D camera, we named the new dataset as HFUT dataset for short. Figure 10 presents samples of the HFUT dataset, which contains 120 images and have many repeated features and repeated structures on the surface of each image. In addition to repeated features, the light for each image is very weak, which pose a new challenge for feature tracking method. In this experiment, we integrated the SPFT feature tracking method into ISFM system [2] to recover the point-cloud model, the results are shown in Figure 11. We can see that the reconstructed point-cloud model has highly geometric consistency with respect to the real scenario. Moreover, we found that the resulting point-cloud model is very dense, which is attributed to the usage of the SPFT feature tracking method. According to our record, the ISFM system with SPFT can recover a high-quality point-cloud model having 338,391 vertices for the HFUT dataset in 5.5 min. As a result, we consider the SPFT has an excellent performance in practice.

Figure 10. Samples from HFUT dataset.

(a) Front of the building

(b) Side of the building

Figure 11. The point-cloud model for HFUT dataset.

5.3. Evaluation of Forensic Dataset

To assess the scalability of the SPFT, we have evaluated it on the Forensic dataset provided by the PIX4D company. The samples of the UAV dataset are provided in Figure 12, which is captured by unmanned aerial vehicle and has large-scale resolution and many repeated features on the surface of each image. In summary, the Forensic dataset is very challenge for feature tracking method and structure from motion.

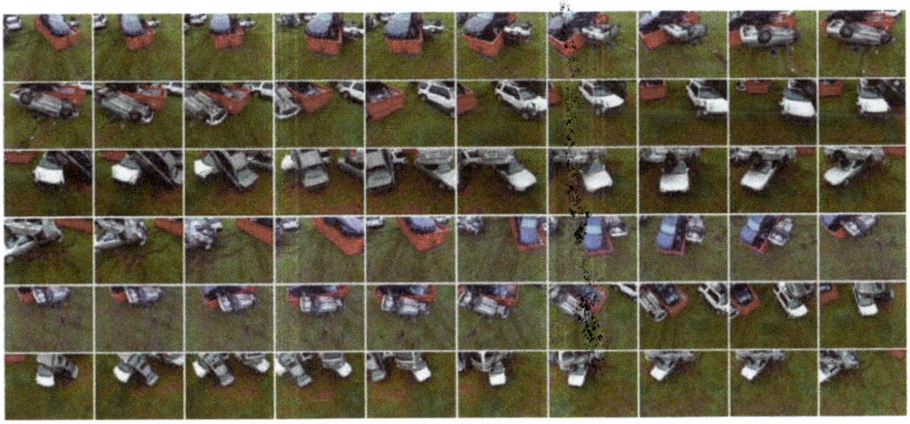

Figure 12. Samples from the Forensic dataset. Note that many repeating features are appeared on the surface of each image.

Figure 13 presents the visual correspondences of each feature tracking method for the Forensic dataset, where the SPFT has obtained the maximum number of feature matches, and has the fastest speed among the compared feature tracking methods. The HPM has the minimum number of visual correspondences, and has the lowest speed. According to our statistic, the HPM has an average of 55 feature correspondences on the Forensic dataset. The number of visual correspondences of the MODS in the second place, and it has lower speed than that of the HPM approach because of views synthesis. Although the ENFT has number of visual correspondences less than that of MODS, which has a cheap computational cost. After a deep analysis for the ENFT method, we found that the ENFT heavily dependents on the segmentation for input video or image sequences to decrease the computational burden. But, the segmented-based approach easily handicaps the quality of the point-cloud model that is constructed by the SFM system. The SuperPoint has more feature correspondences than that of ENFT, but less than that of ours. However, the proposed SPFT method not only has the cheapest computational cost but also has the highest matching precision among these compared feature tracking methods. According to our statistical results in experiment, the SPFT method has an average of 1876 correct feature matches.

In addition to making a comparison with the state-of-the-art method, we have integrated the SPFT into the ISFM system [2], and use the combinational system to estimate the point-cloud model for the Forensic dataset. Figure 14 provides the sparse point-cloud model for the Forensic dataset, which has 2,683,015 vertices and is reconstructed in 10.6 min. We can see that the constructed point-cloud model has good geometric consistency with corresponding to the real scenarios. As a result, we can draw a conclusion that the SPFT has the best performance in both accuracy and efficiency.

Figure 13. Visual correspondences for each method on Forensic dataset.

Figure 14. The sparse point-cloud model for the forensic dataset, and constructed by the ISFM system with the RTFT feature tracking method.

6. Conclusions

In this paper, we proposed an accurate, fast and robust feature tracking method for SFM-based 3D reconstruction, which is based on the superpixel segmentation to increase the number of potentially keypoint and improve the descriptor's quality. In the stage of feature detection, a multiple scale-space analysis and the superpixel-based segmentation technique is used to candidate keypoints, then using non-maximal suppression technique to remove some unstable keypoints from the initial keypoint collection. In the stage of descriptor computing, we use the segment-based binary test to produce a robust descriptor for each keypoints. In the stage of feature matching, the GPU-accelerated KNN method with ratio-test is used to measure the similarity of two descriptors for saving computation time. Finally, we have evaluated the SPFT on the several challenging datasets, and compared it with the state-of-the-arts feature tracking methods. Moreover, the SPFT is integrated into an SFM-based 3D reconstruction system, then resulting high-quality point-cloud models on the challenging datasets. I hold that the SPFT likes a unified framework of feature tracking, in which with different superpixel methods or KNN-like methods, the SPFT may produce a novel feature tracking method. Thus, the SPFT has good ex extendibility.

Besides of promising feature tracking method, we have other valuable findings according to experiments: (1) the number of located keypoints largely depends on multiple scale spaces; (2) the context information is very important to construct a robust descriptor for keypoint; (3) the usage of shared memory in GPU device is also important to accelerate the feature matching speed. In summary, we proposed a promising feature tracking method for SFM-based 3D reconstruction, the quality of point-cloud model is significantly improved when it is used. In the future, we will try to propose a novel feature tracking method based on the proposed SPFT framework for simultaneous localization and mapping.

Author Contributions: Conceptualization, M.C. and L.Z.; methodology, M.C.; software, M.C.; validation, M.C., W.J. and L.Z.; formal analysis, Z.L.; investigation, M.C.; resources, W.J.; data curation, M.C.; writing—original draft preparation, M.C.; writing—review and editing, Z.L., W.J. and X.L.; visualization, M.C.; supervision, X.L.; project administration, L.Z.; funding acquisition, X.L., M.C., W.J. and Z.L.

Funding: This research was funded by [National Science Foundation of China] grant number [61802103, 61877016, and 61673157], and [Postdoctoral Science Foundation] grant number [2018M632522], and [Fundamental Research Funds for the Central Universities] grant number [JZ2018HGBH0280, and PA2018GDQT0014], and [Natural Science Foundation of Shandong Province] grant number [ZR2017QF015], and [Key Research and Development Program in Anhui Province] grant number [1804a09020036].

Acknowledgments: The authors gratefully acknowledge the support of the National Natural Science Foundation (Grant No.: 61802103, 61877016, 61602146 and 61673157), Postdoctoral Science Foundation (Grant No.: 2018M632522), Fundamental Research Funds for the Central Universities (Grant No.: JZ2018HGBH0280 and PA2018GDQT0014), and Natural Science Foundation of Shandong Province (Grant No.: ZR2017QF015), and Key Research and Development Program in Anhui Province (Grant No.: 1804a09020036).

Conflicts of Interest: The authors declare no conflict of interest. The funders had no role in the design of the study; in the collection, analyses, or interpretation of data; in the writing of the manuscript, and in the decision to publish the results.

References

1. Lv, Z.; Li, X.; Li, W. Virtual reality geographical interactive scene semantics research for immersive geography learning. *Neurocomputing* **2017**, *254*, 71–78. [CrossRef]
2. Cao, M.W.; Jia, W.; Zhao, Y.; Li, S.J.; Liu, X.P. Fast and robust absolute camera pose estimation with known focal length. *Neural Comput. Appl.* **2017**, *29*, 1383–1398. [CrossRef]
3. Kong, C.; Lucey, S. Prior-Less Compressible Structure from Motion. In Proceedings of the IEEE Conference on Computer Vision and Pattern Recognition, Las Vegas, NV, USA, 27–30 June 2016; pp. 4123–4131.
4. Cao, M.W.; Li, S.J.; Jia, W.; Li, S.L.; Liu, X.P. Robust bundle adjustment for large-scale structure from motion. *Multimed. Tools Appl.* **2017**, *76*, 21843–21867. [CrossRef]
5. Lu, H.M.; Li, Y.J.; Mu, S.L.; Wang, D.; Kim, H.; Serikawa, S. Motor Anomaly Detection for Unmanned Aerial Vehicles Using Reinforcement Learning. *IEEE Internet Things J.* **2017**, *5*, 2315–2322. [CrossRef]
6. Lu, H.M.; Li, B.; Zhu, J.W.; Li, Y.J.; Li, Y.; Xu, X.; He, L.; Li, X.; Li, J.R.; Serikawa, S. Wound intensity correction and segmentation with convolutional neural networks. *Concurr. Comput. Pract. Exp.* **2017**, *29*, e3927. [CrossRef]
7. Zhang, X.L.; Han, Y.; Hao, D.S.; Lv, Z.H. ARGIS-based Outdoor Underground Pipeline Information System. *J. Vis. Commun. Image Represent.* **2016**, *40*, 779–790. [CrossRef]

8. Serikawa, S.; Lu, H. Underwater image dehazing using joint trilateral filter. *Comput. Electr. Eng.* **2014**, *40*, 41–50. [CrossRef]
9. Ozyesil, O.; Voroninski, V.; Basri, R.; Singer, A. A Survey of Structure from Motion. *Acta Numer.* **2017**, *26*, 305–364. [CrossRef]
10. Snavely, N.; Seitz, S.M.; Szeliski, R. Photo tourism: Exploring photo collections in 3D. *ACM Trans. Graph. (TOG)* **2006**, *25*, 835–846. [CrossRef]
11. Lowe, D.G. Distinctive image features from scale-invariant keypoints. *Int. J. Comput. Vis.* **2004**, *60*, 91–110. [CrossRef]
12. Wu, C. Towards linear-time incremental structure from motion. In Proceedings of the International Conference on 3D Vision-3DV 2013, Seattle, WA, USA, 29 June–1 July 2013.
13. Furukawa, Y.; Ponce, J. Accurate, Dense, and Robust Multi-View Stereopsis. In Proceedings of the IEEE Computer Society Conference on Computer Vision and Pattern Recognition, Minneapolis, MN, USA, 17–22 June 2007.
14. Kazhdan, M.; Bolitho, M.; Hoppe, H. Poisson surface reconstruction. In Proceedings of the Fourth Eurographics Symposium on Geometry Processing, Cagliari, Sardinia, Italy, 26–28 June 2006.
15. Dong, Z.L.; Zhang, G.F.; Jia, J.Y.; Bao, H.J. Keyframe-based real-time camera tracking. In Proceedings of the IEEE 12th International Conference on Computer Vision, Kyoto, Japan, 29 September–2 October 2009.
16. Zhang, G.F.; Liu, H.M.; Dong, Z.L.; Jia, J.Y.; Wong, T.T.; Bao, H.J. ENFT: Efficient Non-Consecutive Feature Tracking for Robust Structure-from-Motion. *arXiv* **2015**, arXiv:1510.08012.
17. Ni, K.; Dellaert, F. HyperSfM. In Proceedings of the 2012 Second International Conference on 3D Imaging, Modeling, Processing, Visualization and Transmission (3DIMPVT), Zurich, Switzerland, 13–15 October 2012.
18. Schönberger, J.L.; Frahm, J.-M. Structure-from-motion revisited. In Proceedings of the 2016 IEEE Conference on Computer Vision and Pattern Recognition (CVPR), Las Vegas, NV, USA, 27–30 June 2016.
19. Zach, C. *ETH-V3D Structure-and-Motion Software.© 2010–2011*; ETH Zurich: Zürich, Switzerland, 2010.
20. Bay, H.; Tuytelaars, T.; van Gool, L. Surf: Speeded up robust features. In Proceedings of the European Conference on Computer Vision, Graz, Austria, 7–13 May 2006; Springer: Berlin/Heidelberg, Germany, 2006; pp. 404–417.
21. Agarwal, S.; Furukawa, Y.; Snavely, N.; Simon, I.; Curless, B.; Seitz, S.M.; Szeliski, R. Building rome in a day. In Proceedings of the IEEE 12th International Conference on Computer Vision, Kyoto, Japan, 27 September–4 October 2009.
22. Zach, C.; Klopschitz, M.; Pollefeys, M. Disambiguating visual relations using loop constraints. In Proceedings of the 2010 IEEE Computer Society Conference on Computer Vision and Pattern Recognition, San Francisco, CA, USA, 13–18 June 2010; pp. 1426–1433.
23. Fan, B.; Wu, F.; Hu, Z. Towards reliable matching of images containing repetitive patterns. *Pattern Recognit. Lett.* **2011**, *32*, 1851–1859. [CrossRef]
24. Roberts, R.; Sinha, S.N.; Szeliski, R.; Steedly, D. Structure from motion for scenes with large duplicate structures. In Proceedings of the 2013 IEEE Conference on Computer Vision and Pattern Recognition, Colorado Springs, CO, USA, 20–25 June 2011; pp. 3137–3144.
25. Wilson, K.; Snavely, N. Network Principles for SfM: Disambiguating Repeated Structures with Local Context. In Proceedings of the IEEE International Conference on Computer Vision, Sydney, NSW, Australia, 1–8 December 2013; pp. 513–520.
26. Ceylan, D.; Mitra, N.J.; Zheng, Y.; Pauly, M. Coupled structure-from-motion and 3D symmetry detection for urban facades. *ACM Trans. Graph.* **2014**, *33*, 57–76. [CrossRef]
27. Saputra, M.R.U.; Markham, A.; Trigoni, N. Visual SLAM and Structure from Motion in Dynamic Environments: A Survey. *ACM Comput. Surv.* **2018**, *51*, 37. [CrossRef]
28. Knapitsch, A.; Park, J.; Zhou, Q.Y.; Koltun, V. Tanks and Temples: Benchmarking Large-Scale Scene Reconstruction. *ACM Trans. Graph.* **2017**, *36*, 78. [CrossRef]
29. Alcantarilla, P.F.; Bartoli, A.; Davison, A.J. KAZE features. In Proceedings of the Computer Vision–ECCV 2012, Florence, Italy, 7–13 October 2012; Springer: Berlin/Heidelberg, Germany, 2012; pp. 214–227.
30. Tombari, F.; Di Stefano, L. *Interest Points via Maximal Self-Dissimilarities*; Springer International Publishing: Cham, Switzerland, 2015.
31. Tomasi, C.; Kanade, T. Detection and tracking of point features. *Int. J. Comput. Vis.* **1991**, *20*, 110–121.

32. Cao, M.; Jia, W.; Lv, Z.; Li, Y.; Xie, W.; Zheng, L.; Liu, X. Fast and robust feature tracking for 3D reconstruction. *Opt. Laser Technol.* **2018**, *110*, 120–128. [CrossRef]
33. Sinha, S.N.; Frahm, J.M.; Pollefeys, M.; Genc, Y. GPU-based video feature tracking and matching. In Proceedings of the EDGE, Workshop on Edge Computing Using New Commodity Architectures, Chapel Hill, NC, USA, 23–24 May 2006.
34. Crandall, D.; Owens, A.; Snavely, N.; Huttenlocher, D. Discrete-continuous optimization for large-scale structure from motion. In Proceedings of the IEEE Conference on Computer Vision and Pattern Recognition, Colorado Springs, CO, USA, 20–25 June 2011; IEEE: Piscataway, NJ, USA, 2011; pp. 3001–3008.
35. Guofeng, Z.; Haomin, L.; Zilong, D.; Jiaya, J.; Tien-Tsin, W.; Hujun, B. Efficient Non-Consecutive Feature Tracking for Robust Structure-From-Motion. *IEEE Trans. Image Process.* **2016**, *25*, 5957–5970.
36. Rublee, E.; Rabaud, V.; Konolige, K.; Bradski, G. ORB: An efficient alternative to SIFT or SURF. In Proceedings of the 2011 IEEE International Conference on Computer Vision (ICCV), Barcelona, Spain, 6–13 November 2011; IEEE: Piscataway, NJ, USA, 2011.
37. Leutenegger, S.; Chli, M.; Siegwart, R.Y. BRISK: Binary robust invariant scalable keypoints. In Proceedings of the 2011 IEEE International Conference on Computer Vision (ICCV), Barcelona, Spain, 6–13 November 2011; IEEE: Piscataway, NJ, USA, 2011.
38. Forssén, P.-E.; Lowe, D.G. Shape descriptors for maximally stable extremal regions. In Proceedings of the IEEE 11th International Conference on Computer Vision (ICCV 2007), Rio de Janeiro, Brazil, 14–21 October 2007; IEEE: Piscataway, NJ, USA, 2007.
39. Rosten, E.; Drummond, T. Machine learning for high-speed corner detection. In Proceedings of the Computer Vision–ECCV 2006, Graz, Austria, 7–13 May 2006; Springer: Berlin, Germany, 2006; pp. 430–443.
40. Mair, E.; Hager, E.M.; Burschka, D.; Suppa, M.; Hirzinger, G. Adaptive and generic corner detection based on the accelerated segment test. In Proceedings of the Computer Vision–ECCV 2010, Crete, Greece, 5–11 September 2010; Springer: Berlin, Germany, 2010; pp. 183–196.
41. Agrawal, M.; Konolige, K.; Blas, M.R. CenSure: Center surround extremas for realtime feature detection and matching. In Proceedings of the Computer Vision–ECCV 2008, Marseille, France, 12–18 October 2008; Springer: Berlin, Germany, 2008; pp. 102–115.
42. Klein, G.; Murray, D. Parallel tracking and mapping for small AR workspaces. In Proceedings of the 6th IEEE and ACM International Symposium on Mixed and Augmented Reality (ISMAR 2007), Nara, Japan, 13–16 November 2007; IEEE: Piscataway, NJ, USA, 2007.
43. Yang, X.; Cheng, K.-T. LDB: An ultra-fast feature for scalable augmented reality on mobile devices. In Proceedings of the 2012 IEEE International Symposium on Mixed and Augmented Reality (ISMAR), Atlanta, GA, USA, 5–8 November 2012; IEEE: Piscataway, NJ, USA, 2012.
44. Yang, X.; Cheng, K.-T. Local difference binary for ultrafast and distinctive feature description. *IEEE Trans. Pattern Anal. Mach. Intell.* **2014**, *36*, 188–194. [CrossRef]
45. Levi, G.; Hassner, T. LATCH: Learned Arrangements of Three Patch Codes. *arXiv* **2015**, arXiv:1501.03719.
46. Trzcinski, T.; Christoudias, M.; Fua, P.; Lepetit, V. Boosting binary keypoint descriptors. In Proceedings of the IEEE Conference on Computer Vision and Pattern Recognition, Portland, OR, USA, 23–28 June 2013.
47. Alahi, A.; Ortiz, R.; Vandergheynst, P. Freak: Fast retina keypoint. In Proceedings of the 2012 IEEE Conference on Computer Vision and Pattern Recognition (CVPR), Providence, RI, USA, 16–21 June 2012; IEEE: Piscataway, NJ, USA, 2012.
48. Wu, C. SiftGPU: A GPU Implementation of Scale Invariant Feature Transform. 2011. Available online: http://cs.unc.edu/~{}ccwu/siftgpu (accessed on 10 November 2018).
49. Graves, A. GPU-accelerated feature tracking. In Proceedings of the 2016 IEEE National Aerospace and Electronics Conference (NAECON) and Ohio Innovation Summit (OIS), Dayton, OH, USA, 25–29 July 2016.
50. Cao, M.; Jia, W.; Li, S.; Li, Y.; Zheng, L.; Liu, X. GPU-accelerated feature tracking for 3D reconstruction. *Opt. Laser Technol.* **2018**, *110*, 165–175. [CrossRef]
51. Xu, T.; Sun, K.; Tao, W. *GPU Accelerated Image Matching with Cascade Hashing*; Springer: Singapore, 2017.
52. Micusik, B.; Wildenauer, H. Structure from Motion with Line Segments Under Relaxed Endpoint Constraints. *Int. J. Comput. Vis.* **2017**, *124*, 65–79. [CrossRef]
53. Sweeney, C.; Fragoso, V.; Hollerer, T.; Turk, M. Large Scale SfM with the Distributed Camera Model. In Proceedings of the 2016 Fourth International Conference on 3D Vision (3DV), Stanford, CA, USA, 25–28 October 2016.

54. Wilson, K.; Snavely, N. Robust global translations with 1dsfm. In Proceedings of the European Conference on Computer Vision, Zurich, Switzerland, 6–12 September 2014; Springer: Berlin, Germany, 2014.
55. Moulon, P.; Monasse, P.; Marlet, R. Global fusion of relative motions for robust, accurate and scalable structure from motion. In Proceedings of the IEEE International Conference on Computer Vision, Sydney, NSW, Australia, 1–8 December 2013.
56. Sweeney, C.; Sattler, T.; Hollerer, T.; Turk, M.; Pollefeys, M. Optimizing the Viewing Graph for Structure-from-Motion. In Proceedings of the IEEE International Conference on Computer Vision, Santiago, Chile, 7–13 December 2015.
57. Goldstein, T.; Hand, P.; Lee, C.; Voroninski, V.; Soatto, S. ShapeFit and ShapeKick for Robust, Scalable Structure from Motion. In Proceedings of the Computer Vision—ECCV 2016: 14th European Conference, Amsterdam, The Netherlands, 11–14 October 2016; pp. 289–304.
58. Cohen, A.; Schonberger, J.; Speciale, P.; Sattler, T.; Frahm, J.; Pollefeys, M. Indoor-Outdoor 3D Reconstruction Alignment. In Proceedings of the Computer Vision—ECCV 2016: 14th European Conference, Amsterdam, The Netherlands, 11–14 October 2016; pp. 285–300.
59. Albl, C.; Sugimoto, A.; Pajdla, T. Degeneracies in Rolling Shutter SfM. In Proceedings of the Computer Vision—ECCV 2016: 14th European Conference, Amsterdam, The Netherlands, 11–14 October 2016; pp. 36–51.
60. Xiao, J.; Owens, A.; Torralba, A. SUN3D: A database of big spaces reconstructed using sfm and object labels. In Proceedings of the IEEE International Conference on Computer Vision, Sydney, NSW, Australia, 1–8 December 2013.
61. Cui, H.; Gao, X.; Shen, S.; Hu, Z. HSfM: Hybrid Structure-from-Motion. In Proceedings of the IEEE Conference on Computer Vision and Pattern Recognition, Honolulu, HI, USA, 21–26 July 2017.
62. Ren, X.; Malik, J. Learning a Classification Model for Segmentation. In Proceedings of the Ninth IEEE International Conference on Computer Vision, Nice, France, 13–16 October2003; IEEE Computer Society: Washington, DC, USA, 2003; p. 10.
63. Achanta, R.; Shaji, A.; Smith, K.; Lucchi, A.; Fua, P.; Susstrunk, S. SLIC Superpixels Compared to State-of-the-Art Superpixel Methods. *IEEE Trans. Pattern Anal. Mach. Intell.* **2012**, *34*, 2274–2282. [CrossRef] [PubMed]
64. Van den Bergh, M.; Boix, X.; Roig, G.; De Capitani, B.; Van Gool, L. SEEDS: Superpixels Extracted Via Energy-Driven Sampling. *Int. J. Comput. Vis.* **2015**, *111*, 298–314. [CrossRef]
65. Moore, A.P.; Prince, S.J.D.; Warrell, J.; Mohammed, U.; Jones, G. Superpixel lattices. In Proceedings of the IEEE Conference on Computer Vision and Pattern Recognition, Anchorage, AK, USA, 23–28 June 2008.
66. Ban, Z.; Liu, J.; Fouriaux, J. GMMSP on GPU. *J. Real-Time Image Process.* **2018**, *13*, 1–13. [CrossRef]
67. Quinlan, J.R. Induction of decision trees. *Mach. Learn.* **1986**, *1*, 81–106. [CrossRef]
68. Salas-Moreno, R.F.; Glocker, B.; Kelly, P.H.J.; Davison, A.J. Dense planar SLAM. In Proceedings of the IEEE International Symposium on Mixed and Augmented Reality, Munich, Germany, 10–12 September 2014.
69. Ge, S.; Li, J.; Ye, Q.; Luo, Z. Detecting Masked Faces in the Wild with LLE-CNNs. In Proceedings of the 2017 IEEE Conference on Computer Vision and Pattern Recognition (CVPR), Honolulu, HI, USA, 21–26 July 2017.
70. Ge, S.; Zhao, S.; Li, C.; Li, J. Low-Resolution Face Recognition in the Wild via Selective Knowledge Distillation. *IEEE Trans. Image Process.* **2019**, *28*, 2051–2062. [CrossRef] [PubMed]
71. Zhu, Z.; Davari, K. Comparison of local visual feature detectors and descriptors for the registration of 3D building scenes. *J. Comput. Civ. Eng.* **2014**, *29*, 04014071. [CrossRef]
72. Aguilera, C.A.; Sappa, A.D.; Toledo, R. LGHD: A feature descriptor for matching across non-linear intensity variations. In Proceedings of the 2015 IEEE International Conference on Image Processing (ICIP), Quebec City, QC, Canada, 27–30 September 2015.
73. Li, S.; Amenta, N. Brute-force k-nearest neighbors search on the GPU. In Proceedings of the International Conference on Similarity Search and Applications, Tokyo, Japan, 24–26 October 2015; Springer: Brelin, Germany, 2015.
74. Roth, L.; Kuhn, A.; Mayer, H. Wide-Baseline Image Matching with Projective View Synthesis and Calibrated Geometric Verification. *PFG J. Photogramm. Remote Sens. Geoinf. Sci.* **2017**, *85*, 85–95. [CrossRef]
75. Lin, W.Y.; Wang, F.; Cheng, M.M.; Yeung, S.K.; Torr, P.H.S.; Do, M.N.; Lu, J. CODE: Coherence Based Decision Boundaries for Feature Correspondence. *IEEE Trans. Pattern Anal. Mach. Intell.* **2018**, *40*, 34–47. [CrossRef]

76. Cheng, J.; Leng, C.; Wu, J.; Cui, H.; Lu, H. Fast and Accurate Image Matching with Cascade Hashing for 3D Reconstruction. In Proceedings of the 2014 IEEE Conference on Computer Vision and Pattern Recognition, Columbus, OH, USA, 23–28 June 2014.
77. Tolias, G.; Avrithis, Y. Speeded-up, relaxed spatial matching. In Proceedings of the 2011 International Conference on Computer Vision, Barcelona, Spain, 6–13 November 2011.
78. Jia, K.; Chan, T.H.; Zeng, Z.; Gao, S.; Wang, G.; Zhang, T.; Ma, Y. ROML: A Robust Feature Correspondence Approach for Matching Objects in A Set of Images. *Int. J. Comput. Vis.* **2016**, *117*, 173–197. [CrossRef]
79. Mishkin, D.; Matas, J.; Perdoch, M. MODS: Fast and robust method for two-view matching. *Comput. Vis. Image Underst.* **2015**, *141*, 81–93. [CrossRef]
80. DeTone, D.; Malisiewicz, T.; Rabinovich, A. Superpoint: Self-supervised interest point detection and description. In Proceedings of the IEEE Conference on Computer Vision and Pattern Recognition, Salt Lake City, UT, USA, 18–22 June 2018.

applied sciences

MDPI

Review

Analysis of the Productive, Structural, and Dynamic Development of Augmented Reality in Higher Education Research on the Web of Science

Jesús López Belmonte, Antonio-José Moreno-Guerrero, Juan Antonio López Núñez and Santiago Pozo Sánchez *

Department of Didactics and School Organization, University of Granada, 18071 Granada, Spain; jesuslopez@ugr.es (J.L.B.); ajmoreno@ugr.es (A.-J.M.-G.); juanlope@ugr.es (J.A.L.N.)

* Correspondence: santiagopozo@correo.ugr.es

Received: 14 November 2019; Accepted: 2 December 2019; Published: 5 December 2019

Abstract: Augmented reality is an emerging technology that has gained great relevance thanks to the benefits of its use in learning spaces. The present study focuses on determining the performance and scientific production of augmented reality in higher education (ARHE). A bibliometric methodology for scientific mapping has been used, based on processes of estimation, quantification, analytical tracking, and evaluation of scientific research, taking as its reference the analysis protocols included in the Preferred Reporting Items for Systematic reviews and Meta-analyses for Protocols (PRISMA-P) matrix. A total of 552 scientific publications on the Web of Science (WoS) have been analyzed. Our results show that scientific productions on ARHE are not abundant, tracing its beginnings to the year 1997, with its most productive period beginning in 2015. The most abundant studies are communications and articles (generally in English), with a wide thematic variety in which the bibliometric indicators "virtual environments" and "higher education" stand out. The main sources of origin are International Technology, Education and Development Conference (INTED) Proceedings and Education and New Learning Technologies (EDULEARN) Proceedings, although Spanish institutions are the most prolific. In conclusion, studies related to ARHE in the WoS have become increasingly abundant since ARHE's research inception in 1997 (and especially since 2009), dealing with a wide thematic variety focused on "virtual environments" and "higher education"; abundant manuscripts are written in English (communications and articles) and originate from Spanish institutions. The main limitation of the study is that the results only reveal the status of this issue in the WoS database.

Keywords: augmented reality; higher education; scientific production; web of science; bibliometric analysis; scientific mapping

---

1. Introduction

Technology is currently in a moment of great development in the field of education, as a result of the continuous advances that are occurring in techno–pedagogical matters that promote its inclusion in learning spaces [1], where technology is increasingly attaining greater use in training activities [2] thanks to its ubiquitous and ergonomic nature [3]. All this has led to new student activities, not only in the way they communicate and collaborate with their teachers and peers but also in the way they interact with contents in a digital way [4].

Educational technology has managed to stimulate the teaching and learning process by enriching interactions with information [5], thereby creating a benefit in the essential aspects of teaching, such as student interests, motivation, and participation [6]. This current educational landscape has conditioned the professional practice of teachers and is visible in the need to carry out innovative practices [7] according to the requirements of an education immersed in this digitalized era [8].

In this sense, the digital competency of teachers has become especially relevant in their daily tasks [9]; indeed, teachers are at the forefront of the education of new generations of students who are highly familiar with technology [10]. Despite this, current students still have certain training lacunae in related professional competences [11]. Therefore, it is required that teachers integrate of knowledge and skills linked to technopedagogy—that is, a new methodological paradigm in which the materials and resources used in the teaching and learning processes are mostly technological and digital. This will improve quality indicators and reproduce innovative learning experiences in the hands of educational technology [12].

One of the technologies with great promise in the field of education is augmented reality (AR) [13,14], which allowing for unique instructional activities to facilitate learning [15]. Experts define this technology as an innovation that "allows the combination of digital information and physical information in real time through different technological devices" [16] (p. 5), thereby promoting access to expanded information about us through mobile devices [17]. The literature shows that AR is a resource that can be used in different educational stages, from the initial stages of school [18] to higher education [19].

Likewise, AR offers a series of potential benefits in the learning process, such as the assumption of a greater role for the student [20], an increase in the student's motivation [21], self-regulation [22], and interest in a task [23], and the exploration of teaching materials and content [24]. AR also encourages digital competition [25] and promotes the development of significant, constructivist, collaborative discovery, and ubiquitous learning [26–28]. These benefits favor both the improvement of teaching results and the environment of training spaces [29].

There has not been a significant number of articles reporting on AR. The main findings of related research have focused on quantifying their scientific productions [30], new technological trends analyzed by researchers [31], and the changes and advances produced by this type of teaching approach [32]. Other studies have focused on the country (Spain) and the time period (2015–2017) for greater scientific production [33]. Likewise, other research has focused on specific fields of education, such as engineering, science [34], the business field [35], and education through applications related to tourism, entertainment, marketing, and transport, among others [36].

Bibliometric studies on AR represent a booming area of study, but much remains to be explored. Bibliometric studies on AR are currently limited, and this has produced a research gap [3] that can only be resolved by enhancing research in this field of study. In a recent study focused on the field of education, a bibliometric analysis carried out on the Web of Science (WoS) has confirmed that the most prolific period is 2015–2017, with Spain being the country with the highest growth production in this field [33], followed by Taiwan [37]. This latest study also found that Taiwan University of Science and Technology is the main institution producing related research, and that C.C. Tsai and G.J. Hwang are the most important authors [37].

In a bibliometric study on AR carried out for the field of business administration, two distinct periods were differentiated in terms of the amount of scholarship produced; the most prolific period corresponds to 2012–2016 [35]. It has been found that quantitative studies predominate [38], especially in communication and presentation formats (moreso than articles [3]), with English being the most widely used language [33]. A combined bibliometric study between AR and M-learning on the WoS found that there is a great variety of main topics being analyzed, with the concept of the phenomenon, the development of new AR methodologies, motivations, special relocations, and the subjects in which AR is implemented being the most popular [32]. A bibliometric study in the specific field of education highlighted that "learning/academic achievement", "motivation", and "attitude" are the most examined variables [38].

The motivation of this study was to investigate the concept of AR in high-impact literature focused on higher education from a novel methodological perspective, with the purpose of achieving new results and deepening existing ones.

The structure of this work follows the methodological process of bibliometric studies. After the presentation of the state of the matter in the analyzed literature, this manuscript continues by drafting the materials and methods used during the investigation, formulating the justification and objectives of the study, and explaining the procedure and data collection. Then, the results related to performance and scientific production, structural and thematic development, the thematic evolution of the terms, and the authors with the highest relevance index are presented. Finally, we present a discussion of the results of the scientific literature we found and offer a set of final conclusions for the entire research process.

This study is limited to analyzing AR in higher education, specifically in the Web of Science, which is the only database we explore. The reason for this study lies in the need to lay the foundations for the scientific development of AR in a university environment, since no precedent has been found in the specialized literature on the state of the question formulated in this investigation. This is the main problem to be solved in this research paper.

Due to the relevance assumed by this emerging technology, and given the benefits of its use in learning spaces, this study is focused on the analysis of scientific productions on augmented reality in higher education (ARHE). The objectives set out in this study are focused on:

1. Evaluating the performance of and scientific productions on ARHE.
2. Establishing the scientific evolution of ARHE in the specialized literature.
3. Discovering the most important topics in the scientific literature on ARHE.
4. Identifying the most relevant authors who study ARHE.

2. Materials and Methods

2.1. Research Design

In order to develop the present study and achieve the formulated objectives, a research methodology of a bibliometric nature has been used, starting from a foundation of previous studies reported in the scientific literature. The use of this research technique lies in the potential reflected by scientometrics, which refers to the quantification, evaluation, and estimation of scientific developments in a specific field of knowledge. This paper examines the evolution of the structure and dynamism of the concept of augmented reality in higher education through an analysis of co-words. In order to do this, the h-index has been taken into account, as well as the citation volume, giving rise to an elaboration of a science map that allows us to observe the yield and locate and determine the terminological subdomains of this field of study, thereby representing the evolution of the subject in specialized literature [39,40]. Also, using the analysis protocols included in the PRISMA-P matrix as a reference, analytical tracking and document measurement techniques have been used through the establishment of different literary control variables. In the same way, the issues concerning AR and its research development have been located through scientific mapping [41,42].

2.2. Procedure, Debugging, and Data Analysis

The present study has been carried out following the structured protocol for different actions. First, the database in which to search for scientific publications was chosen: the Web of Science (WoS), which is a repository that houses a large number of high-impact scientific materials.

The second process is linked to the action of searching for and reporting documents. To carry out this action, the keywords to be used were delimited. These keywords were selected after consulting the ERIC and UNESCO thesauri, in order to obtain the agreed-upon and standardized terms among the scientific community. The main keywords entered in the WoS search field were "augmented reality" and "higher education", which formed the following search equation: (("augmented reality") AND ("higher education") OR ("university") OR ("universities") OR ("colleges") OR ("postdoctoral education")). This algorithm encompassed the entire literary volume (not limited by time period) and focused on the metadata containing titles, abstract, and keywords of publications. The document-reporting process began in May 2019 and ended in August of the same year, resulting in a total of 555 documents—as an analytical unit—that met the inclusion criteria established in Table 1, which have been established to show the most relevant aspects of each. The inclusion criteria for each of the indicators have been developed to show a considerable number of elements, never showing the totality of elements. After checking them (repeated or incorrectly indexed documents), a figure of 552 scientific publications was obtained (Figure 1).

Table 1. Production indicators and inclusion criteria.

Indicators	Criteria
Year of publication	All documents are contemplated
Language	All languages are contemplated
Publication Area	$x \geq 15$
Type of documents	All documents are contemplated
Organizations	$x \geq 5$
Authors	$x \geq 4$
Sources of Origin	$x \geq 6$
Countries	$x \geq 15$
Citation	The five most cited documents

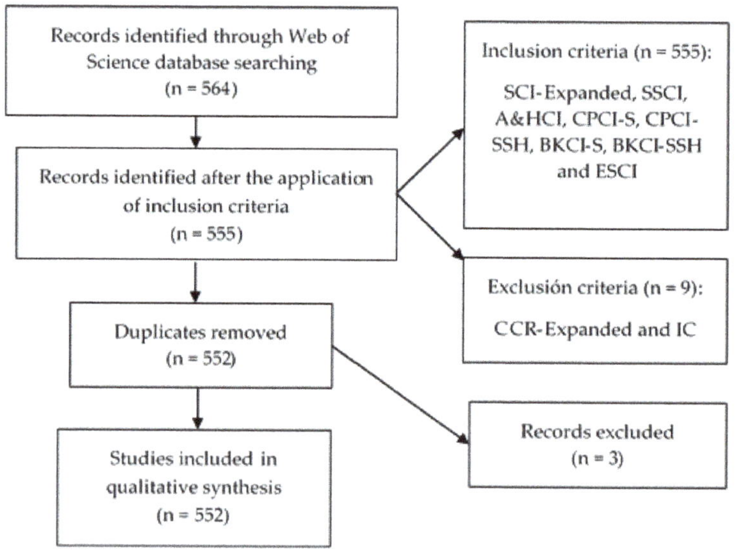

Figure 1. Flowchart according to the PRISMA Declaration.

Co-word analysis has been used to carry out a structural and dynamic study [43], paying special attention to the h-index, among other indicators of scientific quality [44]. This analysis gave rise to a science map that allowed us to study the performance and evolution of AR in the literature and AR's impact on higher education. Likewise, the locations of the subdomains of the concept in this field of research were determined.

Data analysis has been deployed through various programs. Specifically, the analysis of results and creation of the citation report for performance analysis, taking into account the year, type of document, institution, authorship, means of publication, country, language, and document with the most citations. On the other hand, the SciMAT software has been used to develop the structural and dynamic analysis of longitudinal cutting derived from the co-word technique. This program has facilitated the realization of the following processes.

Recognized themes: Based on the 555 references on ARHE, scientific mapping has been carried out to specify the documents that house the state of the art defined in the present study, avoiding non-AR publications in the university field. Thus, a review of the reported literature allowed us to purify the results, eliminating repeated documents and obtaining a figure with 552 works related to ARHE, which served to make co-occurrence connections through nodes, thereby forming a network of co-words through a clustering algorithm.

Reproduced themes were determined through a strategic diagram and a thematic network in two dimensions (centrality and density). This is articulated in four sectors:

1. Upper left: entrenched but isolated issues;
2. Upper right: motor and essential issues;
3. Bottom left: issues that are a priori booming or, on the contrary, are disappearing;
4. Bottom right: poorly developed and transversal issues.

Determined topics: Developed based on an analysis of the evolution of the nodes in different periods of time, configured as follows for the analysis of co-words: P1 = 1997–2015 and P2 = 2016–2019. The reason that they have been limited in such a way is justified by the fact that they cover a minimum of 200 references in temporary spaces. For the authors, a single period (PX) was selected, which compiles all the years of production. Likewise, the strength of association is obtained by the volume of keywords found in common between the different periods.

The assumed performance has been verified through the links established between the keywords and other terms that mark the trend of the node, revealing the use that the scientific community makes of them. A number of aspects have been taken into account. The analysis unit marks the unit of valuation, which in this case refers to the key words marked by the authors in their scientific texts and the key words given by the WoS in relation to those scientific texts, in addition to those of the authors of the various documents. The frequency threshold reflects the minimum frequency threshold for each period, taking into account keywords that appear in at least two documents (for the first period) and three in the second period. The network type reflects the type of network that is going to be built—in this case, a network of the co-occurrence of keywords and authors, or co-word and co-author. The co-occurrence union value threshold establishes the marked periods—in this case, two periods for the keywords and all the years of production for the authors. The normalization measure marks the union threshold, which is the minimum link for that co-occurrence, taking into account unions with a value greater than or equal to 1 in the first and second period (for the keywords) and of 2 (for the authors). The normalization measure marks the measure of similarity used to normalize the network, in this case, the equivalence index *eij* between two entities, *I* and *j*, is calculated in the following way: *eij = cij2/Root (ci – cj)*, where *cij* is the number of co-occurrences of *i* and *j* in the set of documents, *ci* is the number of occurrences of *I*, and *cj* is the number of occurrences of *j*. The clustering algorithm denotes the grouping algorithm used to obtain the map and its associated clusters or themes and subnets. In this case, the simple centers algorithm is used, where the returned clusters are assigned a label that corresponds to the most central node of the group, with no additional processes necessary to assign labels to the group. The evolutionary measure marks the similarity measure needed to construct the evolution map—in this case, the Jaccard Index and the transition map—in this case, the inclusion index, which is reflected in Table 2.

Table 2. Production indicators and inclusion criteria.

Configuration	Values
Analysis unit	Keywords authors, keywords WoS
Frequency threshold	Keywords: P1 = (2), P2 = (3) Authors: PX = (2)
Network type	Co-occurrence
Co-occurrence union value threshold	Keywords: P1 = (1), P2 = (1) Authors: PX = (2)
Normalization measure	Equivalence index
Clustering algorithm	Maximum size: 9; Minimum size: 3
Evolutionary measure	Jaccard index
Overlapping measure	Inclusion Rate

Note: P1: The period from 1997 to 2015; P2: the period from 2016 to 2019; PX: the period from 1997 to 2019.

3. Results

3.1. Performance and Scientific Production

Scientific production on ARHE (n = 552) dates back to 1997 (n = 2) and has remained uninterrupted until the present, albeit with a variable amount of documentation, which is irregular from 1998 (n = 1) until 2014 (n = 23), due to increases and decreases in publications (1999—n = 1; 2000—n = 5; 2001—n = 2; 2002—n = 4; 2003—n = 5; 2004—n = 4; 2005—n = 6; 2006—n = 6; 2007—n = 2; 2008—n = 6; 2009—n = 10; 2010—n = 18; 2011—n = 15; 2012—n = 29; 2013—n = 25). On the other hand, from 2015 (n = 52) onwards, the ascent is prominent (2016—n = 70; 2017—n = 94; 2018—n = 111; 2019—n = 61).

Figure 2 shows the evolutionary development of ARHE throughout its history in the scientific literature. This figure shows three clearly differentiated periods. During the first period, which covers from 1997 to 2007, there is low and uniform production. During the second stage, from 2008 to 2014, the level of production rises irregularly. Finally, in the third period, the production is ascending and more abundant compared to previous periods. The development for 2019 is not significant, since the natural end of the year has not yet occurred, with the literature being open to the publication of new scientific works.

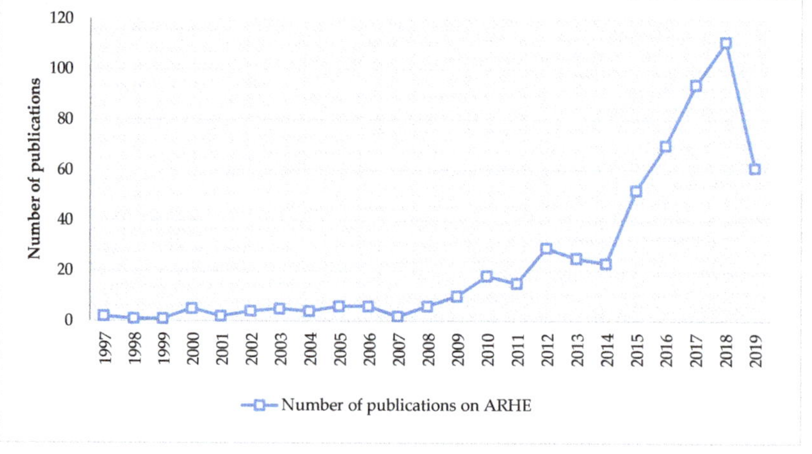

Figure 2. Evolution of scientific production of augmented reality in higher education in the Web of Science (WoS).

English (n = 504) is positioned as the reference language used by various researchers to show their results to the scientific community, followed, to a lesser extent, by Spanish (n = 41), Turkish (n = 3), Portuguese (n = 2), Russian (n = 2), Chinese (n = 1), French (n = 1), and German (n = 1).

Regarding areas of knowledge, a reference was not obtained for studies related to ARHE, since both "education educational research" (n = 220) and "computer science" (n = 208) show similar results. ARHE is also revealed to be a topic of interest for various fields of knowledge, such as "engineering" (n = 144), "telecommunications" (n = 26), "business economics" (n = 20), "social science other topics" (n = 17), optics (n = 16), and "imaging science photographic technology" (n = 15).

For type of document, the scientific community most commonly chooses communications (n = 315) to show the results of their research, followed, to a lesser extent, by articles (n = 235), book chapters (n = 10), literature reviews (n = 9), quick access materials (n = 2), and editorial materials (n = 2). Spanish companies have been verified as reference institutions for ARHE, given their position in Table 3. Of these, the University of la Laguna and the University of Seville are the most common world-wide references on the state of this topic.

Table 3. Institutions of the origin of the manuscripts on the Web of Science (WoS).

Institution	n
Universidad de La Laguna	17
Universidad de Sevilla	15
Universidad de Córdoba	8
Universitat Ramon Llull	8
Polytechnic University of Catalonia	7
Universitat Politécnica de Valencia	7
Universidad de Huelva	6
National Chiao Tung University	5
National Taiwan University of Science Technology	5
RWTH Aachen University	5
State University System of Florida	5
Universitat D'Alacant	5
University of Cambridge	5

The most prolific authors include Martín-Gutiérrez, J. (n = 11), Fonseca, D. (n = 10), Redondo, E. (n = 9), and Cabero, J. (n = 7). Next most productive are Carrera, C.C., Contero, M., Robles, B.F., and Sánchez, A., with five publications, respectively. Finally—complying with the inclusion criteria—Alcaniz, M. has published four works.

"INTED proceedings" is the source of origin with the highest production, followed at a considerable distance by "EDULEARN proceedings" and the other sources listed in Table 4.

Table 4. Source of the origin of manuscripts related to augmented reality in higher education in the WoS.

Source	n
INTED Proceedings	20
EDULEARN Proceedings	13
Lecture Notes in Computer Science	12
Proceedings of SPIE	12
Procedia Computer Science	10
Procedia Social and Behavioral Sciences	8
Advances in Intelligent Systems and Computing	6
Business Horizons	6
Edmetic	6
Iceri Proceedings	6
INTED 2016 10th International Technology Education and Development Conference	6

For countries with greater scientific production, Spain is a worldwide reference for ARHE, since it occupies a high literary volume (n = 103). Spain is accompanied by the United States, with 68 publications. In second place, there are many other countries, including England (n = 36), China (n = 27), Taiwan (n = 23), Australia (n = 22), Italy (n = 22), Germany (n = 20), Romania (n = 20), Turkey (n = 20), Mexico (n = 18), Canada (n = 15), and Malaysia (n = 15), whose production levels are lower.

The scientific reference document for ARHE is an article by Kaufmann and Schmalstieg (2002), due to its high number of recorded citations. The rest, although they are worldwide references, have accumulated lower citation figures (Table 5).

Table 5. Most cited articles.

Reference	Citations
Kaufmann, H.; Schmalstieg, D. [45]	194
Martín-Gutiérrez, J.; Saorín, J.L.; Contero, M.; Alcaniz, M.; Pérez-López, D.C.; Ortega, M. [46]	114
Akcayir, M.; Akcayir, G. [47]	89
Potkinjak, V.; Gardner, M.; Callaghan, V.; Mattila, P.; Guetl, C.; Petrovic, V.M.; Jovanovic, K. [48]	83
Andujar, J.M.; Mejías, A.; Márquez, M.A. [49]	73

3.2. Structural and Thematic Development

The longitudinal view is shown in this case on the transition map. This type of map allows us to detect the evolution of the clusters along different periods, as well as the student, the transient, and new elements of each period. This is reflected in the evolution of key words (Figure 3). By analyzing this figure in depth, two circumferences can be observed. These circumferences represent each of the periods analyzed. From left to right, the first refers to the dates established between 1997 and 2015, and the second refers to the period marked between 2016 and 2019.

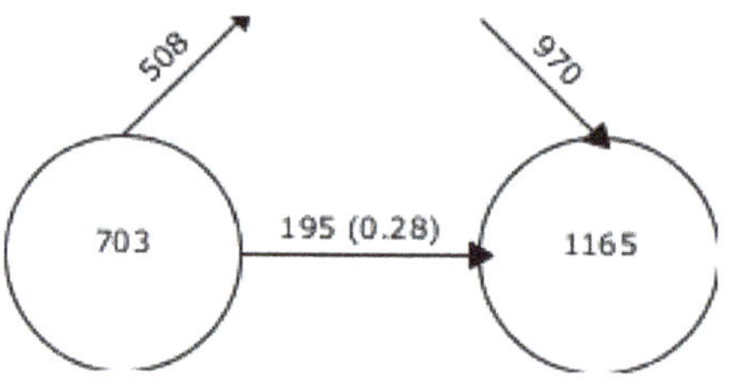

Figure 3. Continuity of keywords between contiguous periods.

In the first circumference, there are 703 keywords. At the top, there is an ascending arrow indicating the number of keywords that will not appear in the second period. The horizontal arrow coming out of the first circumference in the direction of the second one marks the number of coincident keywords in both periods. In this case, a total of 195 represents 28% of the total for both periods. The second circumference has 1165 registered keywords. The descending arrow above reflects the number of new keywords that are incorporated and did not appear in the first period. In this case, there is a total of 970 keywords.

The thematic diversity in the periods established in this study is wide. In the first period, "virtual environments" is the theme with the highest bibliometric values. Other topics offer similar results. In the second period, a pattern similar to the first occurs, highlighting only "higher education", since other topics belong to indicators with a certain degree of similarity (Table 6).

Table 6. Thematic performance.

Denomination	Works	h-Index	g-Index	hg-Index	q2-Index	Citations
Period 1997–2015						
Development	3	3	3	3	7.75	56
Tailored optical fibers	3	3	3	3	7.75	56
Pedagogy	5	2	3	2.45	4.9	24
Teachers	4	4	4	4	9.8	92
Usability	5	3	3	3	7.55	51
Instruction	5	3	3	3	7.35	54
Virtual environments	12	7	9	7.94	18.52	492
Visualization	9	4	7	5.29	7.75	57
Mixed reality	4	1	2	1.41	4.69	23
System	5	3	4	3.46	9.17	102
Social media	3	2	2	2	2.45	8
Online education	2	2	2	2	6.48	30
Gamification	3	1	2	1.41	3.46	13
Context Aware	2	1	1	1	2.24	5
Pattern recognition	2	1	1	1	1	1
3D modeling	3	2	3	2.45	8.12	42
Period 2016–2019						
Technology acceptance	6	2	3	2.45	3.46	12
Improvement	6	2	2	2	2.45	7
Spatial orientation	4	2	3	2.45	5.48	32
Instruction	5	3	3	3	13.64	173
University	6	1	1	1	1	2
Mobile	12	2	3	2.45	3.74	15
Higher education	68	8	16	11.31	11.31	294
Anatomy	5	2	2	2	6.48	23
Usability	5	2	2	2	3.46	8
Framework	6	3	6	4.24	4.58	82
Virtual reality	21	4	5	4.47	8	51
Attitude	4	2	3	2.45	2.83	10
Blended learning	4	0	0	0	0	0
Internet of things	2	0	0	0	0	0

The strategic diagram shows detailed information for each section, through a clustering process, from which a set of interconnected themes are obtained. These topics are obtained thanks to Callon's centrality, which measures the degree of interaction of a network with other networks. Centrality measures the strength of external links to other topics, being the measure of the importance of a topic in the development of the whole field of research analyzed; and to Callon's density, which measures the internal strength of the network, analyzing the internal links between all the key words that describe the research topic, this value being considered as the measure of the degree of development of the topic under study. From both parameters is born the strategic diagram, which is a two-dimensional space constructed through the graphic representation of themes according to their ranges of centrality and density (Figures 4 and 5).

With respect to the analysis of the strategic diagrams of the established periods, the themes of the first period (Figure 4) include "development", which focuses on professional groups, economic issues, the development of control systems, industry, water, electronics, telecommunications, and the information society; "tailored optical fibers", which is oriented toward simulators, reactions, inquiry, higher education, Tesla controllers, digital implementation, and linear accelerators; "mixed reality", which is aimed at educational methodologies, light immersion, maps, mobile augmented reality, telematic presence, and interior design; "pedagogy", which is related to mobile learning, autonomous learning, portable devices, mobile technology, online learning, ubiquitous computing, and architectural design; and "instruction", which is focused on teaching, physics, selection, environments, games, technology, education, and acceptance of the user. This period also highlights "virtual environments", which, although a basic and cross-cutting theme, is of great relevance to the scientific community, due to its high h-index. In the same way, during this period, "3D modeling" themes predominate. "3D modeling" focuses on photography and user interface. This category also includes "gamification", which is aimed at technological, mobile, and gaming information; "context aware", which is focused on image detection and mobile applications; "pattern recognition", focused on 3D and mathematics; and "systems", associated with educational research, open streets, mobile learning, human–computer interactions, and operational and educational engineering. This last topic belongs to an unknown profile, since its location in the diagram defines it as an emerging or disappearing theme.

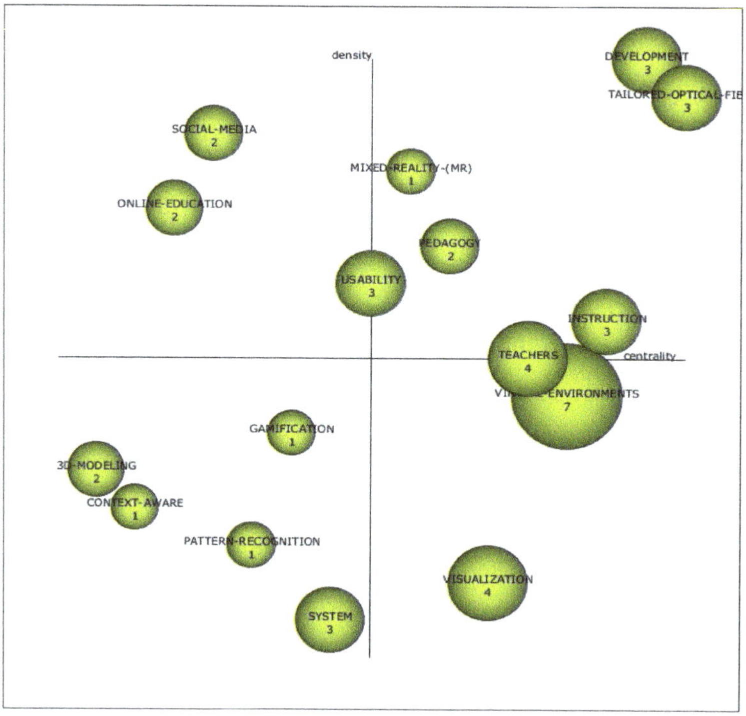

Figure 4. Strategic diagram by h-index: 1997–2015.

In the second period (Figure 5), the thematic engines include "technology acceptance", which refers to supports, experimental learning, sensor networks, remote laboratories, meta-analysis, mobile augmented reality, information, and students; "framework", which relates to building modeling, collaboration, context knowledge, simulation, strategies, industry 4.0, big data, and impact; "university", which is associated with development, skills, applications, English, acceptance, teaching-oriented technology, and gender; "instruction", which is linked to learning systems, designs, spatial capacity, performance, educational technology, education sciences, and cognitive load; "improve", focused on teaching, construction, youth, system, mobile technology, university students, interactive learning environments, and opportunity strategies; "mobile", which is related to architecture, ubiquitous learning, museums, technology, models, technological learning, and tools; and "higher education", which focuses on information and communication technologies, mobile learning, user acceptance, interface, perception, plans, flipped classrooms, and augmented reality. In addition, given its location as an emerging or missed topic, "anatomy", which is related to mathematics, accepted technological models, interactions, devices, visualization, pedagogy, learning, and teacher training, should be kept in mind. "Usability" is focused on interactivity, user experience, location, learning areas, motivation, cultural heritage and science; "attitude" relates to Pokémon Go, portable devices, and difficulties; "blended learning" is focused on gamification, social networks, QR codes, and online learning; and the "internet of things" relates to augmented reality and the training of engineers.

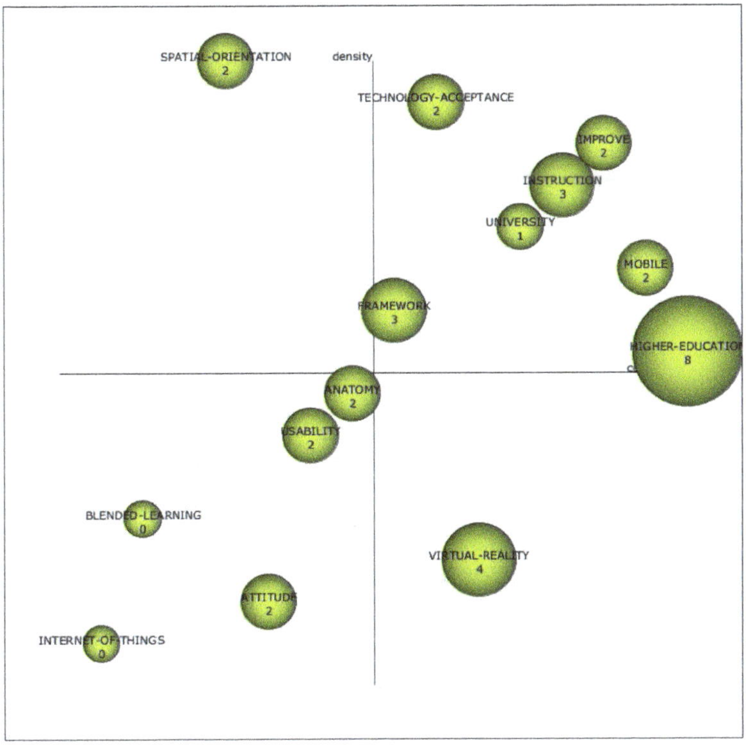

Figure 5. Strategic diagram by h-index: 2016–2019.

3.3. *Thematic Evolution of the Terms*

Considering the thematic evolution, which shows the strength of the evolutions produced in the main thematic areas between consecutive periods from the Jaccard index. Evolution exists if a theme of a period shares keywords with the consecutive theme. The more keywords two clusters of consecutive periods have in common, the more solid their evolution will be. It is necessary to take into account that two types of connections are established: one with a continuous line, whose link is thematic; and one with a dashed line, whose union is based on keywords. Likewise, the thickness of the lines marks the strength of the relationship between themes (Figure 6).

In studies on ARHE, significant thematic variety is observed, with continuity between "usability" and "instruction", since they are repeated in both periods. In the rest of the connections, they show conceptual leaps. There are many connections between the different periods, both conceptual and non-conceptual, but these connections have a weak relationship strength, since the widths of the lines are the lowest. Paying special attention to the topics with the highest h-index of each period, "virtual environments" (first period) connects—in a non-conceptual way—with "spatial orientation" and "higher education" (second period); the latter is conceptually related to "tailored optical fibers" and not conceptually related to "teachers", "instruction", "virtual environments", or "system".

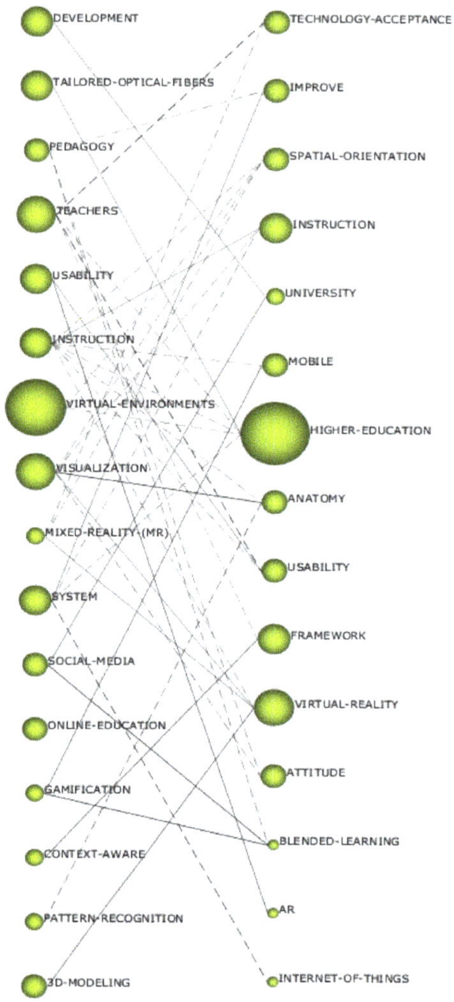

Figure 6. Thematic evolution by h-index.

3.4. Authors with a Higher Relevance Index

Attending to the people who investigate the field of ARHE (Figure 7), the motor authors (by their location in the diagram) most relevant to this field of study are Stoyanova, D., Naves, E.L.M., and Wozniak, P. In addition, Redondo, E., Martín-Gutiérrez, J. and Muñoz-Cristobal, J.A. must be taken into account, since, due to their location in the diagram, they can become motors or disappear. It is noteworthy that the authors with the highest h index are Redondo (h = 3) and Martín-Gutiérrez (h = 4).

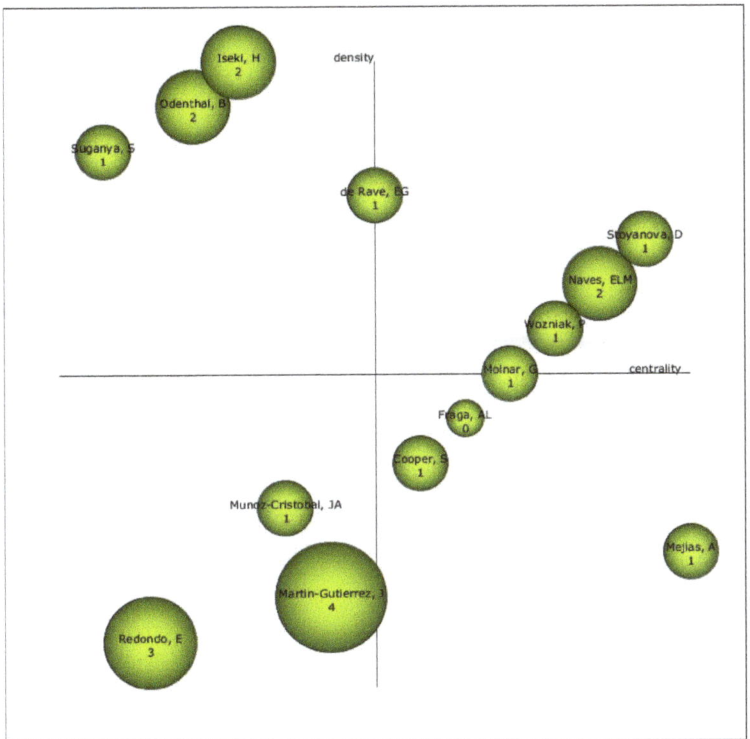

Figure 7. Strategic authoring diagram.

4. Discussion and Conclusions

As has been reflected in previous studies, AR has positioned itself as an emerging technology that yields a set of benefits in training processes. As a result, there is a need to explore its state of affairs in higher education from the perspective of WoS.

With regard to the bibliometric indicators defined in this study, starting with its scientific performance in ARHE, it is shown that ARHE is not abundant, although its production is not recent, since its beginnings date back to 1997. From then until now, its production has been inconstant due to the combination between productive years and recess years, with three different observable moments. In the first moment, dating from 1997 to 2008, production was not high (7.97%), and there was no trend, being instead irregular. During a second period, from 2009 to 2014, production was more extensive (21.74%), albeit with similar trends to the first period. The last period, which appears from 2015 to the present, was the most productive (70.29%), with an upward trend (since the volume of publications is growing every year). These results are similar to those revealed in the literature, although the periods of higher production differ slightly, being established in 2012–2016 [35] and in 2015–2017 [33].

With regard to language, the one used by scientists to present their research is English, as found in other studies [33]. The studies are presented via communication and articles, evenly, with the first being slightly more common—results that are consistent with other studies that also add a paper as a type of relevant study [3], with quantitative studies being the main methodological choice [38]. For sources of origin, "INTED proceedings" (3.62%) and "EDULEARN proceedings" (2.36%) are the most common. The areas of knowledge where research on ARHE is presented are diverse, since there is an even production between "education educational research, computer science, and engineering", which determines the thematic variety of the established field of study.

For the institutions, the Spanish are the pioneers of this type of study, since they take the top positions (highlighting the Universidad de La Laguna (3.08%) and the Universidad de Sevilla (2.72%)). It is worth noting that Martín-Gutiérrez J. (11 works), Redondo E. (9 works), Fonseca D. (10 works), and Cabero, J. (7 works) have provided the most research in this area and even the first two have the highest h-index (four and three, respectively), although they are not the most relevant in this subject. Instead, Stoyanova, D., Naves, E.L.M., and Wozniak, P. are the most relevant by its position in the diagram. The references that have the highest number of citations are [45,46] with 194 and 114 citations, respectively; these scientific texts are the ones that articulate the basis of the current state of research. Studies in the scientific literature have also found that Spain is one of the countries with the highest production in this field (18.66%) [33], but Taiwan is also an important booming country [37]. In contrast, other studies cite the University of Science and Technology of Taiwan as the main institution and C.C. Tsai and G.J. Hwang as the most important authors [37].

Regarding the continuity of keywords between contiguous periods, it is revealed that only 28% of the keywords are repeated between both periods. The increase in the volume of keywords of the second period (n = 1165) with respect to the first (n = 703) stands out. Therefore, 970 new keywords are established in the second period.

The evolution of ARHE has not been regular nor has it settled on a single theme; instead, it has evolved over time and is currently in the process of establishing a solid line of research. This is reflected in the evolution between the periods established in this study, where between 1997 and 2015, the theme with the greatest bibliometric indicators was "virtual environments" (Works = 12; h-index = 7; g-index = 9; hg-index = 7.94; q2-index = 18.52; citations = 492), while between 2016 and 2019, "higher education" (Works = 68; h-index = 8; g-index = 16; hg-index = 11.31; q2-index = 11.31; citations = 294) occupied the top spot. In addition, the same themes are rarely repeated between the two periods.

If the motor themes of both periods are analyzed, the above postulation is confirmed; there is a thematic amalgam in both analyzed periods ("development", "tailored optical fibers", "mixed reality", "pedagogy", "instruction", "virtual environments", "3d modeling", "gamification", "context aware", "pattern recognition", "usability" and "systems" in the first period; "technology acceptance", "framework", "university", "instruction", "improve", "mobile", "higher education", "anatomy", "usability", "attitude", "blended learning", "internet of things" in the second period), whose relationship strength is weak. This shows that ARHE is generating an amplitude both in the field of knowledge and in the various branches of research, as well as conceptual gaps between the established periods. Only, a continuity has been found between "usability" and "instruction" that is repeated in both periods. These results are consistent with those other studies that have found there to be a great variety of research, highlighting the conceptualization of this phenomenon, the development of new RA methodologies, motivation and the attitude, special relocation, academic achievement, and the subjects in which the RA is studied [32,38].

The realization of this study helps to offer the scientific community the most relevant research fields in which ARHE is currently focused, in order to consolidate, in a diachronic manner, the research foundations upon which this emerging technology is based. Therefore, the analysis techniques used in this work provide an expanded and novel vision of the state of ARHE in WoS.

Therefore, this study allows an increase in the knowledge about the use of ARHE because it shows the scientific community where the state of the question about this emerging technology in said educational stage currently. In this way, the research trend so far on this area is shown, enabling researchers who want to study ARHE can select the topics considered relevant in this study and go to the most relevant bibliographic sources on the state of the matter.

The limitations of the study include the location of the references in which the WoS was not determined by keywords, in many cases hindering its localization process. Also, the low volume of scientific papers based on bibliometric analysis made it difficult to discuss the findings obtained in this study with those reported in the literature. This causes the results achieved in this investigation to acquire an exploratory character. This determines the existence of new findings in scientific research on ARHE. In this work, only the main findings obtained from the perspective of the WoS have been presented. For future research, we propose to carry out an investigation with the same structure on other databases, such as Scopus and Google Scholar.

Author Contributions: Conceptualization, S.P.S. and J.A.L.N.; methodology, J.L.B. and A.-J.M.-G; software, A.-J.M.-G.; formal analysis, J.L.B. and A.-J.M.-G.; investigation, J.L.B., A.-J.M.-G., J.A.L.N., and S.P.S.; data curation, S.P.S.; writing—original draft preparation, J.L.B. and S.P.S.; writing—review and editing, J.L.B. and A.-J.M.-G.; visualization, J.A.L.N. and S.P.S.; supervision, J.L.B and J.A.L.N.

Funding: This research received no external funding.

Acknowledgments: We acknowledge the researchers of the research group AREA (HUM-672), which belongs to the Ministry of Education and Science of the Junta de Andalucía and is registered in the Department of Didactics and School Organization of the Faculty of Education Sciences of the University of Granada.

Conflicts of Interest: The authors declare no conflict of interest.

References

1. Rodríguez, A.M.; Cáceres, M.P.; Alonso, S. La competencia digital del futuro docente: Análisis bibliométrico de la productividad científica indexada en Scopus. *Int. J. Innov. Educ. Res.* **2018**, *10*, 317–333.
2. Viñals, A.; Cuenca, J. El rol del docente en la era digital. *Rev. Interuniv. Form. del Profr.* **2016**, *30*, 103–114.
3. Fombona, J.; Pascual, M.Á. La producción científica sobre Realidad Aumentada, un análisis de la situación educativa desde la perspectiva SCOPUS. *EDMETIC* **2017**, *6*, 39–61. [CrossRef]
4. Radu, I. Augmented reality in education: A meta-review and cross-media analysis. *Pers. Ubiquitous Comput.* **2014**, *18*, 1533–1543. [CrossRef]
5. Villalustre, L.; del Moral, M.E. Juegos perceptivos con realidad aumentada para trabajar contenido científico. *Educ. Form. Tecnol.* **2017**, *10*, 36–46.
6. Marín, V.; Muñoz, V.P. Trabajar el cuerpo humano con realidad aumentada en educación infantil. *Rev. Tecnol. Cienc. y Educ.* **2018**, *9*, 148–158.
7. Area, M.; Hernández, V.; Sosa, J.J. Modelos de integración didáctica de las TIC en el aula. *Comunicar* **2016**, *24*, 79–87. [CrossRef]
8. Castañeda, L.; Esteve, F.; Adell, J. ¿Por qué es necesario repensar la competencia docente para el mundo digital? *RED* **2018**, *56*, 1–20. [CrossRef]
9. Fuentes, A.; López, J.; Pozo, S. Analysis of the Digital Teaching Competence: Key Factor in the Performance of Active Pedagogies with Augmented Reality. *REICE* **2019**, *17*, 27–42. [CrossRef]
10. Castellanos, A.; Sánchez, C.; Calderero, J.F. Nuevos modelos tecnopedagógicos. *Competencia digital de los alumnos universitarios. Rev. Electrn. Investig. Educ.* **2017**, *19*, 1–9. [CrossRef]
11. Prendes, M.P.; Gutiérrez, I.; Martínez, F. Competencia digital: Una necesidad del profesorado universitario en el siglo XXI. *RED* **2018**, *56*, 1–22. [CrossRef]
12. Cabero, J.; Barroso, J. Los escenarios tecnológicos en Realidad Aumentada (RA): Posibilidades educativas en estudios universitarios. *Aula Abierta* **2018**, *47*, 327–336. [CrossRef]
13. Cabero, J.; Roig, R. The motivation of technological scenarios in augmented reality (AR): Results of different experiments. *Appl. Sci.* **2019**, *9*, 2907. [CrossRef]
14. Rodríguez, A.M.; Hinojo, F.J.; Ágreda, M. Diseño e implementación de una experiencia para trabajar la interculturalidad en Educación Infantil a través de realidad aumentada y códigos QR. *Educar* **2019**, *55*, 59–77. [CrossRef]
15. Chen, P.; Liu, X.; Cheng, W.; Huang, R. A review of using Augmented Reality in Education from 2011 to 2016. In *Innovations in Smart Learning*; Popescu, E., Kinshuk, M.K., Huang, R., Jemni, M., Chen, N.S., Sampson, D.G., Eds.; Springer: Singapore, 2017; pp. 13–18. [CrossRef]
16. Barroso, J.; Cabero, J.; García, F.; Calle, F.M.; Gallego, Ó.; Casado, I. *Diseño, Producción, Evaluación y Utilización Educativa de la Realidad Aumentada*, 1st ed.; Secretariado de Recursos Audiovisuales y NNTT de la Universidad de Sevilla: Sevilla, Spain, 2017; pp. 55–92. Available online: https://cutt.ly/5etacaZ (accessed on 2 December 2019).
17. Gómez, M.; Trujillo, J.M.; Aznar, I.; Cáceres, M.P. Augment reality and virtual reality for the improvement of spatial competences in Physical Education. *J. Hum. Sport Exerc.* **2018**, *13*, 189–198. [CrossRef]

18. López, J.; Pozo, S.; López, G. La eficacia de la realidad aumentada en las aulas de infantil: Un estudio del aprendizaje de SVB y RCP en discentes de 5 años. *Pixel-Bit* **2019**, *55*, 157–178. [CrossRef]
19. Garay, U.; Tejada, E.; Castaño, C. Percepciones del alumnado hacia el aprendizaje mediante objetos educativos enriquecidos con realidad aumentada. *EDMETIC* **2017**, *6*, 145–164. [CrossRef]
20. Cabero, J.; Llorente, M.C.; Marín, V. Comunidades virtuales de aprendizaje. El Caso del proyecto de realidad aumentada: RAFODIUM. *Perspect. Educ.* **2017**, *56*, 117–138. [CrossRef]
21. Bacca, J.; Baldiris, S.; Fabregat, R.; Graf, S.; Kinshuk. Augmented reality trends in education: A systematic review of research and applications. *Educ. Technol. Soc.* **2014**, *17*, 133–149.
22. Marín, V.; Cabero, J.; Gallego, O.M. Motivación y realidad aumentada: Alumnos como consumidores y productores de objetos de aprendizaje. *Aula Abierta* **2018**, *47*, 337–346. [CrossRef]
23. Cheng, K.H. Reading an augmented reality book: An exploration of learners'cognitive load, motivation, and attitudes. *Australas. J. Educ. Technol.* **2017**, *33*, 53–69. [CrossRef]
24. Fombona, J.; Vázquez, E. Posibilidades de utilización de la Geolocalización y Realidad Aumentada en el ámbito educativo. *Educ. XX1* **2017**, *20*, 319–342. [CrossRef]
25. Toledo, P.; Sánchez, J.M. Realidad Aumentada en Educación Primaria: Efectos sobre el aprendizaje. *RELATEC* **2017**, *16*, 79–92. [CrossRef]
26. Cabero, J.; Llorente, C.; Gutiérrez, J.J. Evaluación por y desde los usuarios: Objetos de aprendizaje con Realidad aumentada. *RED* **2017**, *53*, 1–17. [CrossRef]
27. Kamphuis, C.; Barsom, E.; Schijven, M.; Christoph, N. Augmented reality in medical education? *Perspect. Med. Educ.* **2014**, *3*, 300–311. [CrossRef]
28. Yuen, S.C.; Yaoyuneyong, G.; Johnson, E. *Augmented Reality and Education: Applications and Potentials*, 1st ed.; Springer: Berlin/Heidelberg, Germany, 2013; pp. 77–92. [CrossRef]
29. Prendes, C. Realidad aumentada y educación: Análisis de experiencias prácticas. *Pixel-Bit. Rev. Medios Educ.* **2015**, *46*, 187–203. [CrossRef]
30. Suh, A.; Prophet, J. The state of immersive technology research: A literature analysis. *Comput. Hum. Behave.* **2018**, *86*, 77–90. [CrossRef]
31. Muhamedyev, R.I.; Aliguliyev, R.M.; Shokishalov, Z.M.; Mustakayev, R.R. New Bibliometric Indicators for Prospectivity Estimation of Research Fields. *Ann. Libr. Inf. Stud.* **2018**, *65*, 1–8.
32. Fombona, J.; Pascual, M.Á.; González, M.C. M-learning y realidad aumentada: Revisión de literatura científica en el repositorio WoS. *Comunicar* **2017**, *25*, 63–72. [CrossRef]
33. Lorenzo, G.; Scagliarini, C. Bibliometric review of augmented reality in education. *Rev. Gen. Inf. Doc.* **2018**, *28*, 45–60. [CrossRef]
34. Heradio, R.; de la Torre, L.; Galán, D.; Cabrerizo, F.J.; Herrera, E.; Dormido, S. Virtual and remote labs in education: A bibliometric analysis. *Comput. Educ.* **2016**, *98*, 14–38. [CrossRef]
35. Álvarez, A.; Castillo, M.; Geldes, C. Análisis Bibliométrico de la Realidad Aumentada y su Relación con la Administración de Negocios. *Inf. Tecnol.* **2017**, *28*, 57–66. [CrossRef]
36. Jaramillo, A.M.; Silva, G.J.; Adarve, C.A.; Velásquez, S.M.; Páramo, C.A.; Gómez, L.L. Aplicaciones de Realidad Aumentada en educación para mejorar los procesos de enseñanza-aprendizaje: Una revisión sistemática. *Rev. Espac.* **2018**, *39*, 1–15.
37. Karakus, M.; Ersozlu, A.; Clark, A.C. Augmented Reality Research in Education: A Bibliometric Study. *J. Math. Sci. Technol. Educ.* **2019**, *15*, 1–12. [CrossRef]
38. Arici, F.; Yildirim, P.; Caliklar, Ş.; Yilmaz, R.M. Research trends in the use of augmented reality in science education: Content and bibliometric mapping analysis. *Comput. Educ.* **2019**, *142*, 1–13. [CrossRef]
39. Martínez, M.A.; Cobo, M.J.; Herrera, M.; Herrera, E. Analyzing the scientific evolution of social work using science mapping. *Res. Soc. Work Pract.* **2015**, *25*, 257–277. [CrossRef]
40. Moreno, A.J. Estudio Bibliométrico de la Producción Científica sobre la Inspección Educativa. *REICE. Rev. Iberoam. Sobre Calid. Efic. Cambio Educ.* **2019**, *17*, 23–40. [CrossRef]
41. López-Robles, J.R.; Otegi-Olaso, J.R.; Porto, I.; Cobo, M.J. 30 years of intelligence models in management and business: A bibliometric review. *Int. J. Inf. Manag.* **2019**, *48*, 22–38. [CrossRef]
42. Rodríguez-García, A.M.; López, J.; Agreda, M.; Moreno-Guerrero, A.J. Productive, Structural and Dynamic Study of the Concept of Sustainability in the Educational Field. *Sustainability* **2019**, *11*, 5613. [CrossRef]
43. Hirsch, J.E. An index to quantify an individual's scientific research output. *Proc. Natl. Acad. Sci. USA* **2005**, *102*, 16569–16572. [CrossRef]

44. Cobo, M.J.; López, A.G.; Herrera, E.; Herrera, F. Science mapping software tools: Review, analysis, and cooperative study among tools. *J. Am. Soc. Inf. Sci. Technol.* **2011**, *62*, 1382–1402. [CrossRef]
45. Kaufmann, H.; Schmalstieg, D. Mathematics and geometry education with collaborative augmented reality. *Comput. Graphics* **2002**, *23*, 339–345. [CrossRef]
46. Martín-Gutiérrez, J.; Saorín, J.L.; Contero, M.; Alcaniz, M.; Pérez-López, D.C.; Ortega, M. Design and validation of an augmented book for spatial abilities development in engineering students. *Comput. Graphics* **2010**, *34*, 77–91. [CrossRef]
47. Akcayir, M.; Akcayir, G. Advantages and challenges associated with augmented reality crossMark for education: A systematic review of the literature. *Educ. Res. Rev.* **2017**, *20*, 1–11. [CrossRef]
48. Potkinjak, V.; Gardner, M.; Callaghan, V.; Mattila, P.; Guetl, C.; Petrovic, V.M.; Jovanovic, K. Virtual laboratories for education in science, technology, and engineering: A review. *Comput. Educ.* **2016**, *95*, 309–327. [CrossRef]
49. Andujar, J.M.; Mejías, A.; Márquez, M.A. Augmented Reality for the Improvement of Remote Laboratories: An Augmented Remote Laboratory. *IEEE Trans. Educ.* **2011**, *53*, 492–500. [CrossRef]

MDPI
St. Alban-Anlage 66
4052 Basel
Switzerland
Tel. +41 61 683 77 34
Fax +41 61 302 89 18
www.mdpi.com

Applied Sciences Editorial Office
E-mail: applsci@mdpi.com
www.mdpi.com/journal/applsci

www.ingramcontent.com/pod-product-compliance
Lightning Source LLC
Chambersburg PA
CBHW061808210926
51850CB00006B/646
9783036560618